Winfrid G. Schneeweiss

Boolean Functions

with Engineering Applications and Computer Programs

With 127 Figures

Springer-Verlag Berlin Heidelberg New York
London Paris Tokyo 1989

Prof. Dr. Winfrid G. Schneeweiss
Fachbereich Mathematik und Informatik
Lehrgebiet Technische Informatik
Fernuniversität (Gesamthochschule)
D-5800 Hagen

ISBN 3-540-18892-4 Springer-Verlag Berlin Heidelberg NewYork
ISBN 0-387-18892-4 Springer-Verlag NewYork Berlin Heidelberg

Printed in Germany

Typesetting: Macmillan, India
Printing: Color-Druck, Berlin. Bookbinding: Lüderitz & Bauer, Berlin.
2161/3020-543210 – Printed on acid-free paper

Preface

Modern systems engineering (e.g. switching circuits design) and operations research (e.g. reliability systems theory) use Boolean functions with increasing regularity. For practitioners and students in these fields books written for mathematicians are in several respects not the best source of easy to use information, and standard books, such as, on switching circuits theory and reliability theory, are mostly somewhat narrow as far as Boolean analysis is concerned. Furthermore, in books on switching circuits theory the relevant stochastic theory is not covered. Aspects of the probabilistic theory of Boolean functions are treated in some works on reliability theory, but the results deserve a much broader interpretation. Just as the applied theory (e.g. of the Laplace transform) is useful in control theory, renewal theory, queueing theory, etc., the applied theory of Boolean functions (of indicator variables) can be useful in reliability theory, switching circuits theory, digital diagnostics and communications theory.

This book is aimed at providing a sufficiently deep understanding of useful results both in practical work and in applied research.

Boolean variables are restricted here to indicator or 0/1 variables, i.e. variables whose values, namely 0 and 1, are not free for a wide range of interpretations, e.g. in digital electronics 0 for L ≡ low voltage and 1 for H ≡ high voltage. The use of indicator variables has the advantage that the meaning of 0 and 1 is clear wherever they appear; both remain the well known integers. Compared to this conceptual advantage, the extra effort it may take to define proper binary indicator variables is considered acceptable. Furthermore, indicator variables allow for the use of standard algebra to write Boolean functions; George Boole did this. This certainly does not imply that I think typical Boolean operator symbols like $\vee$ for OR are obsolete; but rather that Boolean analysis with the inclusion of standard operations, like addition, with the operator symbol +, becomes richer. As an example, with X_1, X_2 being indicator variables, i.e. X_1, X_2 can only have the values 0 and 1, as is well known:

$$X_1 \vee X_2 = X_1 \vee \bar{X}_1 X_2$$

with both right-hand terms disjoint. Now, even though

$$X_1 \vee X_2 \neq X_1 + X_2,$$

it is true that

$$X_1 \vee \bar{X}_1 X_2 = X_1 + \bar{X}_1 X_2.$$

Hence, in this case, the + sign can express more detail. Also, in stochastic analysis, with E for expectation,

$$E\{X_1 \vee \bar{X}_1 X_2\}$$

cannot be simplified by standard methods, whereas

$$E\{X_1 + \bar{X}_1 X_2\} = E\{X_1\} + E\{\bar{X}_1 X_2\}.$$

Great care has been taken to use unambiguous language and notation. For instance, the concepts of a minterm of a Boolean function and that of the corresponding binary argument vector are not mixed up. A certain problem exists with the term "contains". If X_2 is said to be contained in $X_1 X_2 X_3$ this appears to be quite clear. However, on the other hand, one might also say $X_1 X_2 X_3$ is contained in X_2 in the sense that the set of the minterms belonging to X_2 will contain the minterms associated with $X_1 X_2 X_3$. When problems of this type could lead to misunderstandings, the text will become slightly more formal.

This text is prepared mainly with first-year graduate students in mind, but can be used for self-study. I know from experience, there is nothing as frustrating as exercises (problems) without solutions (at the back of the book). Therefore, I have given my solutions; alternatives may be correct too. The reader with little time to brood over his or her own solutions should at least quickly look at them, since some contain important additions to the text of the respective chapter (§).

To limit the volume of this text, Boolean vector functions as they appear typically in microprograms (with the address of the relevant VLSI memory unit as the argument (n-tuple)) and in separable binary (k,n) codes for data transfer (with the k net information bits as the argument k-tuple, and the n-k check bits as values of n-k Boolean functions) are not discussed here at any depth.

The following digraph is the (rough) precedence digraph of this book. For a full understanding of a given chapter (§) it is necessary to study much of its preceding chapter(s). The heart of the book are §§ 7, 8, 9.

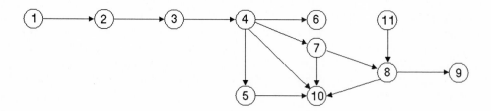

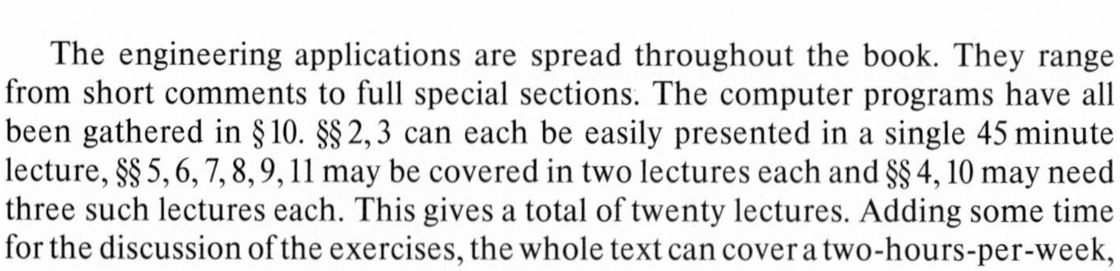

The engineering applications are spread throughout the book. They range from short comments to full special sections. The computer programs have all been gathered in § 10. §§ 2, 3 can each be easily presented in a single 45 minute lecture, §§ 5, 6, 7, 8, 9, 11 may be covered in two lectures each and §§ 4, 10 may need three such lectures each. This gives a total of twenty lectures. Adding some time for the discussion of the exercises, the whole text can cover a two-hours-per-week, one-term course or a three to four days' short course or a seminar.

From my experience, discussing the details of the presentation by the author is not very helpful. The reader should rather look at the table of contents which follows this preface.

It is a great pleasure for me to thank the following persons and institutions who helped me in writing this text: My wife, Dagmar, for her great patience during the last three years; my secretary, E. Lenski, for her expert typing of the manuscript; E. Palmovski for the nice figures; Dr. G. Volta of the European Community's Joint Research Centre at Ispra for offering me funds and discussion partners during part of my sabbatical in 1983; U. Richter, M. Schulte, H. Bähring for many fruitful discussions in connection with proof reading of intermediate or final material; M. Schulte for programming assistance in theory and practice; A. von Hagen and his staff, including linguistic help and anonymous referees, for excellent cooperation in the process of publishing.

Hagen, October 1988 Winfrid G. Schneeweiss

Contents

1 Notation and Glossary of Fundamental Terms and Symbols

Introductory Remark

This text, although mainly written for applications oriented readers, goes to some more depth than for example the usual short introductions to switching algebra found in electrical engineering text books. This necessitates the use of more tools with corresponding further notation. The lengthy notation list is only given for easy reference. New symbols are always explained when used for the first time. For notational simplicity, sometimes the Boolean function symbol φ is used even though the binary variable X_φ of its value would have been a more rigorous notation. But this is common practice in application.

List of Symbols[1]

a_i, b_i, c_i	random events or coefficients, c_i also carry or code bits
$b(x), c(x)$	binary polynomial and code polynomial, respectively
$a, b, c, \alpha, \beta, \gamma$	constants; a, b also random events, α also index
A, B, C	sets in general, A also availability
$\bar{A}$	$A \setminus \Omega$, complement set of set A with respect to Ω
B^n	n-dimensional binary space, $B^1 = \{0, 1\}$
C_i	component i (of a technical system)
$C_N(a)$	number of times event a is found in N trials
$C(\hat{T}_i X_k, \hat{T}_j \bar{X}_k) = \hat{T}_i \hat{T}_j$	concensus of $\hat{T}_i X_k$ and $\hat{T}_j \bar{X}_k$
D_i	duration (time) of state i; specifically downtime i
$D_{i,j}$	duration of $X_i = j$; $j \in \{0, 1\}$; $i \in \{1, \ldots, n, \varphi\}$
d	don't care variable; $d \in \{0, 1\}$ and usual differential symbol
e_i	edge i; error bit i, element i of a certain set
$E\{Y\}$	expectation of the random variable Y
E	set of edges of a graph
$F_Z(z)$	(cumulative) probability distribution function (cdf) of Z (at z); $\bar{F} := 1 - F$
$f_Z(z)$	probability density function (pdf) of Z (at z)

[1] Several local ad hoc notation items are omitted.

$g^*(s)$	$\int_0^\infty g(t)\exp(-st)dt$, Laplace transform of the real function $g(t)$
$\varphi, \varphi(\mathbf{x}), \psi$	Boolean (indicator) functions of Boolean n-tupel (vector) $\mathbf{X}$
$\varphi(X_i=j)$	$\varphi(\mathbf{X})_{/X_i=j} := \varphi(X_1, X_2, \ldots, X_{i-1}, j, X_{i+1}, \ldots, X_n)$
$\bar{\varphi}$	NOT φ (φ can be any Boolean algebra expression)
φ_{si}	sensitivity function of φ with respect to X_i
φ_i', φ_i''	auxiliary functions for switching functions analysis
$g(t), g(x)$	real function and generator polynomial, respectively
$H(t), h(t)$	renewal function; renewal density,
i, j, k, l, m, n	indices
I_j, I_φ	index set j, index set of the minterms of φ
$l_i, l_{i,j}$	indices
L	lifetime (often abbreviated by life)
$\lambda_{i,j}$	state transition rate
λ, μ	rates
$\hat{M}(\varphi)$	set of minterms of φ
$\hat{M}_i, \check{M}_i$	minterm i, maxterm i
m	number of terms of a polynomial or DNF or other type of sum
$\mathbb{N}$	set of cardinal numbers; $\mathbb{N} = \{1, 2, 3, \ldots\}$
N_e, N_v	number of edges resp. vertices of a graph
$N_{\hat{M}}(\varphi)$	number of minterms of Boolean function
$N(t_1, t_2)$	number of random points between t_1 and t_2
n	number of variables of a (general) Boolean function
n_i	number of literals of term i
ν_i	frequency of occurrence of $X_i=0$ or of $X_i=1$; $i \in \{1, 2, \ldots, n, \varphi\}$
$P\{a\}$	probability of (random) event a
$P_i(t)$	$P\{S(t)=S_i\}\dot{P}(t) := dP(t)/dt$, $\ddot{P}(t) := d\dot{P}(t)/dt$
p_i, p_φ	$P\{X_i=1\}$, $P\{X_\varphi=1\}$
$q(\cdot)$	non-negative function (of sets)
R	reliability
$S(t)$	state at time t
S_i	state i (in Markov model)
s_i	elementary system state i and special subsets or sum bits
S, S_φ	(technical) system, described by φ; also a certain set
s	(complex) variable of the Laplace transform
T_i	general term i; also distance of i-th renewal point from time origin
$\hat{T}_i$	conjunction i term
$\check{T}_i$	disjunction term i
$\overset{\diamond}{T}_i$	disjunction of conjunction terms derived from $\hat{T}_i$
t, τ, t'	time
U	unavailability; unreliability, if time-dependent

V	set of vertices of a graph
ω_i	elementary event (of probability theory)
Ω	certain event, reference set
X_i	indicator variable i
x	dummy variable of algebraic coding theory
$Y_{i,j}^{(k)}$	Boolean indicator variable for the existence of a path of length k from vertex i to vertex j
$\mathbf{Y}$	matrix of $Y_{i,j}$; $Y_{i,k} \in \{0, 1\}$
Z	general random variable
ϕ, $\mathbf{0}$, $\mathbf{1}$	the empty set, $\mathbf{0} := (0, \ldots, 0)$, $\mathbf{1} := (1, \ldots, 1)$
$a \in A$	a is an element of set A
$A \subset B$	set A is contained in (is a subset of) set B
$A \cap B$	intersection of the sets A and B, i.e. set of elements contained both in A and in B
$A \cup B$	union of the sets A and B, i.e. set of elements belonging to A or to B (or to both)
$\mathbf{A}$; $\mathbf{A}^T$	matrix, (also vector) $\mathbf{A} := (A_{i,j})_{m,m}$; transposed matrix $\mathbf{A}$
$\mathbf{X}$	vector with a finite number of components $X_1, X_2, \ldots$
$\bar{X}$	negation (complement with respect to 1) of X
$\tilde{X}$	either X or $\bar{X}$
X_φ	(Boolean) value of φ: $B^n \to B$
$X_i \wedge X_j$	conjunction (logical AND) of X_i and X_j
$X_i \not\wedge X_j$	logical NAND = NOT AND of X_i and X_j
$X_i \vee X_j$	disjunction (logical OR) of X_i and X_j
$X_i * X_j$	non-inclusive disjuction (logical EXOR) of X_i and X_j

$$\bigwedge_{i=1}^{m} X_i := X_1 \wedge X_2 \wedge \ldots \wedge X_m$$

$$\bigwedge_{i \in I} X_i = \bigwedge_{i=1}^{m} X_{l_i};\ l_i \in I,\ I = \{l_1, l_2, \ldots, l_m\}$$

$$\bigvee_{i=1}^{m} X_i := X_1 \vee X_2 \vee \ldots \vee X_m$$

$$\mathop{\not\equiv}_{i=1}^{m} X_i = X_1 \not\equiv X_2 \not\equiv \ldots \not\equiv X_m$$

$\mathscr{L}\{g(t)\}$, $\mathscr{L}^{-1}\{g(t)\}$	Laplace transform of $g(t)$ and inverse transform
$A \hat{=} B$	A corresponds to B
$A \Rightarrow B$	A implies B (from statement A there follows statement B)
$A := B$	A is the newly defined name for B
$A \leftarrow B$	A is replaced by B
card $\{A\}$	the number of elements of set A
$\Delta_{X_i} \varphi$	Boolean difference of φ with respect to X_i
$\Delta_{(X_i, X_j)} \varphi$	joint Boolean difference of φ with respect to X_i and X_j
$\Delta^2_{X_i, X_j} \varphi$	second Boolean difference, i.e. $\Delta_{X_j}(\Delta_{X_i} \varphi)$

$\tilde{}, ', ''$	general symbols for indexing with a fixed meaning only for $\tilde{X}$ (see above). (Note that dashes do not mean differentiation here.)
$a(x)(+)b(x)$	polynomial addition, where coefficients are added mod 2
$a(x)(\cdot)b(x)$	polynomial multiplication, where coefficients are added mod 2
$a(x)(:)b(x)$	polynomial division, as inverse operation of $a(x)(\cdot)b(x)$
— —	is the termination symbol for examples, theorems etc.

Abbreviations

cdf,	cumulative distribution function;
pdf	probability density function
DNF†	disjunctive normal form (Boolean polynomial form)
CDNF†	canonical DNF (disjunction of minterms)
DDNF†	DNF of mutually disjoint terms
CNF†; CCNF†	conjunctive normal form; canonical CNF
MTBF (MTTF)	mean time before failure (mean time to failure)
MTTR	mean time to repair
TGF	test generation function
PI	prime implicant (prime term)
iff	if and only if (equivalence)
l.h.s., r.h.s.	left/right hand side
s-independent	stochastically independent
MUX	multiplexer (in digital electronics)
K-map	Karnaugh map
ML	index for multilinear
$\exp(\alpha)$	exponential function at α: e^α

Fundamental Terms Concerning States

A *system* is a well structured, i.e. organized, set of *components*. A component can become a system if it is described in more detail such that sub-components can be identified. If the term "component" is used to identify a vector component, this will be clear from the context.

A component i, when seen as an indivisible unit, can be in either of two *binary states* as described by the two integer values 0 and 1 of its associated indicator variable X_i.

An *elementary state of a system* is a state described by the (ordered) set of the binary states of *all* of its n components. Hence an elementary state corresponds to a binary n-vector.

† Instead of saying "φ is given in . . . form" it is general usage to say "φ is a . . . form".

A *state of a system* is described by the set of the elementary states belonging to it. Typically, a state is defined by the values (0 or 1) of a subset of the components of the binary *state vector* which is the n-vector of the n system components. The non-specified components may be in either of their individual binary states. Hence, every unspecified component increases the number of elementary states belonging to a state by the factor of two. If the *binary state* of a system (e.g. good or bad) is meant, this will be clear from the context, for which the binary indicator variable X_S of the system S is a helpful concept. However, here X_φ is used instead of X_S. This implies that all relevant features of S are expressed by φ.

Basic Terms of Graph Theory [55], [56]

In most practical situations a graph is just a set of "links", called *edges* $e_1, e_2, \ldots, e_{N_e}$, where e_i is described by a pair of "end points", so-called *vertices* or *nodes*:

$$e_i = (v_j, v_k) \ .$$

In the case of a *directed* graph (short: *digraph*) v_j, v_k form an ordered pair and e_i is a *directed* edge. However, there may exist isolated vertices with no *incident* edges. Therefore, it is appropriate to define a graph G as a pair of sets, viz. the set V of its vertices and the set E of its edges, respectively: $G = (V, E)$. An edge-sequence leading from v_{i_1} to v_{im} is the following tuple of edges:

$$[(v_{i_1}, v_{i_2}), (v_{i_2}, v_{i_3}), (v_{i_2}, v_{i_4}), \ldots, (v_{i_{m-1}}, v_{i_m})] \ .$$

If all these edges are different, then this set of edges is a *path* from v_{i_1} to v_{im} and the number of these edges is the *length* of this path. (Note that sometimes an edge sequence is called a path and a path is called a *minimal* path.) The case $v_{i_1} = v_{im}$ describes a cyclic path, i.e. a *cycle*. A graph with paths between all pairs of nodes is *connected*. Trees are connected graphs without cycles. In a directed tree there is one node with out-going edges only; it is its *root*. The end nodes with exactly one link ending in them and none leaving are the *leafs*. Non leaf-nodes are called the *fathers*

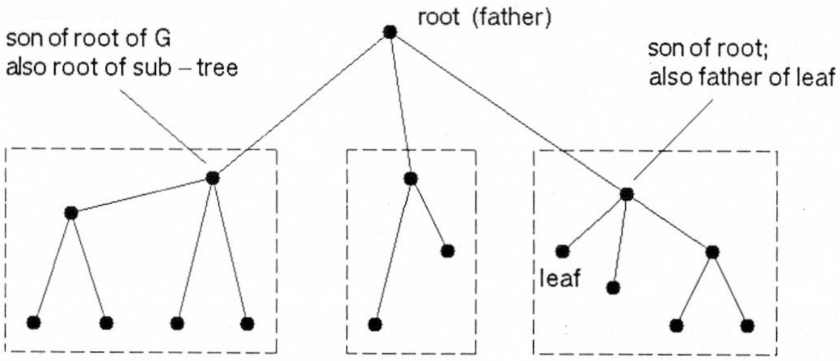

Fig. 1.1. Concept of rooted tree G. Sub-trees of the root in dashed rectangles

of the roots of their sub-trees. The latter roots are the *sons* of the father node; see Fig. 1.1. Non leaf-nodes often have at most two out-going edges. Such trees are *binary* trees. The maximum path length in a rooted tree is the *height* of this tree. A complete binary tree of height c has all the nodes and edges such a tree could possibly have; see e.g. Fig. 3.2.1. A *bipartite* digraph is one with two types of vertices, where edges always connect pairs of vertices of different type.

Fundamental Graphical Symbols

Figure 1.2 shows the symbols for the fundamental Boolean operations of AND, OR, NEGATION, EXOR (the exclusive OR), NAND, and NOR. Negation will be depicted in a shorthand way by a tiny circle as shown for NAND and NOR, the negations of AND and OR. In Fig. 1.2 inputs, i.e. the Boolean argument, value(s) come(s) from below; where two are shown there could be more. The output, i.e. the value of the Boolean function, appears on top of the elementary function generator symbols. The meaning of the outputs is explained in detail in §2.

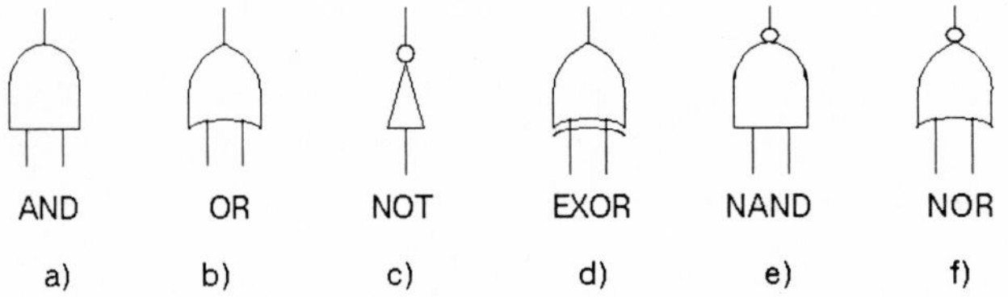

Fig. 1.2a–f. Symbols for logic gates; also used to visualize logic operations

Numbering

All categories of items, such as figures, tables and theorems, are numbered individually in each chapter. For instance, Fig. $X.Y.Z$ means figure Z in chapter (§) $X.Y$. Eq. $(X.Y.Z)$ is formula Z in chapter (§) $X.Y$.

Eq. $(EX.Y)$ is formula Y in the exercises of §X.

Eq. $(SX.Y)$ is formula Y in the solutions of the exercises of §X.

Other (old?) Nomenclature

Readers who are familiar with older literature in the field or who try to read older original papers, e.g. [10], [11] of Quine, may encounter different nomenclature. To help such readers the dictionary of Table 1.1 was compiled.

Table 1.1. Other nomenclature in the field of Boolean functions

clause fundamental formula	$\hat{=}$ (conjunction) term
alternation $\hat{=}$ disjunction	
normal formula sum-of-products form[†]	$\hat{=}$ disjunctive normal form (DNF)
developed $\hat{=}$ expanded (down) to minterms	
term 1 subsumes term 2 $\hat{=}$	all the literals of term 2 are also literals of term 1; term 1 "contains" term 2 (Yet the set of the minterms of term 1 is contained in the set of the minterms of term 2.)
tautology $\hat{=}$ Boolean function $\varphi(\mathbf{X}) = 1$	
affirmed variable (letter) $\hat{=}$ normal variable	

[†] Note that this is not a usual polynomial (of literals) but rather disjunction is written with "+" and conjunction as concatenation (as a product).

2 Fundamental Concepts

In this introductory chapter a number of basic concepts, which are needed in almost all of the following chapters, are presented. Among these concepts are, first of all, those of Boolean variables and Boolean functions of these variables. Next, axioms and further fundamental laws of Boolean algebra are discussed.

Prior to the extensive study of many interesting details, the newcomer to this field of applied mathematics should be told that Boolean functions are, as so-called switching functions, the most important conceptual basis of all computers, from the simple personal computer up to the huge super mainframe computers. All the many control, arithmetic, and memory operations within computers are best described by systems of Boolean functions. Other important applications of Boolean functions can be found in many parts of operations research, typically in reliability theory, where the so-called fault trees are pictures of Boolean functions describing the superposition of component faults to create system faults.

Even though variables with only two possible values and functions with these same two values may look simple at first sight, it is the almost unlimited possibility of combining many functions of many variables through many stages of modern engineering (and social) systems that lends Boolean analysis its own typical complexity in theory and practice.

2.1 Boolean Indicator Variables and All Their Functions of One and Two Variables

This text is devoted to Boolean functions of indicator variables and to an understanding of the behaviour of the systems which are modelled by these functions. It is understood that the n components $C_1, C_2, \ldots, C_n$ of a System S_φ have binary properties which can be modelled by *indicator variables*†

$$X_i = \begin{cases} 0\,, & \text{if } C_i \text{ has a certain property} \\ 1\,, & \text{else} \end{cases}$$

† The concept of indicator variable is a specialization of the concept of *indicator function* [28, p. 81] being 1 if a given random event has occurred, and 0 else. As in [28, pp. 395/6] the term *characteristic function* is retained for the Fourier transform of the pdf of a random variable, which is not used in this book.

and also

$$X_\varphi = \begin{cases} 0 \ , & \text{if } S_\varphi \text{ has a certain property} \\ 1 \ , & \text{else} \ . \end{cases}$$

The above "properties" can be defined freely and individually for the components and for the system[†]. The symbol φ is introduced, because here X_φ will depend on $X_1, \ldots, X_n$ by the Boolean function φ:

$$X_\varphi = \varphi(X_1, \ldots, X_n) \ .$$

(A more formal definition will follow in §4.)

In most of this book a possible dependence of X_i and X_φ on time will be ignored. However, in parts of the stochastic theory of §§8, 9 time will be introduced; see Fig. 2.1.1.

In digital electronics diagrams of the type of Fig. 2.1.1 are called *impulse diagrams* since in many applications, typically in clocked systems, the 1's of computer logic or binary arithmetic are represented by short impulses.

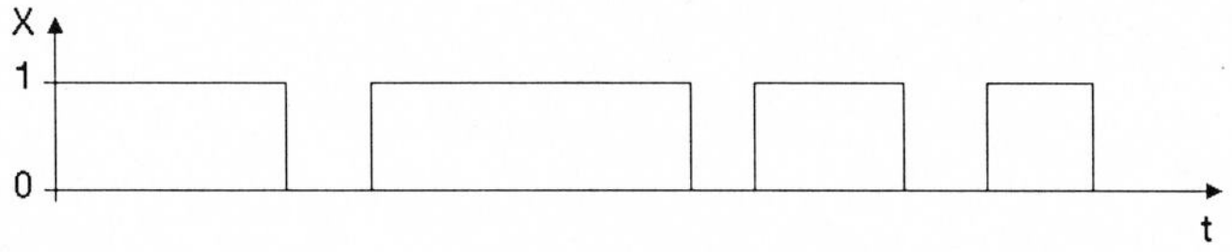

Fig. 2.1.1. Sample function of the binary random process $\{X(t)\}$

2.1.1 The Functions of One Variable

As Table 2.1.1 shows, there are four Boolean functions $\varphi_1, \ldots, \varphi_4$ of 1 variable, namely the constants

$$\varphi_1(X) = 0 \ , \quad \varphi_2(X) = 1 \ ,$$

and (for non-mathematicians the "true" functions)[‡]

$$\varphi_3(X) = X \ , \quad \varphi_4(X) = \bar{X} \ .^{\S}$$

Table 2.1.1. All Boolean functions of 1 variable

X	φ_1	φ_2	φ_3	φ_4
0	0	1	0	1
1	0	1	1	0

[†] Clearly, inconsistent definitions are not allowed. For instance, if $X_i = 1$ means failure of C_i and $\varphi(1, \ldots, 1) = 1$, then, usually, only $X_\varphi = 1$ for system failure will make sense.

[‡] Engineers tend to be reluctant to call a constant value function a "function".

[§] The bar means negation or, rather, complementation with respect to 1, i.e. the values 0 and 1 are replaced by 1 and 0, respectively.

(See Fig. 3.1.1 for a graphical representation of $\varphi_1, \ldots, \varphi_4$.)
Note that the numbering of Boolean functions is non-standard. Any other numbering would be just as good.

2.1.2 The Functions of Two Variables

As Table 2.1.2 shows, there are 16 Boolean functions $\varphi_i(X_1, X_2)$, $i = 1, \ldots, 16$ of two variables. Many of these functions have gained so much practical importance that they have been given plausible names. Specifically, beyond the constant functions

$$\varphi_1(X_1, X_2) = 0 \ , \qquad \varphi_{16}(X_1, X_2) = 1 \ ,$$

one calls

$$\varphi_9 = X_1 \wedge X_2 \qquad \text{(Fig. 1.1a)}$$

the *conjunction* or AND-function, since $\varphi_9 = 1$ iff $X_1 = 1$ and $X_2 = 1$ (where $\wedge$ stands for the latin AUT, meaning "as well as") and

$$\varphi_{15} = X_1 \vee X_2 \qquad \text{(Fig. 1.1b)}$$

the *disjunction* or OR-function, since $\varphi_{15} = 1$ iff $X_1 = 1$ or $X_2 = 1$ or both $X_1 = 1$ and $X_2 = 1$, (where $\vee$ stands for the latin VEL, meaning "or"). In contrast to φ_{15},

$$\varphi_7 = X_1 \not\equiv X_2 \qquad \text{(Fig. 1.1d)}$$

is the *antivalence* or EXOR function, i.e. the *exclusive* OR-function, since $\varphi_7 = 1$ iff either $X_1 = 1$ or $X_2 = 1$, but not both of them. Note that φ_7 is also the 1-bit addition mod 2, and φ_9 is the 1-bit multiplication.

Table 2.1.2. All Boolean functions of two variables

X_1	X_2	φ_1	φ_2	φ_3	φ_4	φ_5	φ_6	φ_7	φ_8
0	0	0	1	0	1	0	1	0	1
0	1	0	0	1	1	0	0	1	1
1	0	0	0	0	0	1	1	1	1
1	1	0	0	0	0	0	0	0	0

X_1	X_2	φ_9	φ_{10}	φ_{11}	φ_{12}	φ_{13}	φ_{14}	φ_{15}	φ_{16}
0	0	0	1	0	1	0	1	0	1
0	1	0	0	1	1	0	0	1	1
1	0	0	0	0	0	1	1	1	1
1	1	1	1	1	1	1	1	1	1

Two more functions of exceptional practical importance are the negations of AND and OR, namely NAND, being

$$\varphi_8 = \overline{X_1 \wedge X_2}\ , \quad \text{(Fig. 1.2e)}\ ,$$

and NOR, being

$$\varphi_2 = \overline{X_1 \vee X_2}\ , \quad \text{(Fig. 1.2f)}\ .$$

Note, that the last five equations are nothing but notation!

Since in most of the applied work final results are noted by using exclusively AND ($\wedge$), OR ($\vee$) and NOT ($\bar{\ }$) we will, as an example, change φ_7 to this form. From Table 2.1.2 it can be verified that

$$X_1 \not\equiv X_2 = (X_1 \wedge \bar{X}_2) \vee (\bar{X}_1 \wedge X_2)\ . \tag{2.1.1}$$

At this point a first glimpse at Boolean functions of n variables is given: Boolean functions φ_i of n binary indicator variables are described mathematically as

$$\varphi_i: B^n \to B^1\ , \quad B^1 = \{0, 1\}\ , \quad B^n = B^{n-1} x B^1 \quad \text{(binary } n\text{-space)}$$

with x for the Cartesian product. Practically, they are simply described by the function table Table 2.1.3. Other non-trivial representations of Boolean functions will follow in §§4, 6, and 7.

Table 2.1.3. Function table of all the Boolean functions of n variables (without details). Such a table (for one function) is also known as a **truth table**

No.	X_1	X_2	... X_n	φ_1	φ_2	... φ_k
1	0	0	... 0	0	1	1
2	0	0	... 1	0	0	1
⋮						
2^n	1	1	... 1	0	0	1

$k := 2^{2^n}$

2.2 Axioms and Elementary Laws of Boolean Algebra

The main interest of this text is focused on Boolean functions. To synthesize these, also for cases of more than two variables (as considered in §2.1), the calculus of Boolean algebra is almost indispensible. This calculus, consisting of axioms and some rather obvious conclusions (so-called elementary laws or rules) is introduced now. Whether or not these axioms are redundant, i.e. a subset of them would suffice for a consistent definition of a Boolean algebra, is not investigated here.

Much deeper treatments of Boolean algebra can be found for example in [2], [3], [5]. Here only a minimum depth is aimed at, even though this text goes further than typical introductory texts on switching circuits theory. A good starting text for deeper applied research may be [2].

Note: The following axioms and laws of Boolean algebra are true for any type of Boolean quantities. It is only for conceptual simplicity that they are noted for single variables here. Even in this book, many of the following results noted for X_i's will be used frequently with X_i's replaced by general Boolean terms T_i's.

2.2.1 Axioms

For Boolean (indicator) variables X_i, X_j, X_k with $i, j, k \in \mathbb{N}$ the following axioms exist which define a Boolean algebra on the above variables by means of the two operations *conjunction* (AND) with the operator $\wedge$ and *disjunction* (OR) with the operator $\vee$:[†]

1. Commutative law of conjunction:

$$X_i \wedge X_j = X_j \wedge X_i \ . \tag{2.2.1}$$

This law is familiar from real algebra with $\wedge$ replaced by $\cdot$ for multiplication. The same is true for the next axiom with $\vee$ replaced by $+$ for addition:

2. Commutative law of disjunction

$$X_i \vee X_j = X_j \vee X_i \ . \tag{2.2.2}$$

3. 1st distributive law (plausible from real algebra):

$$X_i \wedge (X_j \vee X_k) = (X_i \wedge X_j) \vee (X_i \wedge X_k) \ , \tag{2.2.3}$$

4. 2nd distributive law (not so plausible from real algebra):

$$X_i \vee (X_j \wedge X_k) = (X_i \vee X_j) \wedge (X_i \vee X_k) \ . \tag{2.2.4}$$

5. Existence statement of the identity element 1 of conjunction:

$$1 \wedge X_i = X_i \ . \tag{2.2.5}$$

6. Existence statement of the identity element 0 of disjunction;

$$0 \vee X_i = X_i \ . \tag{2.2.6}$$

7. Existence statement of an inverse element $\bar{X}_i$ of X_i such that

$$X_i \wedge \bar{X}_i = 0 \tag{2.2.7}$$

and

$$X_i \vee \bar{X}_i = 1 \ . \tag{2.2.8}$$

[†] In this conceptual frame the operation of *negation* with the operator $^{-}$ is encountered in connection with the existence of an inverse $\bar{X}_i$ of X_i.

Note that the notion of an inverse element is equivalent to a negation operator. As an alternative to presenting these many axioms one can also start with the function tables of AND, OR, and NOT as in §2.1.2 and deduce the laws 1 through 7. In this sense Table 2.2.1 supplies a proof for Eq. (2.2.8). It is frequently practical to have a common name for X_i and $\bar{X}_i$.

Table 2.2.1. Proof of Eq. (2.2.8)

X_i	$\bar{X}_i$	$X_i \vee \bar{X}_i$
0	1	1
1	0	1

Definition 2.2.1. A *literal* is a normal (usual) or a negated (complemented) Boolean variable. — —

Almost needless to say, yet not trivial: Any Boolean algebra is closed under the operations $\wedge$ and $\vee$, i.e. two Boolean quantities connected by these operators yield another (or the same) Boolean quantity. (This is an addendum to the axioms above.)

Theorem 2.2.1. (Principle of duality). Any Boolean algebra theorem is also true if the constants 0 and 1 and the operators $\wedge$ and $\vee$ are interchanged.

Proof. Check all the axioms. This set of assumptions does not change when 0 and 1 and $\wedge$ and $\vee$, respectively, are interchanged, q.e.d.

Comment: Formally, theorem 2.2.1 transforms, e.g., Eq. (2.2.1) to

$$\bar{X}_i \wedge \bar{X}_j = \bar{X}_j \wedge \bar{X}_i \ . \tag{2.2.2a}$$

Yet, since the above axioms are true for any terms T_i, T_j, we can replace in (2.2.2a) X_i by $\bar{X}_i$ and X_j by $\bar{X}_j$ and from Eq. (3.2.9) we find Eq. (2.2.2).

For an elementary treatment of the above see [1]. More advanced material and further literature may be found in [2], [47], [3].

2.2.2 Elementary Laws

For good reasons sets of axioms are usually chosen as small as possible. An obvious consequence of this "purism" is the need for more practical rules, called elementary laws here, to manage Boolean algebra in practice. Such additional laws, including the important idempotence and associativeness laws, are now derived. In introductory texts to "logic" design of digital systems, such as [49] the associative laws of disjunction and conjunction are subsumed under "axioms". However, they can be deduced from the set of axioms given in §2.2.1. Since this deduction is a bit involved, several simpler elementary laws, some of which are needed for the derivation of the associative laws, will be treated next. Most of them are not familiar from elementary algebra.

$$\bar{\bar{X}} := \overline{(\bar{X})} = X \ . \tag{2.2.9}$$

From Table 2.1.2 (see φ_9)

$$0 \wedge X_i = 0 \tag{2.2.10}$$

and (see φ_{15})

$$1 \vee X_i = 1 \ . \tag{2.2.11}$$

Next the *idempotence laws* of disjunction

$$X_i \vee X_i = X_i \tag{2.2.12}$$

and of conjunction

$$X_i \wedge X_i = X_i \tag{2.2.13}$$

are proved [47]: The l.h.s. of Eq. (2.2.12) yields

$$\begin{aligned} X_i \vee X_i &= (X_i \vee X_i) \wedge 1 && \text{by Eq. (2.2.5)} \\ &= (X_i \vee X_i) \wedge (X_i \vee \bar{X}_i) && \text{by Eq. (2.2.8)} \\ &= X_i \vee (X_i \wedge \bar{X}_i) && \text{by Eq. (2.2.4)} \\ &= X_i \vee 0 && \text{by Eq. (2.2.7)} \\ &= X_i && \text{by Eq. (2.2.6).} \end{aligned}$$

Eq. (2.2.13) follows from the principle of duality, q.e.d..

The following *absorption* laws are rather important:

$$X_i \vee (X_i \wedge X_j) = X_i \ , \tag{2.2.14}$$

$$X_i \wedge (X_i \vee X_j) = X_i \ , \tag{2.2.15}$$

$$(X_i \wedge X_j) \vee (X_i \wedge \bar{X}_j) = X_i \ , \tag{2.2.16}$$

$$(X_i \vee \bar{X}_j) \wedge X_j = X_i \wedge X_j \ , \tag{2.2.17}$$

$$(X_i \wedge \bar{X}_j) \vee X_j = X_i \vee X_j \ , \tag{2.2.18}$$

$$(X_i \vee X_j) \wedge (X_i \vee \bar{X}_j) = X_i \ . \tag{2.2.19}$$

Notice that the following pairs of the above absorption laws are duals of each other: Eqs. (2.2.14, 15), Eqs. (2.2.16, 19), and Eqs. (2.2.17, 18). When used with the right-hand side as the starting term and the left-hand side as the result, these formulas can also be interpreted as "*expansion*" *laws*.

Equations (2.2.14) to (2.2.19) can easily be proved using Table 2.1.2. Here we give algebraic proofs:

Equation (2.2.14): By Eqs. (2.2.1), (2.2.3), (2.2.5), and (2.2.11)

$$X_i \vee (X_i \wedge X_j) = (1 \wedge X_i) \vee (X_i \wedge X_j) = (X_i \wedge 1) \vee (X_i \wedge X_j)$$
$$= X_i \wedge (1 \vee X_j) = X_i \wedge 1 = 1 \wedge X_i = X_i \ .$$

Equation (2.2.15): By Eqs. (2.2.3), (2.2.13), (2.2.14)

$$X_i \wedge (X_i \vee X_j) = (X_i \wedge X_i) \vee (X_i \wedge X_j)$$
$$= X_i \vee (X_i \wedge X_j) = X_i \ .$$

Equation (2.2.16): By Eqs. (2.2.1), (2.2.3), (2.2.5), (2.2.8)

$$(X_i \wedge X_j) \vee (X_i \wedge \bar{X}_j) = X_i \wedge (X_j \vee \bar{X}_j)$$
$$= X_i \wedge 1 = 1 \wedge X_i = X_i \ .$$

Equation (2.2.17): By Eqs. (2.2.1), (2.2.3), (2.2.6), (2.2.7)

$$(X_i \vee \bar{X}_j) \wedge X_j = X_j \wedge (X_i \vee \bar{X}_j)$$
$$= (X_j \wedge X_i) \vee (X_j \wedge \bar{X}_j)$$
$$= (X_i \wedge X_j) \vee 0 = X_i \wedge X_j \ .$$

Equation (2.2.18): By Eqs. (2.2.1), (2.2.2), (2.2.4), (2.2.5), (2.2.8)

$$(X_i \wedge \bar{X}_j) \vee X_j = (X_j \vee X_i) \wedge (X_j \vee \bar{X}_j)$$
$$= (X_i \vee X_j) \wedge 1 = 1 \wedge (X_i \vee X_j) = X_i \vee X_j \ .$$

Equation (2.2.19): By Eqs. (2.2.4), (2.2.5), (2.2.8)

$$(X_i \vee X_j) \wedge (X_i \vee \bar{X}_j) = X_i \wedge (X_j \vee \bar{X}_j)$$
$$= X_i \wedge 1 = 1 \wedge X_i = X_i \ .$$

Now the associative laws of disjunction and conjunction, viz.

$$(X_i \vee X_j) \vee X_k = X_i \vee (X_j \vee X_k) \tag{2.2.20}$$

and

$$(X_i \wedge X_j) \wedge X_k = X_i \wedge (X_j \wedge X_k) \tag{2.2.21}$$

are proved. (These laws almost look like axioms. Yet with considerable skill [47] it can be shown that even these fundamental laws follow from the axioms of Boolean algebra.)

Proof of Eq. (2.2.20): Let $L := (X_i \vee X_j) \vee X_k$
By Eqs. (2.2.1), (2.2.5), (2.2.8):

$$L = L \wedge 1 = L \wedge (X_i \vee \bar{X}_i) \ .$$

By Eq. (2.2.3):

$$L = (L \wedge X_i) \vee (L \wedge \bar{X}_i) \ .$$

By Eq. (2.2.1) and on replacing L by $(X_i \vee X_j) \vee X_k$:

$$\begin{aligned} L &= (X_i \wedge L) \vee (\bar{X}_i \wedge L) \\ &= (X_i \wedge [(X_i \vee X_j) \vee X_k]) \vee (\bar{X}_i \wedge [(X_i \vee X_j) \vee X_k]) \ . \end{aligned}$$

By Eq. (2.2.3)

$$L = \{[X_i \wedge (X_i \vee X_j)] \vee (X_i \wedge X_k)\} \vee \{[\bar{X}_i \wedge (X_i \vee X_j)] \vee (\bar{X}_i \wedge X_k)\} \ .$$

By Eqs. (2.2.15) and (2.2.3):

$$L = \{X_i \vee (X_i \wedge X_k)\} \vee \{[(\bar{X}_i \wedge X_i) \vee (\bar{X}_i \wedge X_j)] \vee (\bar{X}_i \wedge X_k)\} \ .$$

By Eqs. (2.2.14) and (2.2.7):

$$L = X_i \vee \{[0 \vee (\bar{X}_i \wedge X_j)] \vee (\bar{X}_i \wedge X_k)\} \ .$$

By Eq. (2.2.6):

$$L = X_i \vee [(\bar{X}_i \wedge X_j) \vee (\bar{X}_i \wedge X_k)] \ .$$

By Eq. (2.2.3):

$$L = X_i \vee [\bar{X}_i \wedge (X_j \vee X_k)] \ .$$

Finally, by (2.2.18):

$$L = X_i \vee (X_j \vee X_k) \ ,$$

which is the r.h.s. of Eq. (2.2.20), q.e.d. The proof of Eq. (2.2.21) follows by the principle of duality.

One of the most frequently used laws is that of *de Morgan*:

$$\overline{X_i \vee X_j} = \bar{X}_i \wedge \bar{X}_j \tag{2.2.22}$$

or, with X_i and X_j replaced by their negations and using Eq. (2.2.9),

$$\overline{X_i \wedge X_j} = \bar{X}_i \vee \bar{X}_j \ . \tag{2.2.23}$$

(see Tables 2.1.1 and 2.1.2 for a simple proof.)

Note: Since many of the above formulas are used so frequently throughout the following text, examples of their applications are omitted here. However, first applications can be found in the exercises at the end of this chapter.

2.2.3 Complete Systems of Operators

As will be shown in detail in §4.3, any Boolean function of n variables can be represented using only the operators $\wedge$, $\vee$ and $\bar{}$ (negation). Hence these three operators are said to form *a complete system.* There are numerous other complete systems using operators out of the set φ_1 to φ_{16} of Boolean operators defining all Boolean functions of two variables. Sometimes – as above – negation is used explicitly; however, in most cases negation is included in at least one of the operators for two-variable functions.

To show (prove) that a set of operators is complete, i.e. is a complete system in the above sense, one simply shows that AND, OR and NOT can be substituted.

Example 2.2.1 NAND ($\bar{\wedge}$) As a complete system

To find $\bar{X}$, use is made of the idempotence law (2.2.13), such that

$$\bar{X} = \overline{X \wedge X} =: X \bar{\wedge} X \ .$$

Further, because of $\bar{\bar{X}} = X$,

$$X_i \wedge X_j = \overline{\overline{X_i \wedge X_j}} = \overline{X_i \bar{\wedge} X_j} = (X_i \bar{\wedge} X_j) \bar{\wedge} (X_i \bar{\wedge} X_j) \ .$$

Finally, by de Morgan's law

$$X_i \vee X_j = \overline{\overline{X_i \vee X_j}} = \overline{\bar{X}_i \wedge \bar{X}_j} = (X_i \bar{\wedge} X_i) \bar{\wedge} (X_j \bar{\wedge} X_j) \ . \text{——}$$

Boolean functions of three or more variables can be synthesized using the operator(s) of a complete system in any combination in a bottom-up (or nested) fashion. For this procedure two notational simplifications are to be recommended:
(I) Omit the AND operator $\wedge$!
(II) Omit brackets around conjunction terms!
For example:

$$(X_1 \wedge X_2) \vee (X_2 \wedge \bar{X}_3 \wedge X_4) =: X_1 X_2 \vee X_2 \bar{X}_3 X_4 \ .$$

Note, that in this context

$$X_1 \wedge X_2 \wedge \ldots \wedge X_m =: \bigwedge_{i=1}^{m} X_i$$

and

$$X_1 \vee X_2 \vee \ldots \vee X_m =: \bigvee_{i=1}^{m} X_i \ .$$

2.3 Polynomials, Vectors, and Matrices with Boolean Elements

Since polynomials, vectors, and matrices belong to the most important multi-component algebraic structures used in Boolean analysis, some basic concepts concerning them and some hints as to their more important engineering applications are in order here. More advanced concepts and application examples will have to be deferred to later chapters. Note that this is not at all a mathematical state of the art report but rather a primer for contemporary applied work, mostly in coding theory and in graph theory.

For deep applications of polynomials with Boolean coefficients (mostly in algebraic coding theory) see [53]. Matrices (and vectors as one-dimensional matrices) are covered in [5].

2.3.1 Polynomials with Boolean Coefficients

Mostly in algebraic coding theory, as far as it is linked with feedback shift registers, polynomials with Boolean coefficients are a well established means for description and algebraic manipulations with coding and decoding. Such polynomials can be of the type

$$b(x) = b_0 + b_1 x + b_2 x^2 + \ldots + b_{n-1} x^{n-1} \; ; \qquad b_i \in \{0, 1\} \tag{2.3.1}$$

corresponding to a binary word (vector)

$$\mathbf{b} = (b_0, b_1, \ldots, b_{n-1}) \; . \tag{2.3.1a}$$

In the framework of the elementary theory of codes using so-called *generator* polynomials (also known as *cyclic* codes) the variable x is not specified. Only the typical power law

$$x^i x^j = x^{i+j}$$

is supposed to hold.

The addition/subtraction $(+)$ of polynomials of the above type is done by mod 2 with respect to the coefficients, which is expressed readily by the EXOR operation:

$$b^{(1)}(x)(+)b^{(2)}(x) = (b_0^{(1)} \not\equiv b_0^{(2)}) + (b_1^{(1)} \not\equiv b_1^{(2)})x + \ldots + (b_{n-1}^{(1)} \not\equiv b_{n-1}^{(2)})x^{n-1} \; . \tag{2.3.2}$$

The multiplication $(\cdot)$ of such polynomials is also done with the addition of products of coefficients mod 2, where product terms equal conjunction terms:

$$\begin{aligned} b^{(1)}(x)(\cdot)b^{(2)}(x) := {} & b_0^{(1)} b_0^{(2)} + (b_0^{(1)} b_1^{(2)} \not\equiv b_1^{(1)} b_0^{(2)})x \\ & + (b_0^{(1)} b_2^{(2)} \not\equiv b_1^{(1)} b_1^{(2)} \not\equiv b_2^{(1)} b_0^{(2)})x^2 + \ldots + b_{n-1}^{(1)} b_{n-1}^{(2)} x^{2n-2} \; . \end{aligned} \tag{2.3.3}$$

The division $(:)$ of such polynomials is the inverse of the above multiplication.

Example 2.3.1 Polynomial division mod 2

$$
\begin{array}{l}
(+)\,\underline{\left|\begin{array}{l}(x^5+x^3+x+1)(:)(x+1) = x^4+x^3+1\\ x^5+x^4\end{array}\right.}\\
\qquad (+)\,\underline{\left|\begin{array}{l}x^4+x^3+x+1\\ x^4+x^3\end{array}\right.}\\
\qquad\qquad (+)\,\underline{\left|\begin{array}{l}x+1\\ x+1\end{array}\right.}\\
\qquad\qquad\qquad 0
\end{array}
$$

Clearly, not all divisions yield the rest 0 as above. — —

In coding theory the addition $(+)$ of polynomials is used to model errors as the $(+)$-addition of an error polynomial $e(x)$ to a codeword $c(x)$. The received word corresponds to the polynomial

$$c'(x) := c(x)(+)e(x) \;. \tag{2.3.4}$$

In the field of error detection codeword polynomials $c(x)$ are used, which are divisible by the generator polynomial $g(x)$:

$$c(x) = i(x)(\cdot)g(x), \quad i(x)\text{: information polynomial} \;, \tag{2.3.5}$$

where the information polynomial contains the net information to be transferred from the sender to the receiver. Hopefully, for many $e(x)$ the received polynomial $c'(x)$ is not divisible by $g(x)$ so that the error can be detected.

For more information on algebraic coding theory, especially on cyclic and convolutional codes, see [25], [53].

2.3.2 Boolean Vector/Matrix Calculus

Boolean operations are readily extended to Boolean vectors. For example, it is quite natural to define for Boolean n-vectors (row vectors) **A** and **B**:

$$\mathbf{A} \bullet \mathbf{B} = (A_1 \bullet B_1, \ldots, A_n \bullet B_n) \;, \qquad A_i, B_i \in \{0, 1\} \;, \tag{2.3.6}$$

where $\bullet$ is a Boolean operation such as $\wedge$, $\vee$, $\not\equiv$. As usual, $\wedge$ will be omitted. The following is simple, but not trivial:

Lemma 2.3.1:

$$\mathbf{A} \not\equiv \mathbf{B} = \mathbf{A}\bar{\mathbf{B}} \vee \bar{\mathbf{A}}\mathbf{B} \;. \tag{2.3.7}$$

Proof. By (2.3.6) for component i of Eq. (2.3.7)

$$A_i \not\equiv B_i = A_i\bar{B}_i \vee \bar{A}_iB_i \;. \tag{2.3.8}$$

which is obviously true, q.e.d.

Note that $\mathbf{A} \not\equiv \mathbf{B}$ corresponds to the $(+)$-addition of binary polynomials discussed in §2.3.1.

The scalar (inner) product of Boolean n-vectors $\mathbf{A}$ and $\mathbf{B}$ is defined to be

$$\mathbf{AB}^{\mathrm{T}} = \bigvee_{i=1}^{n} A_i B_i \; ; \quad \mathbf{A} := (A_1, \ldots, A_n) \; , \quad \mathbf{B} := (B_1, \ldots, B_n) \; . \tag{2.3.9}$$

If A_i and B_i are terms, i.e. conjunctions of literals, then $\mathbf{AB}^{\mathrm{T}}$ can be a (scalar) Boolean function.

Example 2.3.2 2-out-of-3 function as a scalar vector product

Defining $\mathbf{A} = (X_1, X_2, X_3)$, $\mathbf{B} = (X_2, X_3, X_1)$, obviously $\mathbf{AB}^{\mathrm{T}} = X_1X_2 \vee X_2X_3 \vee X_3X_1$. — —

Obviously Eq. (2.3.6) can be readily extended to (m, n)-matrices $\mathbf{A}$ and $\mathbf{B}$, and Eq. (2.3.9) can be extended for the matrix product:

$$\mathbf{C} = \mathbf{AB} \; ; \quad \mathbf{A} = (A_{i,j})_{m,n} \; , \quad \mathbf{B} = (B_{i,j})_{n,l} \; , \tag{2.3.10}$$

where

$$C_{i,k} = \bigvee_{j=1}^{n} A_{i,j} B_{j,k} \; ; \quad A_{i,j}, B_{i,j} \in \{0, 1\} \; . \tag{2.3.10a}$$

[Note that by Eq. (2.3.6) $\mathbf{C} = \mathbf{A} \wedge \mathbf{B}$ would have the meaning $C_{i,j} = A_{i,j}B_{i,j}$].

In linear coding theory also

$$\mathbf{C} = \mathbf{A}(\wedge)\mathbf{B} \; ; \quad C_{i,k} = \mathop{\not\equiv}_{j=1}^{n} A_{i,j} B_{j,k} \tag{2.3.11}$$

is used. Note that $C_{i,k}$ can be interpreted as the *parity* of the n bits $A_{i,j}B_{j,k}$. For more information consult [5].

Exercises

2.1

Show algebraically that $\bar{\bar{X}} = X$.

2.2

Which of the functions of Table 2.1.2 are the functions

$$\varphi = X_1, \bar{X}_1, X_2, \bar{X}_2?$$

2.3

Prove with the aid of Table 2.1.2 the commutative laws

(a) $X_1 \wedge X_2 = X_2 \wedge X_1$, and $X_1 \vee X_2 = X_2 \vee X_1$.

(b) Are there more commutative "true" functions of X_1 and X_2?

2.4

Prove by an argument from the tabular representation (truth table) of Boolean functions that there are 2^{2^n} functions of n variables!

2.5

Show that $\overline{X_1 \not\equiv X_2} = X_1 \not\equiv \bar{X}_2$.

Comment. In the literature, usually the left hand side is written as $X_1 \equiv X_2$ (*equivalence* of X_1 and X_2).

2.6

Prove the correctness of the expansion

$$(X_i \vee X_j)(X_k \vee X_l) = X_i X_k \vee X_i X_l \vee X_j X_k \vee X_j X_l \ . \tag{E2.1}$$

2.7

Prove that (also for $n > 2$)

$$\overline{\bigvee_{i=1}^{n} X_i} = \bigwedge_{i=1}^{n} \bar{X}_i \ . \tag{E2.2}$$

2.8

Prove that NOR is a complete system (of Boolean operators)!

2.9

How many Boolean functions of n variables actually depend on all n variables? Give this number for $n = 1, 2, 3$.

2.10

Calculate

(a) **AB** according to Eq. (2.3.10, 10a),

(b) **A**$(\wedge)$**B** according to Eq. (2.3.11) of

$$\mathbf{A} = \begin{bmatrix} 0 & 1 & 1 \\ 1 & 0 & 1 \\ 0 & 1 & 0 \end{bmatrix}, \quad \mathbf{B} = \begin{bmatrix} 0 & 0 & 1 & 1 \\ 1 & 1 & 1 & 0 \\ 0 & 1 & 0 & 1 \end{bmatrix} .$$

3 Diagrams for Boolean Analysis

In science diagrams are of invaluable help to concentrate the information. Hence one need not be surprised at the enormous amount of work that is being invested into graphical computer science research and development activities all over the world.

There are several typical diagrams used only in Boolean analysis; others are adopted from elsewhere. Some types of diagrams are well known only in special applications. Therefore, the reader is warned not to apply them without proper consideration. Yet any interdisciplinary side effects should be welcome to the scientific community. Of course, only preliminary impressions of the potential use of these graphical aids can be sketched here. Deeper insights must be postponed to later chapters.

Comment. The reader is encouraged to read this book with paper and pencil at hand, and to enhance his/her individual learning process by many additional diagrams. Note that many people have such a strong visual memory that they remember formulae rather as little pictures than by their mathematical contents.

3.1 Standard Graphical Representation of Boolean Functions

Boolean functions of one variable X can be represented by Fig. 3.1.1, with the large dots indicating the possible values of $\varphi(X)$. For two variables X_1 and X_2, the

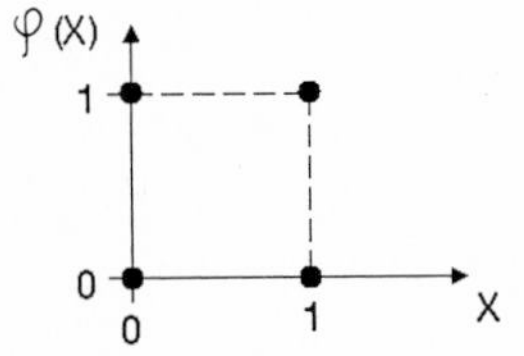

Fig.3.1.1

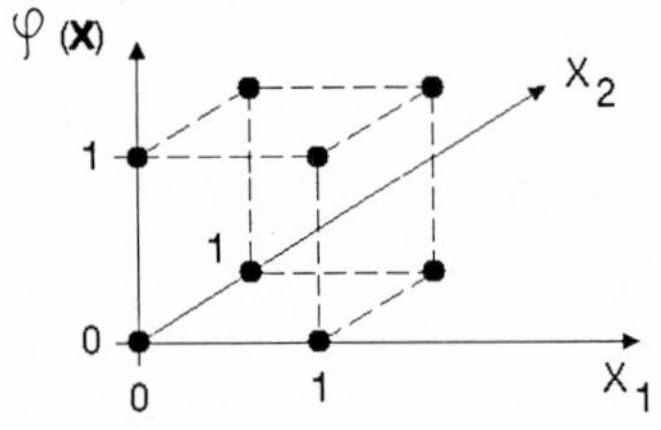

Fig. 3.1.2

Fig. 3.1.1. Standard graph of a Boolean function of 1 variable. (For each value of X only one value of φ is allowed with any specific φ.)

Fig. 3.1.2. Standard graph of a Boolean function of 2 variables. (For each value of $\mathbf{X}$ only one value of φ can be specified.)

relevant graph of φ is contained in Fig. 3.1.2. For higher dimensions a complete standard graphical representation is impossible, for obvious reasons. However, if only the values of **X** are shown together with the values of $\varphi(\mathbf{X})$, it is easy to represent functions of up to the fourth order; see Fig. 3.1.3.

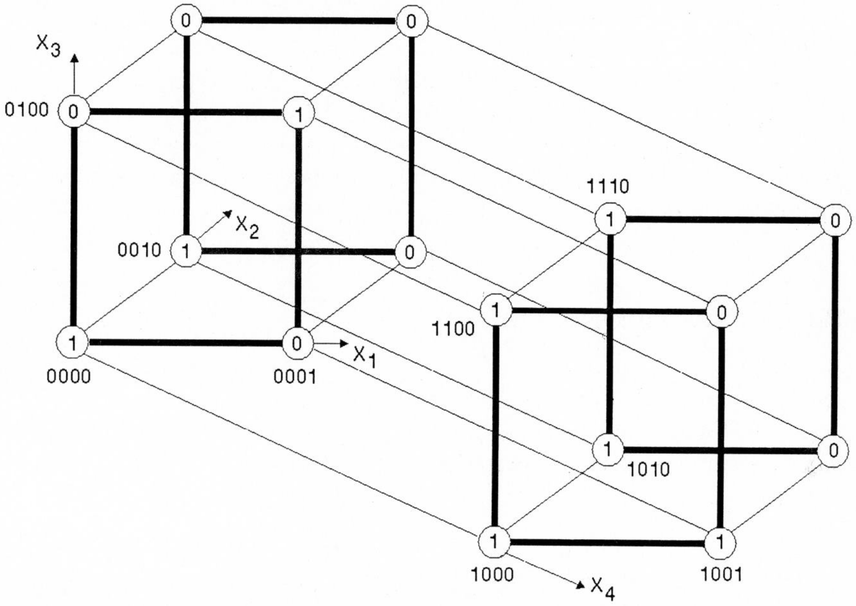

Fig. 3.1.3. Graphical representation of a Boolean function φ of 4 variables. Values of φ are given in the coordinate-quadruple points. Only the fat unit bars are drawn to scale in this hypercube of dimension 4

Note that argument 4-tuples $(X_1, \ldots, X_4)$ are represented by $X_4 \ldots X_1$, which can be used as the binary representation of the vertex number. Figure 3.1.3 shows the way to a simple proof of the formula for the number $N(n)$ of Boolean functions of n vabiables. Since every binary marking of the 2^n nodes of an n-variable φ defines φ uniquely, by a simple argument of induction

$$N(n) = 2^{2^n} . \tag{3.1.1}$$

Definition 3.1.1. The hypercube vertices with $\varphi = i$ are called the i-set; $i = 0, 1$ of φ. — —

Definition 3.1.2. Given two Boolean functions φ_1 and φ_2 we say that φ_1 *implies* φ_2 iff $\varphi_1 = 1 \rightarrow \varphi_2 = 1$. — —

Obviously, φ_1 implies φ_2 if the 1-set of φ_1 is contained in the 1-set of φ_2.

Lemma 3.1.1. The 1-set of $\hat{T} := \tilde{X}_{i_1} \tilde{X}_{i_2} \ldots \tilde{X}_{i_m}$ in binary n-space is the set of vertices of an $(n-m)$-dimensional (unit-)hypercube.

Proof. $\hat{T} = 1$ iff $\tilde{X}_{i_1} = \tilde{X}_{i_2} = \ldots = \tilde{X}_{i_m} = 1$, i.e. $X_{i_m+1}, \ldots, X_{i_n}$ can be 0 or 1. But the vertices with all possible combinations of $X_{i_m+1}, \ldots, X_{i_n} \in \{0, 1\}$ form a B^{n-m}, which is synonymous to an $(n-m)$-dimensional unit hypercube, q.e.d..

Obviously, one of the biggest 1 hypercubes in Fig. 3.1.3 is the square with the vertices **1000, 1010, 1100, 1110**. This 2-dimensional hypercube corresponds to the term $\hat{T} = X_4 \bar{X}_1$. If the right-hand cube of Fig. 3.1.3 were a 1 set it would correspond to $\hat{T} = X_4$.

Definition 3.1.3. The number of bits in which two binary n-vectors (code words) differ is their *Hamming distance.* — —

If a hypercube is taken for a graph, the length of every shortest path between any two nodes is the Hamming distance of the coordinate n-tuples of these nodes. Furthermore, if on a unit-hypercube (of the type of Fig. 3.1.3) movements are only possible along edges then the graph theoretical distance is the actual geometrical (minimal) path length.

3.2 Binary Decision Diagrams

The (truth) table of a Boolean function lends itself in a natural way to a binary tree representation. Figure 3.2.1 shows such a (rooted) tree, called a decision diagram in [37], for three variables, which corresponds to the function table of Table 3.2.1. No

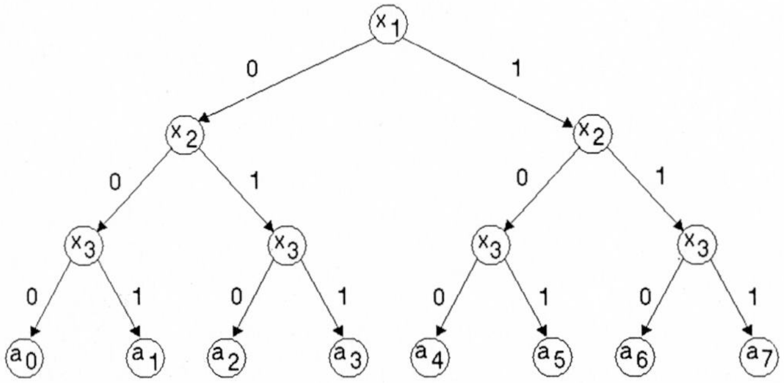

Fig. 3.2.1. Binary tree for any Boolean function of three variables

Table 3.2.1. Function (truth) table of any Boolean function of three variables

X_1	X_2	X_3	φ
0	0	0	a_0
0	0	1	a_1
0	1	0	a_2
0	1	1	a_3
1	0	0	a_4
1	0	1	a_5
1	1	0	a_6
1	1	1	a_7

doubt, only two leaves for the function values 0 and 1 would do. The resulting directed graph would no longer look like a tree, yet it would remain an acyclic digraph.

The marking of the edges means values given to the variables of the nodes they start from. Such binary decision trees, though not necessarily complete ones, can also represent Boolean functions given by algebraic expressions.

Example 3.2.1 Decision tree of a simple function

Let

$$\varphi(X_1, X_2, X_3, X_4) = X_1 X_2 \vee X_1 X_3 \vee X_2 X_3 \vee X_4 \; . \tag{3.2.1}$$

Then Fig. 3.2.2 shows an appropriate decision tree. The leaves are marked with the values of φ found when given the values of the edges to the variables of the nodes where these edges come from. — —

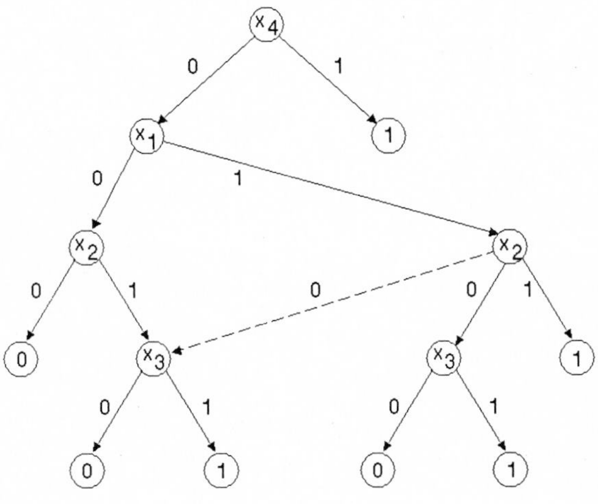

Fig. 3.2.2. Decision tree for Eq. (3.2.1)

Binary decision diagrams need not be trees. For instance, the dotted edge in Fig. 3.2.2 could replace the left hand sub-tree of the node this edge comes from. For further details on decision trees see §4.4.4.

3.3 Karnaugh Maps

The *Karnaugh map* is a special function table which is practical for functions of up to five variables. Its extreme usefulness in the function's minimization will be discussed in §§4 and 5. Here only the systematic way of constructing the $(m+1)$-variable map from the m-variable map is shown.

The Karnaugh (K-)map is for $m = 2k$ like a chess board with every single square corresponding to an argument m-tuple

$$\mathbf{X} = (X_1, \ldots, X_m)$$

of (the Boolean function) $\varphi(\mathbf{X})$, with the value (0 or 1) of φ noted inside the single square. For $m = 2k + 1$ the K-map is doubled by reflection, and it is understood that for the old chess board the new $(m + 1)$-st variable is everywhere 0, and for the new one it is everywhere 1. Figure 3.3.1 shows the details. The values of the variables are noted outside the K-map.

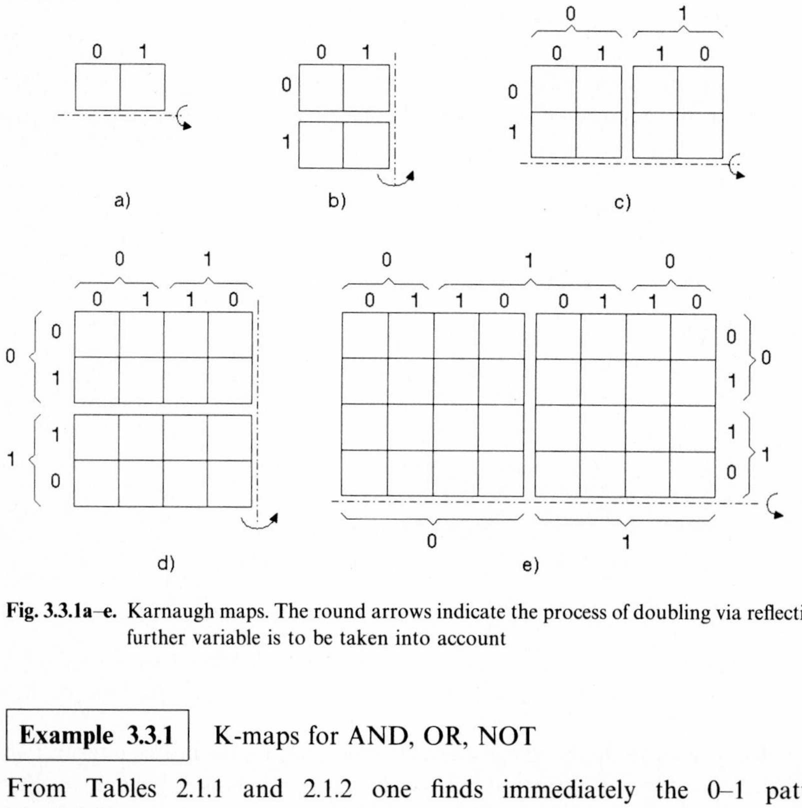

Fig. 3.3.1a–e. Karnaugh maps. The round arrows indicate the process of doubling via reflection when a further variable is to be taken into account

Example 3.3.1 K-maps for AND, OR, NOT

From Tables 2.1.1 and 2.1.2 one finds immediately the 0–1 patterns of Fig. 3.3.2. — —

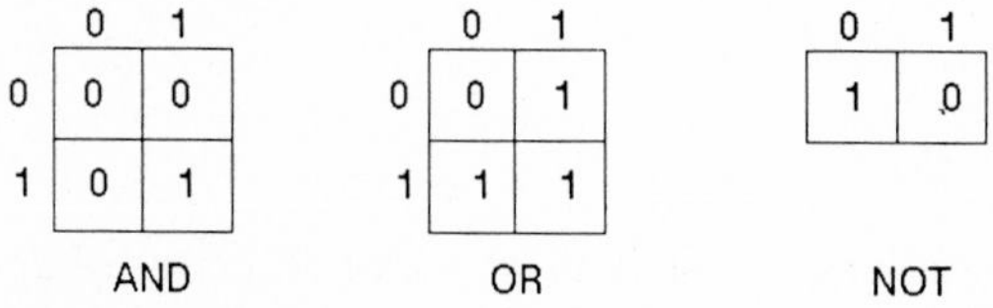

Fig. 3.3.2. K-maps of conjunction, disjunction, and negation

Definition 3.3.1. The small squares with $\varphi = i$ written into them yields the K-diagram (-map) of φ. — —

To avoid ambiguity as to the sequencing of the variables, one often replaces the $X_i = 1$ values by X_i. In other words: X_i denotes the little squares where $X_i = 1$; in the other squares $X_i = 0$. For instance in the case of four variables one gets, instead of Fig. 3.3.1(d), Fig. 3.3.3(a) or (b) or others.

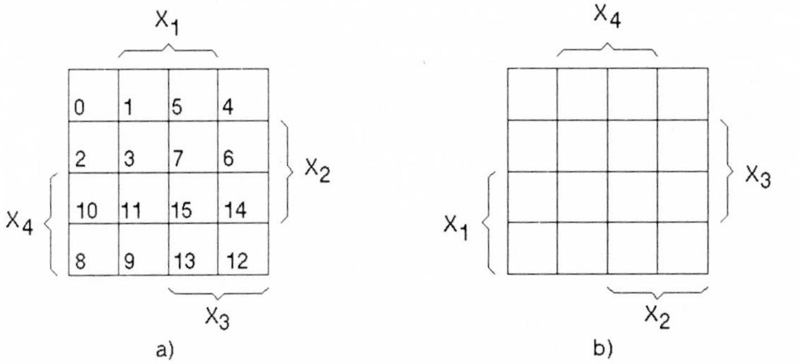

Fig. 3.3.3a, b. Alternative sequencing of variables in a K-map. In (a) the binary form of the number of each small square is $X_4 X_3 X_2 X_1$

Sometimes it is useful to give the small elementary squares of the K-maps names by numbers. The standard numbering is indicated in Fig. 3.3.3(a). The coordinate n-tuple is read as a binary number which is written down in decimal notation. As is easily verified, in every doubling of the map, there corresponds the addition of a further power of 2. For instance in Fig. 3.3.3(a) the numbers in the squares of the lower half are those of the properly reflected upper half increased by $2^3 = 8$.

There is a one-to-one correspondence between the small squares of a K-map and the vertices of a binary hypercube as introduced in Sect. 3.1. Figure 3.3.4 shows the details with the edges and vertices of the four-dimensional hypercube drawn as thick lines and thick points, respectively. The numbers of the little K-map squares

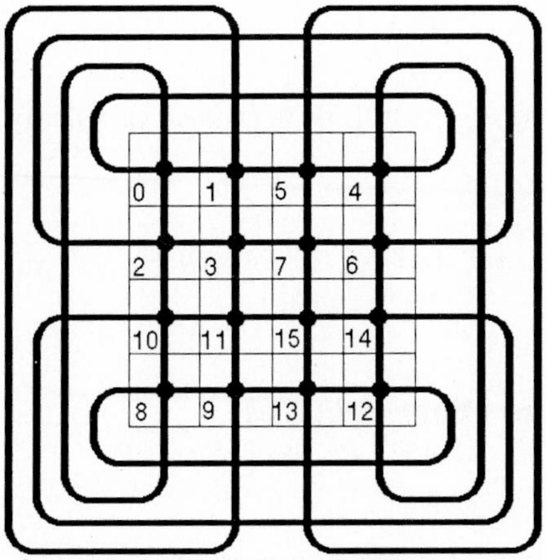

Fig. 3.3.4. Plane equivalent of Fig. 3.1.3 (for any 4-variable φ)

are the numbers of the hypercube vertices. Hence the following alternative for definition 3.1.1 is obvious; see also definition 3.3.1.

Definition 3.3.2. The 1-set of a Boolean function φ corresponds to the set of little squares of the K-diagram of φ which contain the digit 1. We call this set the 1-set of the K-map of φ. — —

Figure 3.3.4 shows very clearly which of the little K-map squares are cyclically neighbours of each other.

K-maps are extremely useful to visualize conjunction terms and disjunctions of such terms, so-called disjunctive normal forms (DNFs), which will be discussed further in §4.

Example 3.3.2 K-map of a four-variables function

Let

$$\begin{aligned}\varphi(X_1, X_2, X_3, X_4) &= X_1 \vee X_2(X_3 \vee \bar{X}_1 X_4) \\ &= X_1 \vee X_2 X_3 \vee \bar{X}_1 X_2 X_4 \,. \end{aligned} \tag{3.3.1}$$

Figure 3.3.5 shows the typical rectangular domains corresponding to the three terms of the DNF, Eq. (3.3.1), of φ. (The domain of $\bar{X}_1 X_2 X_4$ is cyclically rectangular.) — —

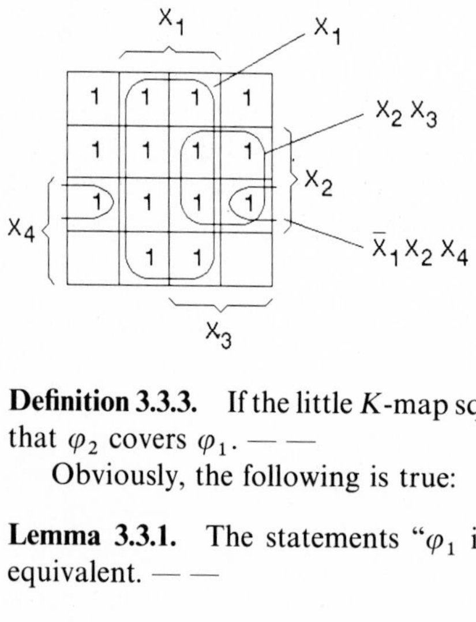

Fig. 3.3.5. K-map of φ of (3.3.1)

Definition 3.3.3. If the little K-map squares of φ_2 cover those of φ_1 then we will say that φ_2 covers φ_1. — —

Obviously, the following is true:

Lemma 3.3.1. The statements "φ_1 implies φ_2" and "φ_1 is covered by φ_2" are equivalent. — —

3.4 Switching Network Graphs (Logic Diagrams) and Syntax Diagrams

Boolean functions can be nicely depicted using the symbols shown in Fig. 1.1. This way one immediately understands the plans which electronics engineers use to document their switching networks. The circuits for AND, OR and NOT are called *gates* and the lines between gates represent wiring, i.e. either thin layers of tinned copper on insulating material or copper wires. (With VLSI, things are a bit more complex and will not be discussed here.)

Example 3.4.1 2-out-of-3-majority voter

The direct translation of

$$X_\varphi = \varphi(X_1, X_2, X_3) = X_1X_2 \vee X_1X_3 \vee X_2X_3 \tag{3.4.1}$$

into this type of directed graph, often called a logic diagram, yields Fig. 3.4.1. Switching networks are directed graphs even though, usually, no arrows indicate the directions of the edges. The gates are nodes with several inputs (except for the NOT-gate) and one output. The small thick dots are nodes with one input edge and at least two output edges.

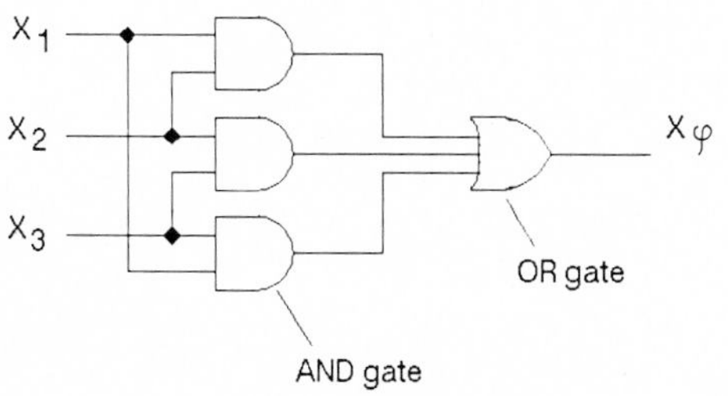

Fig. 3.4.1

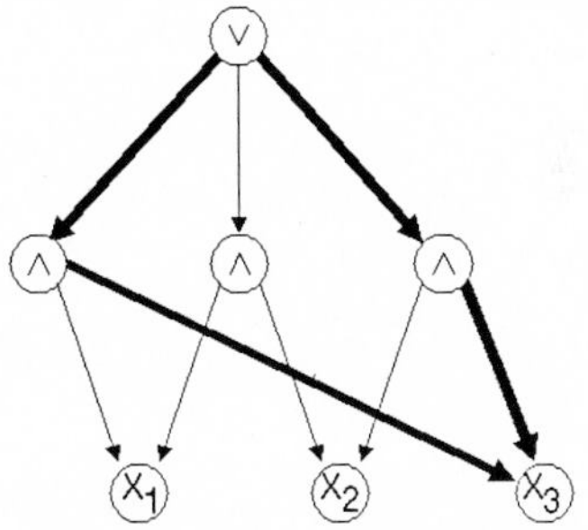

Fig. 3.4.2

Fig. 3.4.1. Electronics switching circuits picture of Eq. (3.4.1). The symbols are explained in Fig. 1.2. $X_\varphi = 1$, if at least $X_i = 1$, $X_j = 1$; $i, j \in \{1, 2, 3\}$

Fig. 3.4.2. Syntax diagram corresponding to 2-out-of-3 majority function of Fig. 3.4.1

Comment. Switching networks as modelling tools can sometimes be replaced by Petri nets; see §3.10.

There is a one-to-one correspondence between the above switching circuit graphs and syntax diagrams of Boolean functions. The latter are found from the former simply by replacing the gate symbols by node symbols which are marked by the appropriate operators. For example, Fig. 3.4.2 corresponds to Fig. 3.4.1. Note that syntax diagrams for (high-level) computer languages are completely different from the above type of digraph.

The extension of graphs of the types of Figs. 3.4.1 and 3.4.2 to cases of several functions of the same variables, i.e. to Boolean vector functions, is obvious. Both types of graphs remain acyclic, yet reconverging branches (paths) as in Figs. 3.4.1 and 3.4.2 are allowed. For instance, the two thick paths in Fig. 3.4.2 are *reconverging* since they have two non-adjacent nodes in common.

Note that in some cases, especially in coding technology, the EXOR gate is replaced by a little circle with a plus sign within. This yields easier-to-"read" pictures and emphasizes the arithmetic interpretation of EXOR as addition mod 2; see Fig. 3.4.3.

For bigger networks "black" boxes are often used, to hide too much detail and ease a higher-level overview. Figure 3.4.4(a) shows the black box of a half-adder whose details are shown in Fig. 3.4.4(b).

Fig. 3.4.3. EXOR symbols; right: usual EXOR gate symbol; left: special symbol as used in coding technology.

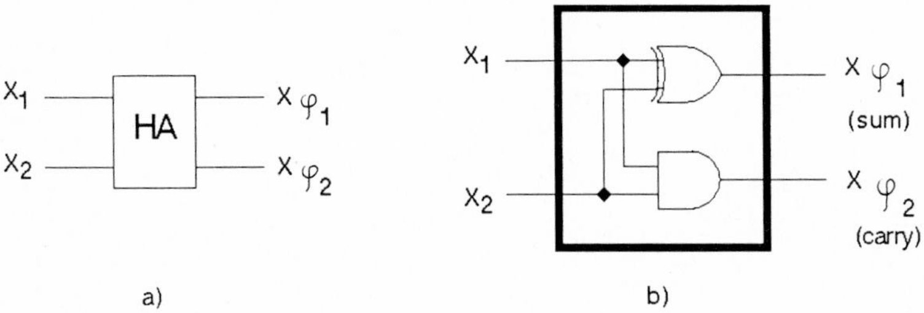

Fig. 3.4.4a, b. Block (black box) of a block diagram, here that of a half-adder

3.5 Venn Diagrams

Venn diagrams are used whenever non-trivial sets which are "composed" of more elementary sets are to be visualized. Figure 3.5.1 shows, as the probably best known example, a Venn diagram for the union and the intersection of two sets A_1 and A_2. The elements of these sets (in the diagram) are points or regions within the contours.

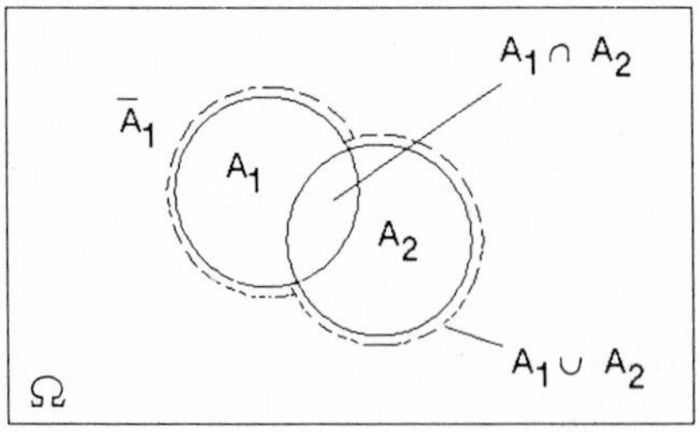

Fig. 3.5.1. Venn diagram with two "basic" sets A_1 and A_2. All sets of this figure are subsets of Ω

$\bar{A}_1$ is the set of elements (of Ω) "outside" of A_1. The extension to cases with more than two basic sets $A_1, A_2, \ldots$ is obvious.

Ω is the reference set. All sets that are regarded here are parts of Ω, and the complement (set) $\bar{A}_i$ of A_i is defined with respect to Ω, i.e. $A_i \cup \bar{A}_i = \Omega$. As to set "subtraction", it is interesting to note that

$$A_1 \backslash A_2 = A_1 \cap \bar{A}_2 \ .$$

In fact, on both sides of this equation all the elements belonging to A_1 but not to A_2 are involved.

3.6 State Transition Graphs

In the *state transition graphs* (diagrams) the states of a system are modelled to become the vertices (nodes) of a directed graph. The directed edges of this graph indicate the possible changes of the system state.

Figure 3.6.1 shows the typical graph of a 1-out-of-2: G system. (Notation according to Trans. IEEE-R.) It is a system that is good iff at least one of its two components is good.

The corresponding elementary states are depicted in Fig. 3.6.2. The first and the second binary digit corresponds to the state of the first and the second component of the system respectively. (State 2 in Fig. 3.6.1 is composed of elementary states 01

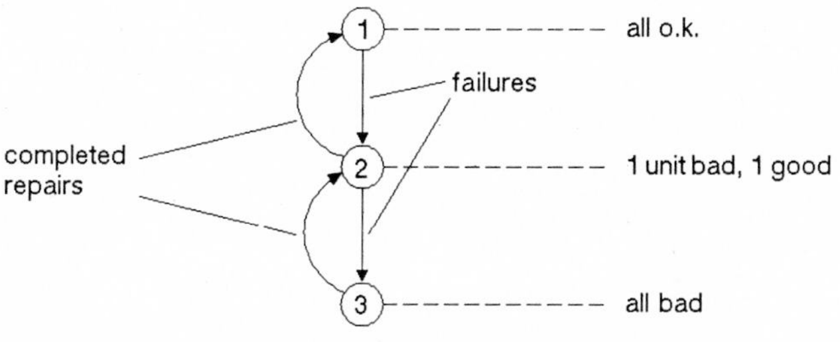

Fig. 3.6.1. State transition graph of a 1-out-of-2 system. Typically, vertices 1 and 3 have no direct connections with each other

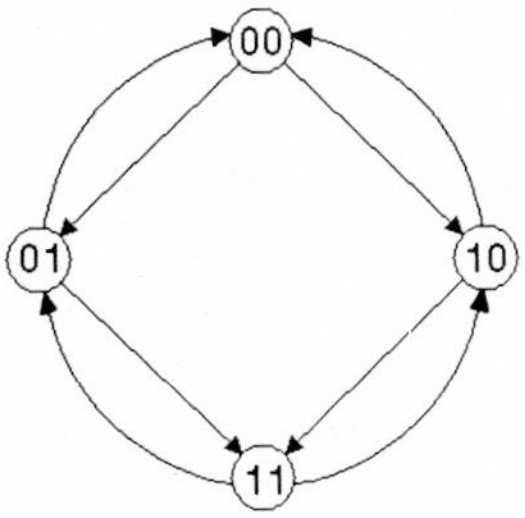

Fig. 3.6.2. Transition possibilities between (all) elementary states of a 2-components system corresponding to the transitions in Fig. 3.6.1

and 10.) Note that only in clocked systems (synchronous automata) transitions of a state to itself make sense; but these are not discussed in this text. Hence transition graphs of this text will be free of loops, i.e. edges beginning and ending in the same vertex. (This should be kept in mind when modelling Markov chains in §§8 and 9.)

3.7 Communications Graphs

In communications (in the widest sense) just simple non-directed or directed graphs are used to show the nodes of information processing as vertices and the communication links as edges. Figure 3.7.1 shows a typical example – (a) for the undirectional case, (b) for a special directional case.

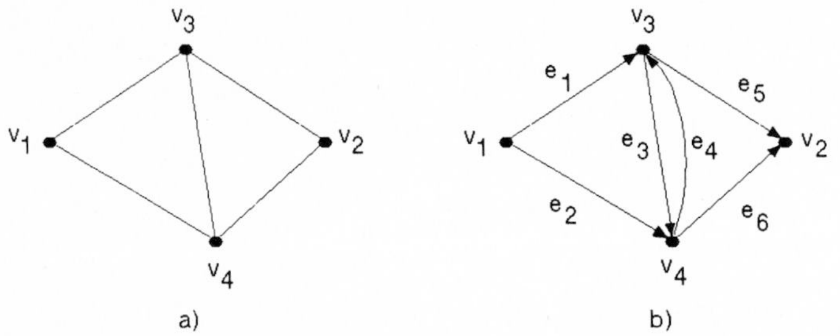

Fig. 3.7.1a, b. Graph of a simple "bridge" network: (**a**) bidirectional flow, (**b**) with special flow directions for each edge

Comment. It can be advantageous to replace the thick dots for the vertices by circles into which the numbers of the vertices can be written, as in Fig. 3.6.1. Note that for communications graphs loops make little sense. However, cycles (rings) of length $l > 1$ are implemented frequently.

Sometimes it is advantageous to model a non-directed graph by a symmetrical digraph, i.e. by a digraph, where with each edge (v_i, v_j) there also exists edge (v_j, v_i). Examples will follow.

3.8 Reliability Block Diagrams

In reliability theory a special kind of pseudo graph, *the reliability block diagram*, is used for good reasons. The "pseudo" is due to the fact that different links (edges) may have the same name. The system is "good" so long as at least one path from left to right through the graph touches only good components. To illustrate this, Fig. 3.8.1 shows two graphs of the well-known 2-out-of-3 system. The corresponding so-called *fault trees* (not trees in the strict sense of graph theory) are shown in Fig. 3.8.2. In an exercise for §3 it will be shown that both pictures of Fig. 3.8.2 model the same Boolean function.

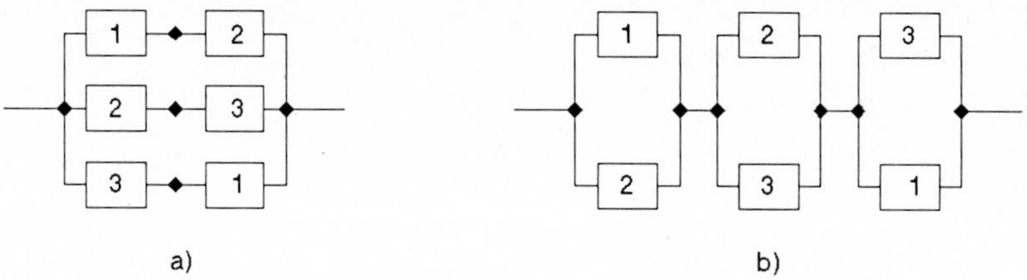

Fig. 3.8.1a, b. Two possible reliability block diagrams of a 2-out-of-3 system. The "blocks" contain names for the edges of the graph

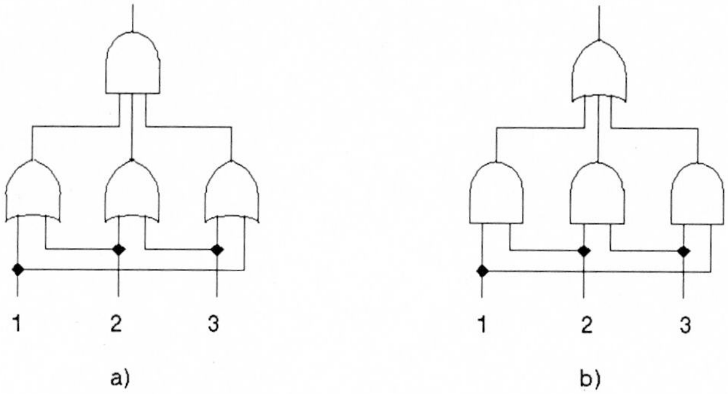

Fig. 3.8.2a, b. Fault trees of the 2-out-of-3 system

A point of some recent interest is the general relation between a fault tree and a corresponding reliability block diagram. Basically, since the block diagram is "success" oriented, a "serial" structure, as in Fig. 3.8.3(a) corresponds to an OR-gate in the fault tree, and a "parallel" structure, as in Fig. 3.8.3(b), corresponds to an AND gate. Furthermore, from the above, if in a fault tree OR- and AND-gates follow each other on any path from any input to the output, every further layer of gates (i.e. in graph theoretical terminology: every increase of the height of the tree by one) corresponds to a further step in nesting depth; see Fig. 3.8.4. The corresponding Boolean function, which is sometimes called the (redundancy) structure function, is

$$X_\varphi = X_3 \vee (X_1 \vee X_2)(X_4 \vee X_5 X_8 \vee X_6) X_9 \vee X_7 \ . \tag{3.8.1}$$

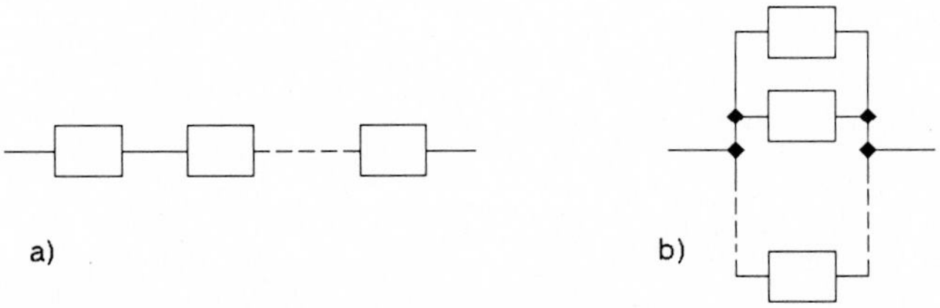

Fig. 3.8.3. Series structure (**a**) and parallel structure (**b**)

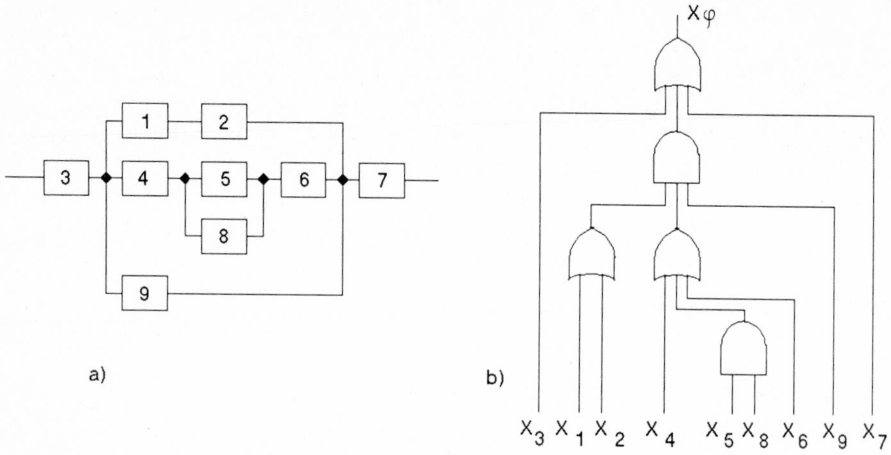

Fig. 3.8.4. (**a**) Reliability block diagram, (**b**) corresponding fault tree

3.9 Flowcharts

Flowcharts, i.e. control flow graphs, are of great help in visualizing the structure of algorithms at different levels. They are directed graphs consisting of one type of edge and two types of vertices.

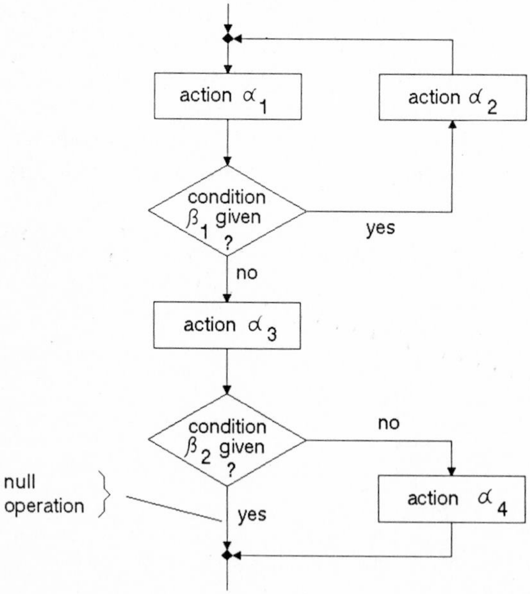

Fig. 3.9.1. Typical simple flowchart

On the edges the actions, specifically data manipulations (including transfers), are noted in rectangular "boxes". The null operation is not mentioned. (Special boxes for inputs and outputs are not discussed here.)

The vertices consist of "decision" nodes (for binary decisions) with one input and two outputs, and of "merging" nodes with several inputs and one output; Fig. 3.9.1. (Special nodes for start and stop etc. are not discussed here.) Flowcharts will be used extensively in §10.

3.10 Petri Nets

Petri nets are special bipartite digraphs, i.e. directed graphs with two types of nodes called *places* and *transitions*, respectively, where such places are never adjacent to, i.e. direct neighbours of, transitions and vice versa. Places can "contain" *tokens*. Once all the predecessor places of a given transition contain at least one token this transition "fires", sending one token to each following place; see Fig. 3.10.1. In order to model delays one can assume that tokens are delayed in the transitions.

Petri nets have lots of applications, especially wherever sequential switching operations are involved. Since Boolean functions are primarily concerned with instantaneous switching operations, the usefulness of Petri nets in this field is not quite obvious. However, whenever the electronic details of implementing certain representations of Boolean functions are discussed, e.g. the hazards of §4.7, then

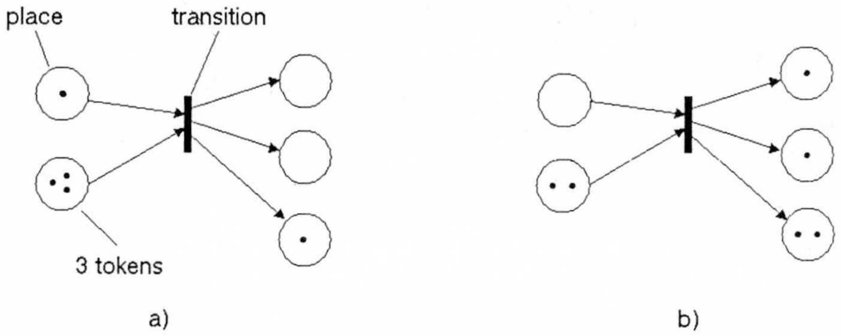

Fig. 3.10.1. Firing of a transition; (**a**) before, (**b**) after firing. (The number of tokens can vary with time.)

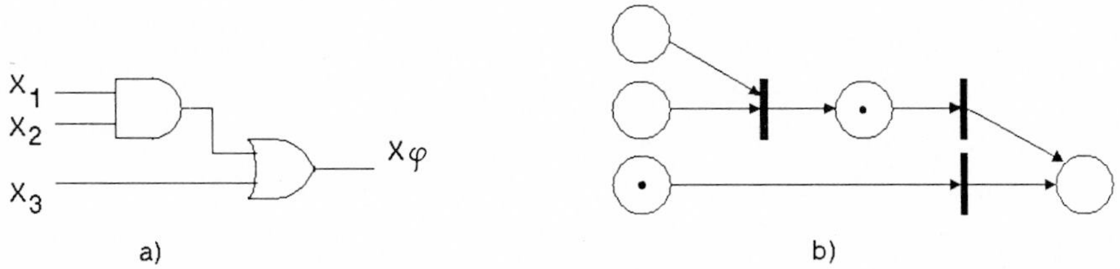

Fig. 3.10.2. Petri net of an AND gate whose output is one input of an OR gate; (**a**) logic diagram, (**b**) Petri net

Petri nets are in place. Figure 3.10.2 shows the modelling of information transfer through a very simple switching network, where the tokens represent the advent of the voltage corresponding to logical (Boolean) 1. Figure 3.10.2b shows the situation, where X_3 became 1 after X_1 and X_2 did. But X_φ is not yet 1, due to switching delays.

Exercises

3.1

Prove that in a K-map of four variables neighbouring elementary squares (those with one common side) differ in only one variable, being normal for one of them and negated for the other.

3.2

Usually, in switching circuits, feedback produces useless circuits. Explain that and why the circuit of Fig. E3.1 is an exception! (It is the famous SR-flip-flop for storing 1 bit of information.)

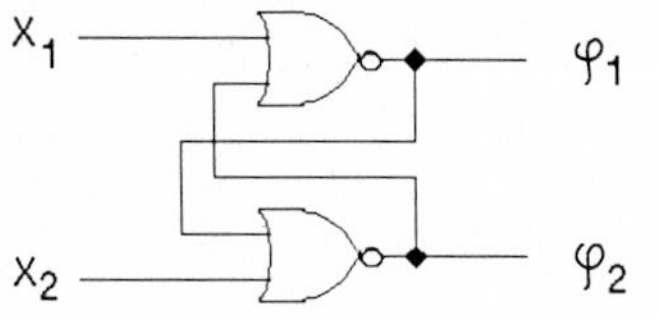

Fig. E3.1. *SR*-flip-flop; *S* for set, *R* for reset

Hint. Assume a change of X_1 and/or X_2 at t and tell what can be observed at the outputs φ_1 and φ_2 at $t+\Delta t$! The case $X_1(t)=X_2(t)=1$ is supposed to be impossible; it can lead to instability, viz. non deterministic behaviour, when being changed.

3.3

Explain that for two sets A_1 and A_2

$$A_1 \cup A_2 = A_1 \cup (\bar{A}_1 \cap A_2) .$$

3.4

Draw a state transition graph for the elementary states of a three-component system in which at most one component can change its (binary) state at a time!

3.5

Show by algebraic manipulations that the two switching function representations of Fig. 3.8.2 belong to the same Boolean function !

3.6

Draw two decision diagrams for the function $\varphi = X_1 \not\equiv X_2 \not\equiv X_3$, viz.

(a) a binary tree,
(b) a minimal graph as to numbers of vertices and edges.

4 Representations (Forms) and Types of Boolean Functions

One of the typical features of Boolean functions is their non-uniqueness of representation, which is obvious from §2.2. This has strongly influenced the search for canonical representations, which are unique. Hence two different representations can be shown to be the same function by transforming them to the same canonical form. (The function table would be a trivial ultima ratio to this end.)

Let it be stressed right here: There is no need to represent Boolean functions by using Boolean operators (§7 will give details) but in most situations this is very practical because of the great number of nice simplification rules of standard Boolean algebra; see §2.2.2.

The biggest sub-section, namely §4.4, is devoted to three algorithms for the production of fairly short polynomial-type forms which are extremely useful in the stochastic theory of Boolean functions (§§8, 9). Probably, this material appears here for the first time in a text book. The main applications area for this material is reliability theory, i.e. fault-tree analysis.

Further sub-sections of this §4 are devoted to various types of Boolean functions and special properties of them, respectively. Numerous small applications in digital electronics are included.

4.1 Negation (Complement)

Before turning to details of representing Boolean functions, it seems appropriate to elaborate quickly on a very useful general result for negating a given φ. It is an extension of the de Morgan laws (Eqs. (2.2.22, 2.2.23)).

Theorem 4.1.1 (Shannon's inversion rule). In order to negate (invert) a Boolean function given in the AND/OR/NOT system of operators, just interchange all AND and OR operators and negate all literals. In an obvious symbolism:

$$\bar{\varphi}(\mathbf{X};\wedge,\vee,\bar{\ })=\varphi(\bar{\mathbf{X}},\vee,\wedge,\bar{\ })^{\dagger}\ .\text{— —}\tag{4.1.1}$$

Comment. The case $\varphi=\bar{\varphi}_0$, where one would, usually, not apply Eq. (4.1.1), but, rather, Eq. (2.2.9) to get $\bar{\varphi}=\varphi_0$, need not be excluded. Since the order of two successive negations can be inverted, one can first apply Eq. (4.1.1) to φ_0, yielding

† The bar means negations except that of literals; see definition 2.2.1.

$\bar{\varphi}_0$ and then negate this result, since the function to be processed by Eq. (4.1.1) was not φ_0 but $\bar{\varphi}_0$. By Eq. (2.2.9) the final result would again be $\bar{\varphi}=\varphi_0$ with a negation bar over the r.h.s. For instance, by Eq. (4.1.1)

$$\varphi=\overline{X_1 \vee X_2} \tag{4.1.2}$$

is transformed to

$$\bar{\varphi}=\overline{\bar{X}_1\bar{X}_2}\ . \tag{4.1.3}$$

This is correct, since, by de Morgan's law, Eq. (2.2.22), the r.h.s. of Eq. (4.1.3) equals $X_1 \vee X_2$. Consequently, in the following proof of Eq. (4.1.1) the case of $\varphi=\bar{\varphi}_0$, i.e. the case of a function, whose whole r.h.s. is the formal negation of another function will be disregarded.

Proof of Eq. (4.1.1): Excepting the case $\varphi=\bar{\varphi}_0$, either

$$\varphi=\varphi_1 \vee \varphi_2\ , \tag{4.1.4}$$

or

$$\varphi=\varphi_1\varphi_2\ , \tag{4.1.5}$$

where both φ_1 and φ_2 are shorter than φ. By de Morgan's laws either

$$\bar{\varphi}=\bar{\varphi}_1\bar{\varphi}_2\ ,$$

or

$$\bar{\varphi}=\bar{\varphi}_1 \vee \bar{\varphi}_2\ .$$

In any case, the AND and OR operators of Eqs. (4.1.4) and (4.1.5) have been interchanged. Using the above approach repeatedly yields, after the next step, (for $i=1, 2$) either

$$\bar{\varphi}_{\mathrm{i}}=\bar{\varphi}_{\mathrm{i},1}\bar{\varphi}_{\mathrm{i},2} \quad \text{or} \quad \bar{\varphi}_{\mathrm{i}}=\bar{\varphi}_{\mathrm{i},1} \vee \bar{\varphi}_{\mathrm{i},2}\ ,$$

and, after a finite number of steps, functions $\bar{\varphi}_{\mathrm{i,j},\ldots,\mathrm{m}}$, which are literals, that were negated in the last expansion step, q.e.d..

Example 4.1.1 Negating a deeply nested function

Let

$$\varphi(X_1,\ldots,X_4)=X_1[X_2 \vee \bar{X}_3 \vee X_4(X_1 \vee X_2 \vee \bar{X}_3)] \vee X_2(\bar{X}_3 \vee X_4)\ .$$

By theorem 4.1.1

$$\bar{\varphi}(X_1,\ldots,X_4)=\bar{X}_1 \vee [\bar{X}_2X_3(\bar{X}_4 \vee \bar{X}_1\bar{X}_2X_3)](\bar{X}_2 \vee X_3\bar{X}_4)\ . \tag{4.1.6}$$

Using only Eqs. (2.2.22, 2.2.23) would have resulted in the following awkward sequence of steps:

$$
\begin{aligned}
\bar{\varphi} &= \overline{X_1[X_2 \vee \bar{X}_3 \vee X_4(X_1 \vee X_2 \vee \bar{X}_3)] \vee X_2(\bar{X}_3 \vee X_4)} \\
&= \overline{X_1[X_2 \vee \bar{X}_3 \vee X_4(X_1 \vee X_2 \vee \bar{X}_3)]}\ \overline{X_2(\bar{X}_3 \vee X_4)} \\
&= \bar{X}_1 \vee \overline{[X_2 \vee \bar{X}_3 \vee X_4(X_1 \vee X_2 \vee \bar{X}_3)]}[\bar{X}_2 \vee \overline{(\bar{X}_3 \vee X_4)}] \\
&= \bar{X}_1 \vee \bar{X}_2\overline{[\bar{X}_3 \vee X_4(X_1 \vee X_2 \vee \bar{X}_3)]}(\bar{X}_2 \vee X_3\bar{X}_4) \\
&= \bar{X}_1 \vee \bar{X}_2 X_3\overline{[X_4(X_1 \vee X_2 \vee \bar{X}_3)]}(\bar{X}_2 \vee X_3\bar{X}_4) \\
&= \bar{X}_1 \vee \bar{X}_2 X_3[\bar{X}_4 \vee \overline{(X_1 \vee X_2 \vee \bar{X}_3)}](\bar{X}_2 \vee X_3\bar{X}_4) \\
&= \bar{X}_1 \vee \bar{X}_2 X_3[\bar{X}_4 \vee \bar{X}_1\overline{(X_2 \vee \bar{X}_3)}](\bar{X}_2 \vee X_3\bar{X}_4) \\
&= \bar{X}_1 \vee \bar{X}_2 X_3(\bar{X}_4 \vee \bar{X}_1\bar{X}_2 X_3)(\bar{X}_2 \vee X_3\bar{X}_4)
\end{aligned}
$$

as in Eq. (4.1.6). — —

Note that a very simple possibility of inverting φ is by using (see Table 2.1.2)

$$\bar{\varphi} = \varphi \not\equiv 1 \ . \tag{4.1.7}$$

In closing, the uniqueness of Boolean negation is demonstrated.

Lemma 4.1.1: If for Boolean φ and φ'

$$\varphi\varphi' = 0 \ , \qquad \varphi \vee \varphi' = 1 \ ,$$

then $\varphi' = \bar{\varphi}$.

Proof. By Eqs. (2.2.7), (2.2.8)

$$\varphi\bar{\varphi} = 0 \ , \qquad \varphi \vee \bar{\varphi} = 1 \ .$$

$\varphi \vee \varphi' = 1$ and $\varphi \vee \bar{\varphi} = 1$ AND-ed by $\bar{\varphi}$ and φ' respectively, yield

$$\bar{\varphi}\varphi \vee \bar{\varphi}\varphi' = \bar{\varphi} \ , \quad \text{and} \quad \varphi'\varphi \vee \varphi'\bar{\varphi} = \varphi' \ ,$$

and because of $\bar{\varphi}\varphi = \varphi'\varphi = 0$:

$$\varphi' = \bar{\varphi} \ ,$$

q.e.d.

4.2 Special Terms

The term "term" is generally used to describe a "part" of a function, the notation of which forms a certain unit; typically a product (term) in a sum in "usual" algebra.

Definition 4.2.1. Unless otherwise stated, in this text a *term* will be understood to be the conjunction of different literals, excluding pairs of the type X_i, $\bar{X}_i$.[†]

In Boolean analysis the most frequently used special terms are minterms, prime implicants (terms) and maxterms. (Maxterms are not terms in the sense of definition 4.2.1.)

Definition 4.2.2. A *minterm* $\hat{M}_i$ is a maximum length conjunction term of a Boolean function φ, i.e. it contains all variables normal (unchanged) or negated:

$$\hat{M}_i = \bigwedge_{k=1}^{n} \tilde{X}_k \; ; \quad \tilde{X}_k \in \{X_k, \bar{X}_k\} \; . \text{——} \tag{4.2.1}$$

Which $\tilde{X}_k = X_k$ and which $\tilde{X}_k = \bar{X}_k$, specifies $\hat{M}_i$ uniquely.

By Eq. (2.2.11): $\hat{M}_i = 1 \Rightarrow X_\varphi = 1$.

Example 4.2.1 The minterms with one, two and three variables

One variable: $X, \bar{X}$. (4.2.2)

Two variables: $X_1X_2, \bar{X}_1X_2, X_1\bar{X}_2, \bar{X}_1\bar{X}_2$. (4.2.2a)

Three variables: $X_1X_2X_3, \bar{X}_1X_2X_3, X_1\bar{X}_2X_3, \bar{X}_1\bar{X}_2X_3, X_1X_2\bar{X}_3,$

$\bar{X}_1X_2\bar{X}_3, X_1\bar{X}_2\bar{X}_3, \bar{X}_1\bar{X}_2\bar{X}_3$. —— (4.2.2b)

A minterm gives a minimal amount of information on φ since its value is 1 for precisely one binary argument n-tuple $\mathbf{X}$ of φ and 0 else. Also, a minterm corresponds to exactly one (little) 1-square of the Karnaugh map of φ (see § 3.3) as well as to one node of the 1-set of definition 3.1.1.

Definition 4.2.3. Let us call $\hat{M}(\varphi)$ the set of minterms of $\varphi := \varphi(X_1, \ldots, X_n)$. ——

Clearly, $\hat{M}(\hat{M}_i) = \{\hat{M}_i\}$, card $\{\hat{M}(\varphi)\} \leqq 2^n$.

To find the minterms of a term $\hat{T}_i$ of φ one can use successively the expansion (2.2.16) or one can "AND" $\hat{T}_i$ with the 1-function which is the disjunction of all minterms for those variables of φ, which do not show up in $\hat{T}_i$; see example 4.2.1. As to the first alternative, let $X_{i_1}, X_{i_2}, \ldots$ not be contained (not show up) in $\hat{T}_i$, then

$$\begin{aligned} \hat{T}_i &= \hat{T}_i(X_{i_1} \vee \bar{X}_{i_1}) = \hat{T}_iX_{i_1} \vee \hat{T}_i\bar{X}_{i_1} \, , \\ \hat{T}_iX_{i_1} &= \hat{T}_iX_{i_1}(X_{i_2} \vee \bar{X}_{i_2}) = \hat{T}_iX_{i_1}X_{i_2} \vee \hat{T}_iX_{i_1}\bar{X}_{i_2} \, , \\ \hat{T}_i\bar{X}_{i_1} &= \hat{T}_i\bar{X}_{i_1}(X_{i_2} \vee \bar{X}_{i_2}) = \hat{T}_i\bar{X}_{i_1}X_{i_2} \vee \hat{T}_i\bar{X}_{i_1}\bar{X}_{i_2} \end{aligned} \tag{4.2.3}$$

etc. This introduces the next theorem, where the phrase "φ has the minterm $\hat{M}_j$" means "$\hat{M}_j$ is a minterm of φ".

[†] This way a term is never identically 0. Also the term 1 is excluded.

Theorem 4.2.1. Prolonged terms have less minterms. More precisely: If $\tilde{X}_j$ is no literal of the term $\hat{T}_i$, then

$$\hat{M}(\hat{T}_i\tilde{X}_j) \subset \hat{M}(\hat{T}_i) , \quad \text{card } \{\hat{M}(\hat{T}_i\tilde{X}_j)\} = \tfrac{1}{2} \text{ card } \{\hat{M}(\hat{T}_i)\} . \tag{4.2.4}$$

Proof. If $\hat{T}_i$ contains m literals and minterms contain n, then $\hat{T}_i$ can be expanded in $n-m$ steps according to the above to yield 2^{n-m} minterms, whereas $\hat{T}_i\tilde{X}_j$ will yield only half as many since the starting term contains $m+1$ literals, thus allowing for only $n-m-1$ expansion steps, q.e.d..

Definition 4.2.4. Two Boolean functions φ_1 and φ_2 are *disjoint* if their conjunction is zero:

$$\varphi_1\varphi_2 = 0 \ . \text{ — —} \tag{4.2.5}$$

(In §4.4 it will be proved that this is equivalent to

$$\hat{M}(\varphi_1) \cap \hat{M}(\varphi_2) = \phi \ ,$$

which gives the term “disjoint” a proper meaning.)

Theorem 4.2.2. All the minterms of the same set of indicator variables are disjoint, i.e. their conjunctions are zero.

Proof. Any two minterms $\hat{M}_i$, $\hat{M}_j$ must have at least one literal $\tilde{X}_k$ in one of them to which corresponds $\bar{\tilde{X}}_k$ in the other. Hence, by Eq. (2.2.7)

$$\hat{M}_i\hat{M}_j = X_k\bar{X}_k \ldots \text{(further literals)} = 0 \ ,$$

q.e.d.

Definition 4.2.5. A term $\hat{T}_i$ which *implies* φ, i.e. for which $\hat{T}_i = 1 \Rightarrow X_\varphi = 1$ is an *implicant* of φ. — —

Definition 4.2.6. A *prime implicant* (*prime term*) of φ is an “indivisable” conjunction term which implies φ but which could not be (formally) divided by one of its literals without loosing the property of being an implicant. In other words: a prime implicant is a shortest implicant. — —

For small numbers of variables prime implicants (PI’s) are easily found via the Karnaugh map.

Example 4.2.2 Prime terms via the Karnaugh map

Let $\varphi(X_1, X_2, X_3, X_4)$ be given by the Karnaugh map of Fig. 4.2.1 (hereinafter called K-map). Clearly, all the prime terms of this function φ are

$$X_1X_2,\ X_1X_3,\ X_1X_4,\ \bar{X}_2X_4 \ .$$

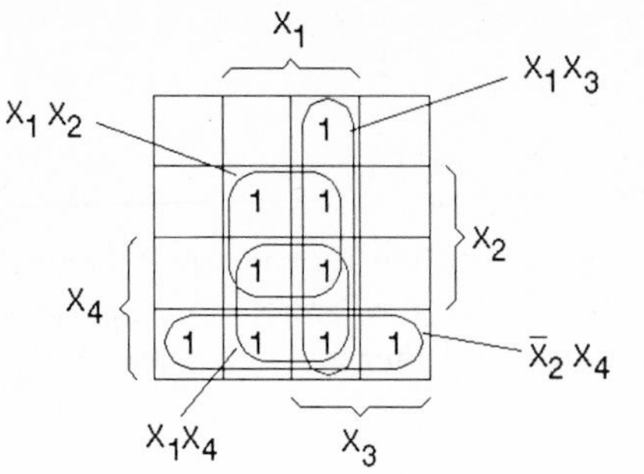

Fig. 4.2.1. Karnaugh map with prime implicants

Hint. The problem of finding all the prime implicants of φ will be discussed at length in § 5.2.

Definition 4.2.7. A *disjunction term* $\check{T}$ is a disjunction of literals. For practical reasons we exclude cases of repeated literals, of pairs X_i and $\bar{X}_i$ and the constants 0 and 1 in any disjunction term. Otherwise we will call a term a "raw" term. — —

Definition 4.2.8. The "dual" of a minterm $\hat{M}_i$, where

$$\hat{M}_i = \bigwedge_{j=1}^{n} \tilde{X}_j ,$$

is a *maxterm*. It is the maximum length disjunction term

$$\check{M}_i := \bigvee_{j=1}^{n} \tilde{X}_j . — — \tag{4.2.6}$$

What does "dual" mean here ? By Shannon's inversion rule (Eq. (4.1.1))

$$\bar{\hat{M}}_i = \bigvee_{j=1}^{n} \bar{\tilde{X}}_j . \tag{4.2.7}$$

Hence, with $\hat{M}_i$ being a special Boolean function, i.e.

$$\hat{M}_i = \varphi_i(\mathbf{X}) , \tag{4.2.8}$$

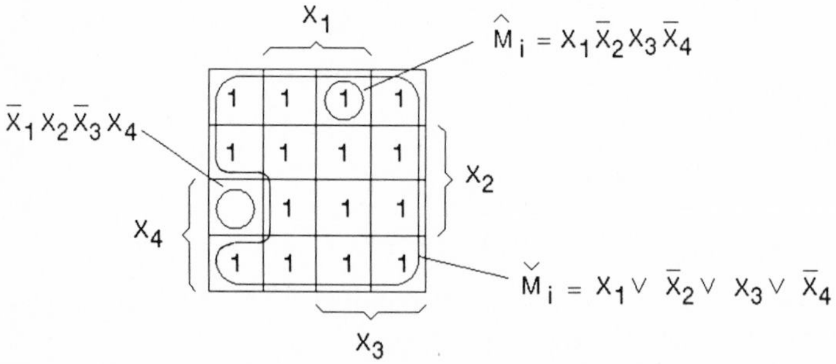

Fig. 4.2.2. Minterm $\hat{M}_i$ and dual maxterm $\check{M}_i$ with the same literals

we have

$$\check{M}_i = \overline{\varphi_i(\bar{\mathbf{X}})} \ . \tag{4.2.9}$$

In K-map language, $\check{M}_i$ is the function with 1's everywhere except for the small square corresponding to a minterm, in which all $\tilde{X}_k$ of $\hat{M}_i$ are replaced by $\bar{\tilde{X}}_k$; see Fig. 4.2.2.

Example 4.2.3 Minterm and dual maxterm

A special minterm called $\hat{M}_i$ and its associated maxterm $\check{M}_i$ are depicted in Fig. 4.2.2.

Remark. It should be mentioned that for minterms there exists a standard type of indexing, namely by choosing for i of $\hat{M}_i$ the number of the corresponding small square of the K-map, viz. the number one gets when interpreting $\hat{M}_i$ with $X_j := 1$ and $\bar{X}_j := 0$ as a binary number where X_j corresponds to 2^{j-1}. For instance, in case of three variables

$$X_1 X_2 X_3 = \hat{M}_7 \ , \quad X_1 X_2 \bar{X}_3 = \hat{M}_3 \ ;$$

see Fig. 3.3.3(a). Note that here the least significant bit is the leftmost bit.

4.3 Normal Forms and Canonical Normal Forms

The possible representations (forms) of Boolean functions are so numerous that even so-called *normal forms*, mostly such with a twofold nested structure, depicted in the tree of height two of Fig. 4.3.1, are not unique.

For instance, a syntax tree of the type in Fig. 4.3.1 may represent

$$X_\varphi = X_1 X_2 X_3 \vee X_1 X_2 \bar{X}_3 \vee X_4 \vee \bar{X}_4 X_5 = X_1 X_2 \vee X_4 \vee X_5 \ .$$

Both forms are of the disjunctive normal type.

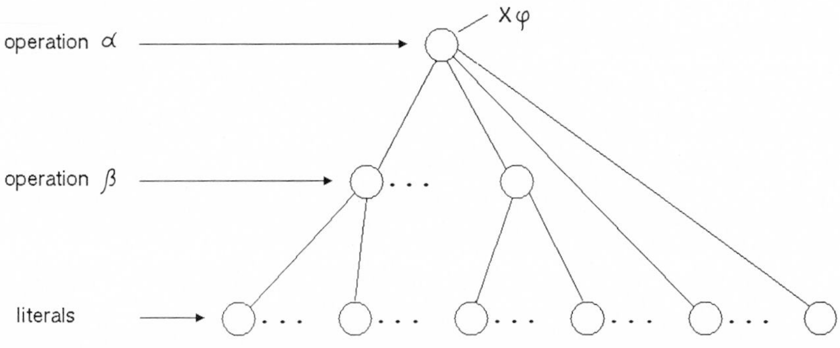

Fig. 4.3.1. Syntax diagram of a typical normal form of a Boolean function

Therefore, there is a need for normal forms, which are unique, so-called *canonical normal forms*.

In sub-section 4.3.1 the elementary theory will be developed with several examples from digital electronics. The following § 4.3.2 will contain graph theoretical applications which are of some importance in various fields of operations research, including reliability theory.

4.3.1 Elementary Theory

Theorem 4.3.1. Any Boolean function† can be represented as a disjunction of conjunction terms, i.e. as

$$X_\varphi = \varphi(X_1, \ldots, X_n) = \bigwedge_{i=1}^{m} \hat{T}_i \; ; \quad \hat{T}_i = \bigwedge_{j=1}^{n_i} \tilde{X}_{l_{i,j}} \tag{4.3.1}$$

with

$$\tilde{X}_k \in \{X_k, \bar{X}_k] \; ; \quad l_{i,j} \in \{1, 2, \ldots, n\} \; ; \quad n_i \leqq n, \quad m \leqq 2^n \; . \tag{4.3.1a}$$

(A proof follows later.)

To ease certain discussions it is understood that all $\hat{T}_i$ are terms according to definition 4.2.1.

Definition 4.3.1. Eq. (4.3.1) defines a *disjunctive normal form* (DNF). — — Note that every term of a DNF of φ implies φ.

Example 4.3.1 Two-out-of-three majority function

In case $X_\varphi = 1$ iff at least two of the three variables X_1, X_2, and X_3 assume the value 1, obviously (see also Fig. 3.4.1)

$$X_\varphi = X_1 X_2 \vee X_1 X_3 \vee X_2 X_3 \; . \tag{4.3.2}$$

Here $m=3$, $\hat{T}_1 = X_1 X_2$, $\hat{T}_2 = X_1 X_3$, $\hat{T}_3 = X_2 X_3$, $n_1 = n_2 = n_3 = 2$,

$$\tilde{X}_{l_{1,1}} = X_1 \; , \; \tilde{X}_{l_{1,2}} = X_2 \; , \; \tilde{X}_{l_{2,1}} = X_1 \; , \; \tilde{X}_{l_{2,2}} = X_3 \; , \; \tilde{X}_{l_{3,1}} = X_2 \; , \; \tilde{X}_{l_{3,2}} = X_3 \; .$$

One proof of theorem 4.3.1 is a straight forward consequence of the following theorem.

Theorem 4.3.2. (The Shannon decomposition theorem). Given a Boolean function φ, for every variable X_i, on which φ depends,

$$\begin{aligned} \varphi(X_1, \ldots, X_n) = {} & X_i \varphi(X_1, \ldots, X_{i-1}, 1, X_{i+1}, \ldots, X_n) \\ & \vee \bar{X}_i \varphi(X_1, \ldots, X_{i-1}, 0, X_{i+1}, \ldots, X_n) \; . \end{aligned} \tag{4.3.3}$$

† Except for the constant function 0.

Proof: For $X_i=1$ the left-hand side of Eq. (4.3.3) trivially becomes

$$\varphi(X_1,\ldots,X_{i-1},1,X_{i+1},\ldots,X_n)\ . \tag{4.3.4}$$

Because of $1 \wedge X=X$, $\bar{1}=0$ and $0 \wedge X=0$ the right-hand side equals expression (4.3.4). An analogous result is easily derived for $X_i=0$. Hence, both sides of Eq. (4.3.3) are equal for all **X**, q.e.d. (Examples of applications will follow.)

To prove Eq. (4.3.1) via Eq. (4.3.3) just consider the fact that after at most n decomposition steps one has a nested (binary tree) structure of the type

$$\varphi(\mathbf{X})=X_i(X_j[\ldots] \vee \bar{X}_j[\ldots]) \vee \bar{X}_i(X_k[\ldots] \vee \bar{X}_k[\ldots]) \tag{4.3.5}$$

which is, in principle, easily transformed via repeated applications of Eq. (2.2.3) to a DNF, q.e.d.

Since there is much free choice in the variable X_i of Eq. (4.3.3) there is, in general, more than one DNF for a given φ. However, if every term of Eq. (4.3.1) is "blown up" to an individual DNF of minterms only, and if on composing the $\hat{T}_i$'s of Eq. (4.3.1) duplicates of minterms are deleted because of the idempotence rule (Eq. (2.2.12)), then the resulting DNF, the so-called *canonical DNF* (CDNF), is unique.

In general we have the following theorem.

Theorem 4.3.3. The canonical DNF of a Boolean function is unique.

Proof. Every (binary) argument vector **X** results in exactly one minterm becoming 1 and all the others becoming 0. Hence, there is a one-to-one correspondence between the **X**'s for which $\varphi(\mathbf{X})=1$ and the minterms of φ. Since this set of **X**'s of φ is unique, so is $\hat{M}(\varphi)$, the set of minterms of φ, q.e.d.

Corollary 4.3.1. Moreover, for the canonical DNF one has with $X_j^0 := \bar{X}_j$, $X_j^1 := X_j$ (from [2]),

$$\varphi(\mathbf{X}) = \bigvee_{\mathbf{X}_i \in B^n} \left[\varphi(\mathbf{X}_i) \bigwedge_{j=1}^{n} X_j^{X_{i,j}} \right], \qquad \mathbf{X}_i := (X_{i,1},\ldots,X_{i,n})\ , \tag{4.3.6}$$

which means that one can get the canonical DNF as the disjunction of those minterms one gets from the argument n-tuples, for which $\varphi=1$, on replacing $X_{i,j}=0$ by $\bar{X}_j$ and $X_{i,j}=1$ by X_j. (An example precedes the proof.)

Example 4.3.2 Minterms of two-out-of-three majority function

Table 4.3.1 is the function table. From this table, i.e. its 1 set:

$$\varphi=\bar{X}_1X_2X_3 \vee X_1\bar{X}_2X_3 \vee X_1X_2\bar{X}_3 \vee X_1X_2X_3\ . \tag{4.3.7}$$

Proof of Eq. (4.3.6): The minterm

$$\bigwedge_{j=1}^{n} X_j^{X_{i,j}} = \begin{cases} 1\ , & \mathbf{X}=\mathbf{X}_i; \quad \mathbf{X}_i := (X_{i,1},\ldots,X_{i,n}) \\ 0\ , & \text{else}\ . \end{cases} \tag{4.3.8}$$

Table 4.3.1. The two-out-of-three majority function

X_1	X_2	X_3	φ
0	0	0	0
0	0	1	0
0	1	0	0
0	1	1	1
1	0	0	0
1	0	1	1
1	1	0	1
1	1	1	1

Hence, for every (binary) **X** all the terms of the disjunction $\bigvee_{\mathbf{X}_i \in B^n} \ldots$ vanish except for the term with $\mathbf{X}_i = \mathbf{X}$, such that

$$\varphi(\mathbf{X})_{/\mathbf{X}_i} = \varphi(\mathbf{X}_i) \wedge 1 = \varphi(\mathbf{X}_i) \ ,$$

q.e.d.
A less sophisticated form of Eq. (4.3.6) is

$$\varphi(\mathbf{X}) = \bigvee_{i \in I_\varphi} \hat{M}_i \ , \tag{4.3.9}$$

where I_φ is the set of the indices of the minterms of φ.

Definition 4.3.2. To ease the description of implicant terms $\hat{T}_i$ of a Boolean function φ we say "$\hat{T}_i$ implies φ" instead of "$\hat{T}_i$ is contained in φ". The meaning is clear from the fact that the set of the minterms of $\hat{T}_i$ is contained in the set of the minterms of φ, which can easily be visualized by the K-map. This is readily generalized as follows.

Lemma 4.3.1. If φ_1 implies φ_2, i.e. if $\varphi_1 = 1 \Rightarrow \varphi_2 = 1$ holds, then

$$\hat{M}(\varphi_1) \subset \hat{M}(\varphi_2) \ . \text{— —}$$

(See lemma 3.3.1.)

It is interesting to note that for minterms the disjunction (OR) is the same as the exclusive disjunction (EXOR).

Lemma 4.3.2. For minterms $\hat{M}_i$ and $\hat{M}_j$ (of the same variables)

$$\hat{M}_i \vee \hat{M}_j = \begin{cases} \hat{M}_i \not\equiv \hat{M}_j \ ; & i \neq j \\ \hat{M}_i \ ; & i = j \ . \end{cases} \tag{4.3.10}$$

Proof: From Eq. (2.1.1)

$$\hat{M}_i \not\equiv \hat{M}_j = \hat{M}_i \bar{\hat{M}}_j \vee \bar{\hat{M}}_i \hat{M}_j \ .$$

Now, by (4.2.7)

$$\hat{M}_i\bar{\hat{M}}_j=\hat{M}_i\left(\bigvee_{k=1}^{n}\bar{\bar{\tilde{X}}}_{l_k}\right)=\bigvee_{k=1}^{n}(\hat{M}_i\bar{\bar{\tilde{X}}}_{l_k})=\begin{cases}\hat{M}_i\ , & i\neq j\\ 0\ , & i=j\end{cases}$$

since, if $\hat{M}_i$ contains $\bar{\bar{\tilde{X}}}_l$, then $\hat{M}_i\bar{\bar{\tilde{X}}}_l=\hat{M}_i$; if $\hat{M}_i$ contains $\tilde{X}_l$, then $\hat{M}_i\bar{\bar{\tilde{X}}}_l=0$, and, unless $i=j$, there is at least one k for which $\bar{\bar{\tilde{X}}}_{l_k}$ is a literal of $\hat{M}_i$. Hence

$$\hat{M}_i\bar{\hat{M}}_j \vee \hat{M}_j\bar{\hat{M}}_i=\begin{cases}\hat{M}_i \vee \hat{M}_j\ , & i\neq j\\ 0\vee 0=0\ , & i=j\ ,\end{cases}$$

q.e.d. (A much more elementary proof is given in Exercise 4.7.)

Vectors with Minterms as Components

In order to represent a CDNF as a scalar vector product, one can try to achieve a certain amount of "canonicalism" by choosing **A** or **B** in Eq. (2.3.9) as the vector of all the minterms of a certain subset of the variables of

$$\varphi=\mathbf{AB}^{\mathrm{T}}=\mathbf{BA}^{\mathrm{T}}\ . \tag{4.3.11}$$

For instance, for three variables X_1,X_2,X_3 one could choose

$$\begin{aligned}\mathbf{A}=(&\bar{X}_3\bar{X}_2\bar{X}_1,\bar{X}_3\bar{X}_2X_1,\bar{X}_3X_2\bar{X}_1,\bar{X}_3X_2X_1\ ,\\ &X_3\bar{X}_2\bar{X}_1,X_3\bar{X}_2X_1,X_3X_2\bar{X}_1,X_3X_2X_1)\ .\end{aligned} \tag{4.3.12}$$

Clearly, with such a choice of **A**, the components of **B** would not always be minterms of the remaining variables of φ.

Example 4.3.3

Let

$$\varphi=X_1X_2X_3 \vee X_1\bar{X}_2\bar{X}_3\bar{X}_4 \vee \bar{X}_2X_3\ . \tag{4.3.13}$$

For

$$\mathbf{A}=(X_1X_2\ ,\bar{X}_1X_2\ ,X_1\bar{X}_2\ ,\bar{X}_1\bar{X}_2) \tag{4.3.14}$$

one expands $\bar{X}_2X_3$ as follows:

$$\bar{X}_2X_3=X_1\bar{X}_2X_3 \vee \bar{X}_1\bar{X}_2X_3\ ,$$

which yields

$$\begin{aligned}\varphi&=X_1X_2X_3 \vee X_1\bar{X}_2(\bar{X}_3\bar{X}_4 \vee X_3) \vee \bar{X}_1\bar{X}_2X_3\\ &=X_1X_2X_3 \vee X_1\bar{X}_2(X_3 \vee \bar{X}_4) \vee \bar{X}_1\bar{X}_2X_3\\ &=(X_1X_2,\bar{X}_1X_2,X_1\bar{X}_2,\bar{X}_1\bar{X}_2)(X_3,0,X_3 \vee \bar{X}_4,X_3)^{\mathrm{T}}\ .\end{aligned} \tag{4.3.15}$$

Obviously, here none of the components of **B** is a minterm (of X_3 and X_4). — —

Hence, one may call Eq. (4.3.11) with **A** or **B** of the type of Eq. (4.3.12) a *semi canonical* form of φ. A fully canonical form is found if the components of the factor (**A** or $\mathbf{B}^T$), whose components are not all the minterms of a subset $\{X_s\}$ of $\{X_1, \ldots, X_n\}$, are CDNFs of the other components of **X**. For instance, in example 4.3.3 **B** would then become [see Eq. (4.3.15)]:

$$(X_3X_4 \vee X_3\bar{X}_4, 0, X_3X_4 \vee X_3\bar{X}_4 \vee \bar{X}_3\bar{X}_4, X_3X_4 \vee X_3\bar{X}_4) .$$

In this context a non-trivial minimization problem is obvious: Which choice of **A** (or **B**) according to Eq. (4.3.12) yields a (the) shortest φ of Eq. (4.3.11)?

Returning to the non-canonical DNF, one is sometimes interested in the negated function, i.e. in

$$\bar{\varphi}(X) := \overline{\varphi(X)} .$$

Here we can use the following special case of theorem 4.1.1.

Corollary 4.3.3. (Shannon's Inversion Rule for the DNF). Let φ be as in Eq. (4.3.1), i.e. let

$$X_\varphi = \bigvee_{i=1}^{m} \bigwedge_{j=1}^{n_i} \tilde{X}_{l_{i,j}} , \tag{4.3.16}$$

then

$$\bar{X}_\varphi = \bigwedge_{i=1}^{m} \bigvee_{j=1}^{n_i} \bar{\tilde{X}}_{l_{i,j}} . \tag{4.3.17}$$

In other words: Inversion of variables together with the exchange of all AND's and OR's results in the inversion of a DNF.

Proof. By de Morgan's law

$$X_\varphi = \bigvee_i \hat{T}_i$$

results in

$$\bar{X}_\varphi = \bigwedge_i \bar{\hat{T}}_i = \overline{\bigwedge_i \bigwedge_j \tilde{X}_{l_{i,j}}} = \bigwedge_i \bigvee_j \bar{\tilde{X}}_{l_{i,j}} , \tag{4.3.18}$$

q.e.d.

Definition 4.3.3. A Boolean function form of the type of the r.h.s. of Eq. (4.3.17), i.e. a conjunction of disjunctions, is called a *conjunctive normal form* (CNF).

Lemma 4.3.3. To find a CNF of a function φ one can take a DNF of $\bar{\varphi}$ and then apply Shannon's inversion rule.

Proof. Replace X_φ by $\bar{X}_\varphi$ in Eqs. (4.3.16, 4.3.17)! Then Eq. (4.3.17) is a CNF of $\bar{\bar{\varphi}} = \varphi$, q.e.d.

Remark. Often φ is given as a DNF as in Eq. (4.3.16). Then the r.h.s. of Eq. (4.3.17) has to be "factored out".

Example 4.3.4 Finding a CNF (from [4])

Let

$$\varphi = X_1 X_2 \vee \bar{X}_2 X_3 X_4 \vee X_3 \bar{X}_4 \ . \tag{4.3.19}$$

By Shannon's inversion rule

$$\bar{\varphi} = (\bar{X}_1 \vee \bar{X}_2)(X_2 \vee \bar{X}_3 \vee \bar{X}_4)(\bar{X}_3 \vee X_4) \ . \tag{4.3.20}$$

When first factoring out the last two disjunctions it should be observed that $\bar{X}_3 \bar{X}_3 = \bar{X}_3$ and that $\bar{X}_3$ will absorb other terms containing $\bar{X}_3$. By this and other obvious simplifications

$$\bar{\varphi} = (\bar{X}_1 \vee \bar{X}_2)(\bar{X}_3 \vee X_2 X_4) = \bar{X}_1 \bar{X}_3 \vee \bar{X}_1 X_2 X_4 \vee \bar{X}_2 \bar{X}_3 \ . \tag{4.3.21}$$

Finally, again by Shannon's inversion rule, we have the desired CNF

$$\varphi = (X_1 \vee X_3)(X_1 \vee \bar{X}_2 \vee \bar{X}_4)(X_2 \vee X_3) \ . \tag{4.3.22}$$

In case a non-DNF is to be transformed to a CNF a dual version of the Shannon expansion theorem 4.3.2 can be used:

Theorem 4.3.4 (Dual Shannon's decomposition). Given the Boolean function φ: $B^n \to B$, for every variable X_i of φ:

$$\begin{aligned} \varphi(X_1, \ldots, X_n) = {} & [X_i \vee \varphi(X_1, \ldots, X_{i-1}, 0, X_{i+1}, \ldots, X_n)] \\ & \wedge [\bar{X}_i \vee \varphi(X_1, \ldots, X_{i-1}, 1, X_{i+1}, \ldots, X_n)] \ . \end{aligned} \tag{4.3.23}$$

Proof. The equality of both sides of Eq. (4.3.23) is obvious for $X_i = 0$ and $X_i = 1$, q.e.d.

Comment. On applying Eq. (4.3.23) recursively one can choose between looking at $X_i \vee \varphi(X_i = 0)$ and $\bar{X}_i \vee \varphi(X_i = 1)$ or only at $\varphi(X_i = 0)$ and $\varphi(X_i = 1)$, respectively, as the functions to be developed further. The first choice leads directly to CNFs.

Example 4.3.5 (Example 4.3.4 resumed)

Let

$$\varphi = X_1 X_2 \vee \bar{X}_2 X_3 X_4 \vee X_3 \bar{X}_4 \ . \tag{4.3.24}$$

Then, by Eq. (4.3.23) with $i = 4$

$$\varphi = (X_4 \vee X_3 \vee X_1 X_2)(\bar{X}_4 \vee X_1 X_2 \vee \bar{X}_2 X_3) \ .$$

Furthermore, by Eq. (4.3.23) with $i=1$

$$X_4 \vee X_3 \vee X_1X_2=(X_1 \vee X_4 \vee X_3)(\bar{X}_1 \vee X_4 \vee X_3 \vee X_2) ,$$

and with $i=2$

$$\bar{X}_4 \vee X_1X_2 \vee \bar{X}_2X_3=(X_2 \vee \bar{X}_4 \vee X_3)(\bar{X}_2 \vee \bar{X}_4 \vee X_1) .$$

Composing these results in Eq. (4.3.24) yields – after a trivial reordering of literals –

$$\varphi=(X_1 \vee X_3 \vee X_4)(\bar{X}_1 \vee X_2 \vee X_3 \vee X_4) \wedge (X_1 \vee \bar{X}_2 \vee \bar{X}_4)(X_2 \vee X_3 \vee \bar{X}_4) . \text{——} \qquad (4.3.25)$$

Notice that in a recursive binary tree type application of theorem 4.3.4 the number of variables is not necessarily decreased in every step. Rather, the number of terms connected by conjunction operators is doubled, and in each such term the number of single literals connected disjunctively (ORed) is increased.

Clearly, as a dual (see the principle of duality in §2.2) of theorem 4.3.1 the following is easily proved via theorem 4.3.4.

Theorem 4.3.5. Any Boolean function can be represented as a CNF. ——

Another proof is the following. Use the procedure of lemma 4.3.3. The DNF of $\bar{\varphi}$ exists because of theorem 4.3.1, q.e.d.

Definition 4.3.4. A *canonical conjunctive normal form* (CCNF) is a conjunction of maxterms.

Lemma 4.3.4. If the CDNF of a Boolean function φ is

$$\varphi = \bigvee_{i \in I_\varphi} \hat{M}_i \, ; \quad \hat{M}_i := \bigwedge_{j=1}^{n} \tilde{X}_j \, ,$$

then its CCNF is

$$\varphi = \bigwedge_{i \in I_{\bar{\varphi}}^{-1}} \check{M}_i \, , \quad \check{M}_i := \bigvee_{j=1}^{n} \tilde{X}_j \, , \qquad (4.3.26)$$

where $I_{\bar{\varphi}}^{-1}$ is found from the set of the indices of the minterms of $\bar{\varphi}$ by exchanging 0 and 1 in the binary representation of that index.

Proof. Since minterms not belonging to φ belong to $\bar{\varphi}$, we have

$$\bar{\varphi} = \bigvee_{i \in I_{\bar{\varphi}}} \hat{M}_i \, ,$$

where $I_{\bar{\varphi}}$ is the set of the indices of the minterms of $\bar{\varphi}$. Obviously

$$I_{\bar{\varphi}} := \{0, 1, \ldots, 2^n-1\} \setminus I_\varphi \, . \qquad (4.3.27)$$

By de Morgan's law

$$\bar{\bar{\varphi}} = \varphi = \bigwedge_{i \in I_{\bar{\varphi}}} \bar{\hat{M}}_i = \bigwedge_{i \in I_{\bar{\varphi}}^{-1}} \check{M}_i \ ,$$

q.e.d. (Notice the similarity with lemma 4.3.3.)

Example 4.3.6 Two-out-of-three function

As shown in example 4.3.2, the CDNF is here

$$X_\varphi = X_1 X_2 X_3 \vee X_1 X_2 \bar{X}_3 \vee X_1 \bar{X}_2 X_3 \vee \bar{X}_1 X_2 X_3 \ .$$

$$= \hat{M}_7 \vee \hat{M}_3 \vee \hat{M}_5 \vee \hat{M}_6 \ .$$

Hence, using the above standard indexing of minterms, by Eq. (4.3.27)

$I_\varphi = \{3, 5, 6, 7\}$, $I_{\bar{\varphi}} = \{0, 1, 2, 4\}$, $I_{\bar{\varphi}}^{-1} = \{7, 6, 5, 3\}$. Hence

$X_\varphi = \check{M}_7 \check{M}_3 \check{M}_5 \check{M}_6$. — —

Note that in general $I_{\bar{\varphi}}^{-1} \neq I_{\bar{\varphi}}$.
Now we come to an obvious dual of theorem 4.3.2.

Theorem 4.3.6. The CCNF is unique.

Proof. Since for a given φ the index set I_φ of its minterms is unique, also the index set $I_{\bar{\varphi}}^{-1}$ of the CCNF is unique, q.e.d.

Prior to turning to some applications in electronics, remember the following: In practice quite often normal forms (DNF and CNF) are expanded in an ad-hoc fashion for various reasons. For a DNF term $\hat{T}_i$ this is done simply via Eq. (2.2.16):

$$\hat{T}_i = \hat{T}_i(X_j \vee \bar{X}_j) = \hat{T}_i X_j \vee \hat{T}_i \bar{X}_j \ . \tag{4.3.28}$$

For a CNF term $\check{T}_i$ this is done via (2.2.19):

$$\check{T}_i = (\check{T}_i \vee X_i)(\check{T}_i \vee \bar{X}_i) \ . \tag{4.3.29}$$

Implementing DNFs in Electronics

Nowadays DNFs are usually "programmed" VLSI circuits, mostly PLAs (programmable logic arrays) or ROMs (read only memories). In the latter case the function table is put in the memory; often only the 1-set, even though this can mean a poor use of memory space.

In closing this section, two interesting ways to implement an electronic CDNF network are mentioned, a *multiplexer* and an *address decoder* with an OR gate. A multiplexer is a circuit for the selection of one of several data streams; see Fig. 4.3.2, where $X_1, \ldots, X_n$ are the (binary) control inputs.

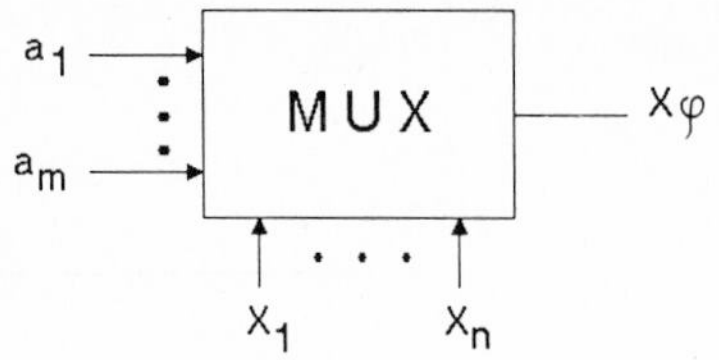

Fig. 4.3.2. Multiplexer; a_i: input i, X_j: control j; X_φ: output

Setting the indicators

$$a_i = \begin{cases} 1\,, & \text{if } \hat{M}_i \in \hat{M}(\varphi) \\ 0\,, & \text{if } \hat{M}_i \notin \hat{M}(\varphi)\,, \end{cases}$$

one has

$$X_\varphi = \varphi(X_1, \ldots, X_n)\;.$$

Let us look briefly at an example with $m=4$ and $n=2$.

Example 4.3.7 Four-inputs multiplexer to implement EXOR

Figure 4.3.3 shows the details of a multiplexer which can choose one out of 4 data streams. In order to implement

$$\varphi = X_1 \not\equiv X_2 = X_1\bar{X}_2 \vee \bar{X}_1 X_2$$

on has to set

$$a_1 = 0\,, \quad a_2 = 1\,, \quad a_3 = 1\,, \quad a_4 = 0\;.$$

Then by $a_2 = 1$, $X_\varphi = 1$ if $\bar{X}_1 X_2 = 1$, by $a_3 = 1$, $X_\varphi = 1$ if $X_1 \bar{X}_2 = 1$:

$$X_\varphi = \bar{X}_1 X_2 \vee X_1 \bar{X}_2\;. \text{——}$$

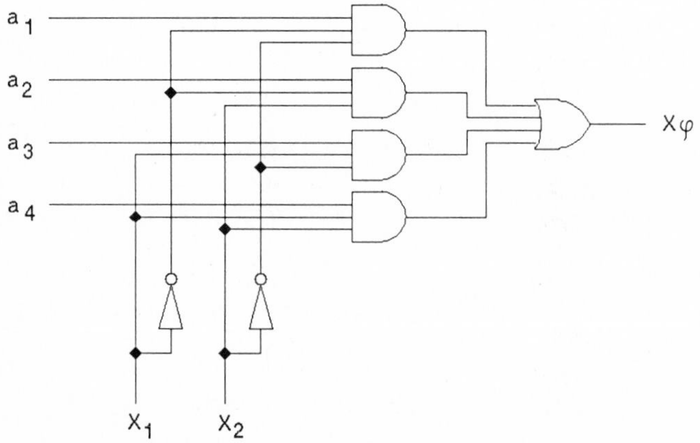

Fig. 4.3.3. Logic diagram of an electronic multiplexer

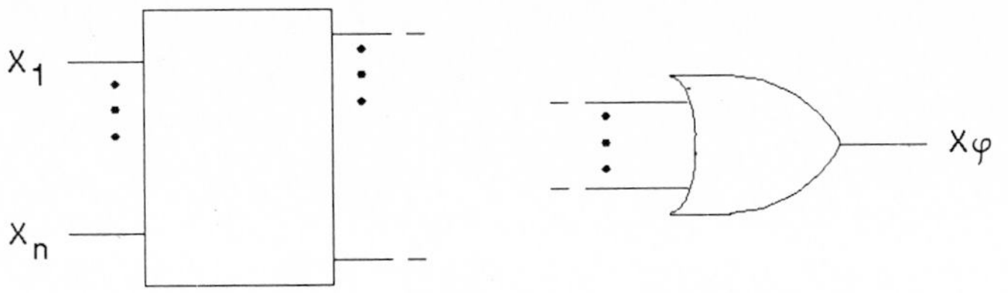

Fig. 4.3.4. Address decoder with a selection of outputs ORed together

An address decoder is a circuit with n inputs and 2^n outputs. For each input vector $\mathbf{X}$ exactly one output bit is 1, the others are 0. For a given $\varphi(\mathbf{X})$ the outputs corresponding to $\mathbf{X}$s with $\varphi(\mathbf{X}) = 1$ are ORed; see Fig. 4.3.4.

Example 4.3.8 Four-outputs address decoder

Figure 4.3.5 shows a (trivially small) address decoder. The ORed outputs are chosen to yield the EXOR function

$$X_\varphi = X_1 \bar{X}_2 \vee \bar{X}_1 X_2 \ .$$

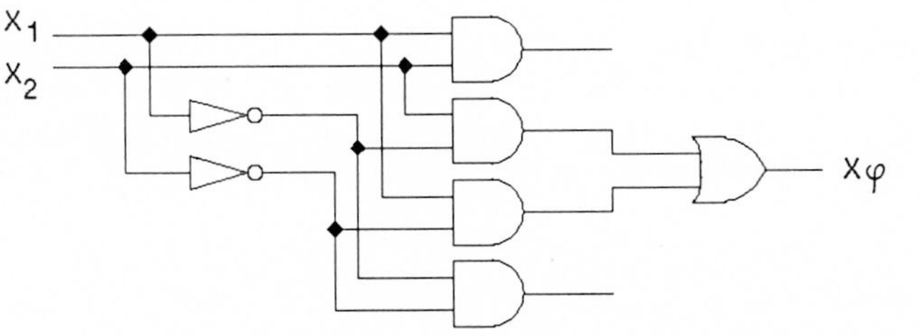

Fig. 4.3.5. Decoder for 2-bits addresses with 2 outputs ORed

4.3.2 Description of Connectivity Properties of Graphs

Boolean functions, and especially DNFs, are extremely useful for describing the existence of a connection between two vertices of a graph (directed or not) and for the description of the connectedness of the graph, i.e. the existence of at least one path between any two vertices. In this context, the basic two terms of graph theory are the *path (tie) set* and the *cut set.*

Definition 4.3.5. A *path (tie) set* of a graph is a set of its edges forming a path from a starting point (input point) to an end point.

Definition 4.3.6. A *cut set* of a graph is a set of edges whose deletion would interrupt every path between two given vertices.

Example 4.3.9 The simplified ARPA computer network (in the USA)

From Fig. 4.3.6 there follows directly the path[†] sets (of edges e_i) between the vertices v_1 and v_2: $\{e_1, e_2, e_3\}$, $\{e_1, e_2, e_5, e_7\}$, $\{e_1, e_3, e_4, e_5\}$, $\{e_1, e_4, e_7\}$, $\{e_6, e_7\}$, $\{e_2, e_3, e_4, e_6\}$, $\{e_3, e_5, e_6\}$. The corresponding cut sets are (see Fig. 4.3.6): $\{e_1, e_6\}$, $\{e_2, e_4, e_6\}$, $\{e_3, e_4, e_5, e_6\}$, $\{e_1, e_4, e_5, e_7\}$, $\{e_2, e_5, e_7\}$, $\{e_3, e_7\}$. — —

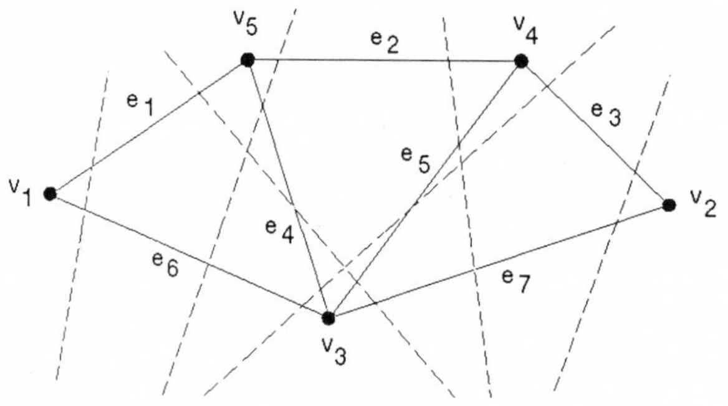

Fig. 4.3.6 Simplified ARPA net with cuts between v_1 and v_2

Now we introduce the indicator variables

$$X_i = \begin{cases} 0, & e_i \text{ is deleted} \\ 1, & \text{else.} \end{cases} \quad , \quad X_\varphi = \begin{cases} 0 & \text{if there is no path} \\ 1 & \text{if there is at least one path} \end{cases} \text{ between } v_1 \text{ and } v_2$$

(Clearly, in real life the deletion of e_i would mean the non-viability of the corresponding path, whatever this may mean in detail. It could mean the destruction of a bridge on road i or a failure state of component i.)

Obviously, a path refers to a conjunction of the associated variables, e.g.

$$\{e_1, e_2, e_3\} \hat{=} X_1 X_2 X_3 \ , \quad \text{since} \quad X_1 X_2 X_3 = 1 \Rightarrow X_\varphi = 1 \ ,$$

and the connectedness of v_1 and v_2 is described by the DNF of the terms corresponding to all the paths between v_1 and v_2. Likewise, a cut[‡] refers to a conjunction of the associated negated (complemented) variables, e.g.

$$\{e_1, e_6\} \hat{=} \bar{X}_1 \bar{X}_6 \ , \quad \text{since} \quad \bar{X}_1 \bar{X}_6 = 1 \Rightarrow \bar{X}_\varphi = 1 \Leftrightarrow X_\varphi = 0 \ ,$$

and the disconnectedness of v_1 and v_2 is described by the DNF of the terms corresponding to all the cuts between v_1 and v_2.

[†] Usually, one is only interested in minimal paths, i.e. paths without cycles. All the paths discussed here are minimal paths.

[‡] All cuts discussed here are minimal cuts, i.e. no edge in these sets could be deleted.

In the above example we have thus for the indicator variable X_φ of a connection between v_1 and v_2

$$X_\varphi = X_1X_2X_3 \vee X_1X_2X_5X_7 \vee X_1X_3X_4X_5 \vee X_1X_4X_7 \vee X_6X_7 \vee X_2X_3X_4X_6 \vee X_3X_5X_6 \ , \tag{4.3.30}$$

and

$$\bar{X}_\varphi = \bar{X}_1\bar{X}_6 \vee \bar{X}_2\bar{X}_4\bar{X}_6 \vee \bar{X}_3\bar{X}_4\bar{X}_5\bar{X}_6 \vee \bar{X}_1\bar{X}_4\bar{X}_5\bar{X}_7 \vee \bar{X}_2\bar{X}_5\bar{X}_7 \vee \bar{X}_3\bar{X}_7 \ . \text{——} \tag{4.3.31}$$

Obviously, both sets, cuts and paths can be calculated from each other or rather minimal cuts from minimal paths. The procedure is given by Shannon's inversion rule (theorem 4.1.1). The details are:

(1) Negate the given DNF using de Morgan's theorem to get a CNF (conjunction of disjunctions)!
(2) Transform this CNF to a DNF by using

$$(\hat{T}_1 \vee \hat{T}_2 \vee \ldots)(\hat{T}_1^* \vee \hat{T}_2^* \vee \ldots) = \hat{T}_1\hat{T}_1^* \vee \hat{T}_1\hat{T}_2^* \vee \ldots \hat{T}_2\hat{T}_1^* \vee \hat{T}_2\hat{T}_2^* \vee \ldots$$

and by simplifying the result, mainly via §2.2.2!

Finding Minpaths by Backtracking

For directed graphs with more than, say, 10 edges the visual approach used in connection with Fig. 4.3.6 will easily lead to errors because human pattern recognition capability is limited. In such cases the following backtracking procedure to find minpaths is better:

(I) Define auxiliary indicator variables X'_i, where

$$X'_1 = \begin{cases} 1, & \text{if there is a connection between the source node} \\ & v_s \text{ and the end node of edge } i \ (e_i) \\ 0, & \text{else.} \end{cases} \tag{4.3.32}$$

Clearly, if edges $k_1, \ldots, k_j$ end in node $k(V_k)$, then

$$\psi_k := \bigvee_{i=1}^{j} X'_{k_i} \tag{4.3.33}$$

is the Boolean indicator that a path exists from the source node to V_k. (In Fig. 4.3.6 v_1 is the source node.)

If in a stochastic graph only edges can vanish (at random), then, if e_i starts from v_k, then

$$X'_i = X_i \psi_k \ . \tag{4.3.34}$$

This expresses the obvious fact that, if the end point of e_i is connected with v_s, then e_i exists (is "good") and the starting node of e_i is connected with v_s. Note the subtle difference between the end point of e_i and its end node. This differentiation looks less artificial in the main field of applications of this procedure, namely communications engineering. There simply

$X_i'=1$ means: A signal comes through e_i.

$X_i=1$ means: Link i is "good".

$\psi_k=1$ means: Link i gets correct input signals.

(II) Start with the receiver node v_r. (In Fig. 4.3.6 v_2 is the receiver node.) Obviously, in equations like (4.3.31)

$$\varphi=\psi_r \; . \tag{4.3.35}$$

Then, "simply", successively replace in φ all X_i' variables by X_j variables!

Example 4.3.10 Simple bridge network

From Fig. 3.7.1b for shortest, i.e. acyclic connections between v_1 and v_2

$$\varphi:=\psi_2=X_5' \vee X_6' \; , \tag{4.3.36}$$

where

$$X_5'=X_5(X_1' \vee X_4') \; , \quad X_1'=X_1 \; , \quad X_4'=X_4X_2' \; , \quad X_2'=X_2 \; ,$$
$$X_6'=X_6(X_2' \vee X_3') \; , \quad X_3'=X_3X_1' \; . \tag{4.3.37}$$

Hence

$$\varphi=X_5(X_1 \vee X_2X_4) \vee X_6(X_2 \vee X_1X_3)$$
$$=X_1X_5 \vee X_2X_6 \vee X_1X_3X_6 \vee X_2X_4X_5 \; . \tag{4.3.38}$$

(Of course, this would have followed more easily from the minpaths.)

Comment: In the above derivation cyclic interdependencies of X_i's were avoided since they yield inconsistent results. For instance, using (4.3.33, 4.3.34) "verbally" would mean

$$X_4'=X_4(X_2 \vee X_3') \; , \quad X_3'=X_3(X_1 \vee X_4') \; ,$$

i.e.

$$X_4'=X_4(X_2 \vee X_1X_3 \vee X_3X_4') \; . \tag{4.3.39}$$

However, for independent X_1, X_2, X_3, X_4 neither $X_4'=0$ nor $X_4'=1$ make sense: $X_4'=0$ would mean

$$0=X_4(X_2 \vee X_1X_3) \; ,$$

and $X_4'=1$ would mean

$$1=X_4(X_2 \vee X_3) \ .$$

Also with cuts (as in paths as above) cycles are a problem. This problem is solved satisfactorily by the following addendum to the backtracking procedure discussed:

(III) If the directed graph under consideration contains cycles in paths from v_s to v_r, then, in order to get minimal paths, any such cycle must be broken by at least one added assumption $X_i=0$.
For instance, Figs. 4.3.7(a) and (b) show what was tacitly assumed in the first r.h.s. of Eq. (4.3.38).

Fig. 4.3.7. Dissolving cycles in backtracking

The following matrix methodology shows a differend way of managing cycles in graphs.

Finding Paths by Matrix Manipulations

The paths between two vertices can be found in a more systematic way by Boolean matrix operations. Therefore, the concept of the adjacency matrix is very useful.

Definition 4.3.7. Given a *graph* $G(V, E)$ with V and E the sets of its N_v vertices v_i and N_e edges $e_{ij}=(v_i, v_j)$, the *adjacency matrix* is the square matrix

$$\mathbf{Y}=(Y_{i,j})_{N_v,N_v}$$

with

$$Y_{i,j}=\begin{cases}1 \ , & \text{if } e_{i,j} \text{ is "operable"}^{\dagger} \\ 0 \ , & \text{else} \ ,\end{cases} \qquad (4.3.40)$$

where "operable" includes the existence of an edge, but may give more detailed information.

Even if there is no edge starting and ending in v_i it makes sense to set

$$Y_{i,i}=1 \ . \ —— \qquad (4.3.40a)$$

Definition 4.3.8. The "length" of a path in a graph is the number of edges it consists of. ——

† The conceptually more appropriate notation $e_{l_{i,j}}$; $l_{i,j}\in\{1, \ldots, N_e\}$ was simplified to $e_{i,j}$.

For matrix conjunction we define as follows; see Eqs. (2.3.10, 2.3.10a):

Definition 4.3.9.

$$\mathbf{YY}=\left(\bigvee_{k=1}^{N_v} Y_{i,k}\,Y_{k,j}\right)_{N_v,N_v} \quad ; \qquad \mathbf{Y}^{(l)}:=\mathbf{Y}^{(l-1)}\mathbf{Y}\ , \qquad \mathbf{Y}^{(1)}:=\mathbf{Y}\ . \tag{4.3.41}$$

Theorem 4.3.7. The indicator function of the connectiveness of vertices *i* and *j* is

$$\psi_{i,j}:=\bigvee_{l=1}^{N_v} Y_{i,j}^{(l)}\ . \tag{4.3.42}$$

Proof. By induction (shortened),

$$Y_{i,j}^{(l)}=\bigvee_{k=1}^{N_v} Y_{i,k}^{(l-1)}\,Y_{k,j}$$

is obviously the indicator function of all the paths of length l from vertex i to vertex j. Typically, the last vertex before v_j, being v_k, is then reached from v_i on paths of length $l-1$. Initially

$$Y_{i,j}^{(2)}:=\bigvee_{k=1}^{N_e} Y_{i,k}^{(1)}\,Y_{k,j}\ , \qquad Y_{i,j}^{(1)}:=Y_{i,j}\ ,$$

q.e.d.

Comment. Since the example of Fig. 4.3.6 is already a bit too complicated for a first demonstration of theorem 4.3.7 we use the simpler bridge network example of Fig. 3.7.1.

Example 4.3.11 Simple bridge network

We are asking for the "logics" of having a connection between v_1 and v_2. From Fig. 3.7.1(a) the adjacency matrix is

$$\mathbf{Y}=\begin{bmatrix} 1 & 0 & Y_{1,3} & Y_{1,4} \\ 0 & 1 & Y_{2,3} & Y_{2,4} \\ Y_{3,1} & Y_{3,2} & 1 & Y_{3,4} \\ Y_{4,1} & Y_{4,2} & Y_{4,3} & 1 \end{bmatrix}.$$

Specifically $Y_{1,2}=0$.

By Eq. (4.3.41), taking into account that, for an undirected graph,

$$Y_{i,j}=Y_{j,i}\ , \tag{4.3.43}$$

$$\mathbf{Y}^{(2)} = \begin{bmatrix} 1 & Y_{1,3}Y_{3,2} \vee Y_{1,4}Y_{4,2} & Y_{1,3} \vee Y_{1,4}Y_{4,3} & Y_{1,4} \vee Y_{1,3}Y_{3,4} \\ Y_{1,3}Y_{3,2} \vee Y_{1,4}Y_{4,2} & 1 & Y_{2,3} \vee Y_{2,4}Y_{4,3} & Y_{2,4} \vee Y_{2,3}Y_{3,4} \\ Y_{1,3} \vee Y_{1,4}Y_{4,3} & Y_{2,3} \vee Y_{2,4}Y_{4,3} & 1 & Y_{3,1}Y_{1,4} \vee Y_{3,2}Y_{2,4} \vee Y_{3,4} \\ Y_{1,4} \vee Y_{1,3}Y_{3,4} & Y_{2,4} \vee Y_{2,3}Y_{3,4} & Y_{3,1}Y_{1,4} \vee Y_{3,2}Y_{2,4} \vee Y_{3,4} & 1 \end{bmatrix} \tag{4.3.44}$$

Specifically

$$Y_{1,2}^{(2)} = Y_{1,3}Y_{3,2} \vee Y_{1,4}Y_{4,2} \; .$$

Since here (acyclic) paths cannot have a length exceeding 3, $Y^{(3)}$ is not needed, only the element $Y_{1,2}^{(3)}$ of $\mathbf{Y}^{(2)}\,\mathbf{Y}$:

$$\begin{aligned} Y_{1,2}^{(3)} &= Y_{1,3}Y_{3,2} \vee Y_{1,4}Y_{4,2} \vee (Y_{1,3} \vee Y_{1,4}Y_{4,3})Y_{3,2} \\ &\quad \vee (Y_{1,4} \vee Y_{1,3}Y_{3,4})Y_{4,2} \; . \\ &= Y_{1,3}Y_{3,2} \vee Y_{1,4}Y_{4,2} \vee Y_{1,4}Y_{4,3}Y_{3,2} \vee Y_{1,3}Y_{3,4}Y_{4,2} \; . \end{aligned} \tag{4.3.45}$$

Finally, from Eq. (4.3.42), absorbing multiple terms, the v_1–v_2-connection indicator function is $\varphi = \psi_{1,2}$, where

$$\begin{aligned} \varphi &= Y_{1,2} \vee Y_{1,2}^{(2)} \vee Y_{1,2}^{(3)} \\ &= Y_{1,3}Y_{3,2} \vee Y_{1,4}Y_{4,2} \vee Y_{1,4}Y_{4,3}Y_{3,2} \vee Y_{1,3}Y_{3,4}Y_{4,2} \; . \end{aligned} \tag{4.3.46}$$

This result is easily checked by means of Fig. 3.7.1a. It is Eq. (4.3.38) in new notation. — —

As with the backtracking discussed above, cycles must be weeded out. They are easy to find, since they lead to conjunctions $Y_{i,j}Y_{j,k} \ldots Y_{l,i}$, where the second index of an $Y_{i,j}$ equals the first one of another.

It should be obvious by now how to formulate the *connectiveness* of a graph which means that there exists a path between any two edges. With ψ_G the indicator function of the fact that G is a *connected graph*

$$\psi_G = \bigwedge_{i,j:\, v_i,\, v_j \in V} \psi_{i,j} \; , \tag{4.3.47}$$

where $\psi_{i,j}$ is defined in Eq. (4.3.42).

More on Boolean matrices may be found in [5].

4.4 Disjunctive Normal Forms of Disjoint Terms

First, let us remember definition 4.2.4 for the disjointness of two Boolean functions, since terms are also functions.

The interest in disjoint (conjunction) terms is strongly motivated from the field of probabilistic Boolean analysis. As will be explained in detail in § 8, terms $\hat{T}_i$ can be used to model events a_i, for which, in general (see the appendix § 11),

$$P\left\{\bigcup_i a_i\right\}=\sum_i P\{a_i\}-\sum_{i<j} P\{a_i\cap a_j\}$$

$$+\sum_{i<j<k} P\{a_i\cap a_j\cap a_k\}-+\ldots\pm P\left\{\bigcap_i a_i\right\} . \quad (4.4.1)$$

Because of $P\{\phi\}=0$, disjoint a_i are highly desirable since with them Eq. (4.4.1) becomes simply

$$P\left\{\bigcup_i a_i\right\}=\sum_i P\{a_i\} ,$$

and disjoint a_i imply disjoint corresponding terms $\hat{T}_i$.

A trivial example for making two terms of a DNF disjoint is given by

$$X_1\vee X_2=X_1\vee \bar{X}_1X_2 , \quad (4.4.2)$$

which is true by Eq. (2.2.18). An alternate proof is by theorem 4.3.2 (with $i=1$). A further proof starts with the production of the CDNF (assuming $n=2$):

$$X_1\vee X_2=X_1X_2\vee X_1\bar{X}_2\vee \bar{X}_1X_2 , \quad (4.4.3)$$

and continues with merging the first two minterms to obtain X_1, q.e.d.

After giving some fundamentals (§4.4.1) in §§4.4.2, 4.4.3 and 4.4.4 three modern algorithms for the production of a DDNF, i.e. a DNF with pairwise disjoint terms, will be presented, which will be compared with each other in §4.4.5.

4.4.1 Fundamentals

Definition 4.4.1. A DNF of disjoint terms is abbreviated by DDNF. — —

A trivial solution to the problem of finding a DDNF is to find the proper CDNF because by theorem 4.2.2 the CDNF is a DDNF. However, in general the CDNF is of unmanageable length. The following theorem gives a necessary and sufficient condition for the disjointness of conjunction terms.

Theorem 4.4.1. Two terms are disjoint iff one of them contains at least one negated variable of the other.

Proof.

(a) If

$$\hat{T}_i=X_k\hat{T}'_i , \quad \hat{T}_j=\bar{X}_k\hat{T}'_j ,$$

then by Eq. (2.2.7)

$$\hat{T}_i\hat{T}_j=0 . \quad (4.4.4)$$

(b) If no X_k of the type just mentioned exists, then

$$\hat{T}_i \hat{T}_j = \bigwedge_{l \in I_{i,j}} \tilde{X}_l = \begin{cases} 1 , & \tilde{X}_l = 1, \forall l \in I_{i,j} \\ 0 , & \text{else} , \end{cases} \tag{4.4.5}$$

where $I_{i,j}$ is the index set of the literals appearing in $\hat{T}_i$ or $\hat{T}_j$, i.e. $\hat{T}_i$ and $\hat{T}_j$ are not disjoint, q.e.d.

A motivation for the concept of disjoint terms is given by the following theorem.

Theorem 4.4.2. Disjoint terms have disjoint sets of minterms.

Proof. (By contradiction):

Let $\hat{M}(\hat{T}_1) \cap \hat{M}(\hat{T}_2) \neq \phi$. Specifically define

$$\hat{T}_1 = \bigvee_{i \in I_1} \hat{M}_i , \quad \hat{T}_2 = \bigvee_{k \in I_2} \hat{M}_k \tag{4.4.6}$$

and let $j \in I_1 \cap I_2$, i.e. let $\hat{M}_j$ be a common minterm. Then, because of [see theorem 4.2.2 and (2.2.13)]

$$\hat{M}_i \hat{M}_k = \begin{cases} \hat{M}_i , & i = k \\ 0 , & \text{else} \end{cases} \tag{4.4.7}$$

one would have

$$\hat{T}_1 \hat{T}_2 = M_j \vee \ldots \neq 0 ,$$

q.e.d.

Corollary 4.4.1. Disjoint terms have disjoint areas on the K-map.

Proof. Notice the one-to-one correspondence of minterms and squares on the K-map.

Theorem 4.4.3. Let $\varphi_1, \varphi_2, \varphi_3, \varphi_4$ be Boolean functions $\neq 0$. Then the disjointness of φ_1 and φ_2 implies the same for the pair $\varphi_1 \varphi_3$ and $\varphi_2 \varphi_4$.

Proof. By the laws of commutation and association $\varphi_1 \varphi_2 = 0$ results in

$$(\varphi_1 \varphi_3)(\varphi_2 \varphi_4) = (\varphi_1 \varphi_2) \varphi_3 \varphi_4 = 0 , \tag{4.4.8}$$

q.e.d.

The main topic of this §4.4 is the transformation of the DNF

$$\varphi = \bigvee_i \hat{T}_i \tag{4.4.9}$$

to a DNF of mutually disjoint terms $\hat{T}'_j$:

$$\varphi = \bigvee_j \hat{T}'_j , \quad \hat{T}'_i \hat{T}'_k = 0 , \quad i \neq k . \tag{4.4.10}$$

Notice that the r.h.s. of Eq. (4.4.10) can be considerably longer than that of Eq. (4.4.9). This can be the "price" of mutual disjointness of all terms.

4.4.2 A Set-addition Algorithm

We begin with the derivation of an important auxiliary result concerning an optimal partition of the set of all the minterms of m variables.

As is obvious from Eq. (4.4.7), the canonical DNF of any $\varphi: B^n \to B$ is a DDNF. However, it is the longest one and, generally, the short ones are of more interest. Therefore, we introduce what we call the minimal DDNF of the constant 1 function.

Minimal Disjoint DNF of $\varphi = 1$

Definition 4.4.2. Call 1_m any shortest DDNF with at least one minterm of the Boolean function

$$X_\varphi = \varphi(X_1, \ldots, X_m) = 1. \text{ — —}$$

How can one find short DNFs or even DDNFs of $\varphi = 1$?

From Eq. (2.2.8), trivially

$$X_1 \vee \bar{X}_1 = 1_1 \ . \tag{4.4.11}$$

Hence, by an incomplete expansion (factoring); see also Fig. 4.4.1(a):

$$(X_1 \vee \bar{X}_1)(X_2 \vee \bar{X}_2) = X_1(X_2 \vee \bar{X}_2) \vee \bar{X}_1 = X_1 X_2 \vee \bar{X}_1 \vee X_1 \bar{X}_2 = 1_2 \ .$$

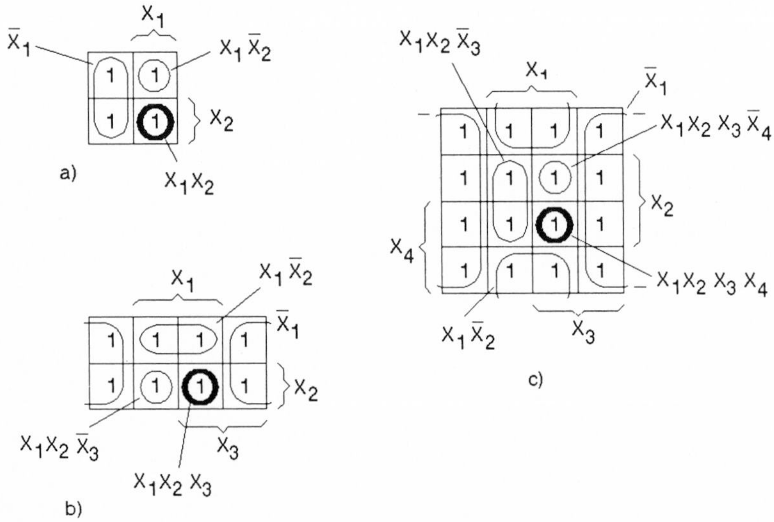

Fig. 4.4.1. K-maps of 1_2, 1_3, and 1_4 in (**a**), (**b**) and (**c**), respectively

Thus, by the same type of expansion; see also Fig. 4.4.1(b):

$$(X_1X_2 \vee \bar{X}_1 \vee X_1\bar{X}_2)(X_3 \vee \bar{X}_3) = X_1X_2X_3 \vee \bar{X}_1 \vee X_1\bar{X}_2 \vee X_1X_2\bar{X}_3 = 1_3 \ .$$

Hence, similarly, see Fig. 4.4.1c:

$$(X_1X_2X_3 \vee \bar{X}_1 \vee X_1\bar{X}_2 \vee X_1X_2\bar{X}_3)(X_4 \vee \bar{X}_4)$$
$$= X_1X_2X_3X_4 \vee \bar{X}_1 \vee X_1\bar{X}_2 \vee X_1X_2\bar{X}_3 \vee X_1X_2X_3\bar{X}_4 = 1_4 \ .$$

The generalization of these results will be done in several steps.

In general we have

Lemma 4.4.1.

$$1'_m := X_1 \ldots X_m \vee \bar{X}_1 \vee X_1\bar{X}_2 \vee X_1X_2\bar{X}_3 \vee \ \ldots \ \vee X_1 \ldots X_{m-1}\bar{X}_m = 1 \ . \tag{4.4.12}$$

Hint. This equation is also correct for X_i replaced by $\tilde{X}_i$.

Proof (by induction).
Equation (4.4.12) is true for $m = 1, 2, 3, 4$. If it were true for $m = k - 1$, then for $m = k$

$$1'_k = 1'_{k-1}(X_k \vee \bar{X}_k) = X_1 \ldots X_k \vee \bar{X}_1 \vee \ \ldots$$
$$\vee X_1 \ldots X_{k-2}\bar{X}_{k-1} \vee X_1 \ldots X_{k-1}\bar{X}_k = 1 \ ,$$

where only the first term of Eq. (4.4.12) was expanded and all the others (since essentially multiplied by 1) remained unchanged. Obviously, Eq. (4.4.12) is also true for $m = k$ and therefore for any integer m, q.e.d.

Corollary 4.4.2. The function

$$\check{M}^* := \bar{X}_1 \vee X_1\bar{X}_2 \vee X_1X_2\bar{X}_3 \vee \ \ldots \vee X_1 \ldots X_{m-1}\bar{X}_m \tag{4.4.13}$$

is a DDNF of the maxterm $\bar{X}_1 \vee \ \ldots \ \vee \bar{X}_m$.

Proof. (4.4.13) is the right-hand side of Eq. (4.4.12) except for the first term. By Eq. (4.4.12)

$$\check{M}^* \vee X_1 \ldots X_m = 1 \ .$$

By theorem 4.4.1 each term of $\check{M}^*$ is disjoint with $X_1 \ldots X_m$. Hence, by lemma 4.1.1 and by de Morgan's law

$$\check{M}^* = \overline{X_1 \ldots X_m} = \bar{X}_1 \vee \ \ldots \ \vee \bar{X}_m \ , \tag{4.4.13a}$$

q.e.d.

More details on disjointness within $\check{M}^*$ gives the following simple lemma.

Lemma 4.4.2. The terms of $1'_m$ are all pairwise disjoint.

Proof. $X_1 \dots X_m$ is disjoint with all other terms because each contains a negated variable of $X_1 \dots X_m$. $\bar{X}_1$ is disjoint with all the following terms because all of them contain X_1. $X_1\bar{X}_2$ is disjoint with all the following terms, since all of them contain X_2 etc., q.e.d.

We are on the way to showing that $1'_m$ of Eq. (4.4.12) is an 1_m of definition 4.4.2. Yet the minimality of the length of $1'_m$ remains to be shown.

Theorem 4.4.4. The Boolean function representation (of a maxterm)

$$\check{M}^* := \bar{X}_1 \vee X_1\bar{X}_2 \vee X_1X_2\bar{X}_3 \vee \dots \vee X_1 \dots X_{m-1}\bar{X}_m$$

is a minimal DDNF in the sense that it consists of a minimum number of terms whose added total length is minimal.

Proof. A term of length 1 (as $\bar{X}_1$) is the shortest possible. It corresponds to 2^{m-1} minterms of (the above) $\check{M}^*$. Since $\check{M}^*$ consists of $2^m - 1$ minterms, a DDNF of $\check{M}^*$ cannot contain more than one term of length 1. Beyond one term of length 1, $\check{M}^*$ can contain at most one term of length 2 (as $X_1\bar{X}_2$), which consists of 2^{m-2} minterms, since beyond the 2^{m-1} minterms for the term of length 1, only

$$2^m - 1 - 2^{m-1} = 2^{m-1} - 1$$

have to be covered by the rest of a (minimal) DDNF. By exactly the same argument it can be shown that $\check{M}^*$ can contain at most one term of any length between 1 and m, q.e.d.

Comment. The above results concerning 1_m are true for any permutation of $(X_1, \dots, X_m)$. However, the 1_m of (4.4.12) is probably one of the easiest to remember.

Corollary 4.4.3. $\check{M}^*$ of theorem 4.4.4 consists of m terms whose added total length is

$$1 + 2 + \dots + m = \frac{m}{2}(m+1) = \binom{m+1}{2} . \text{——}$$

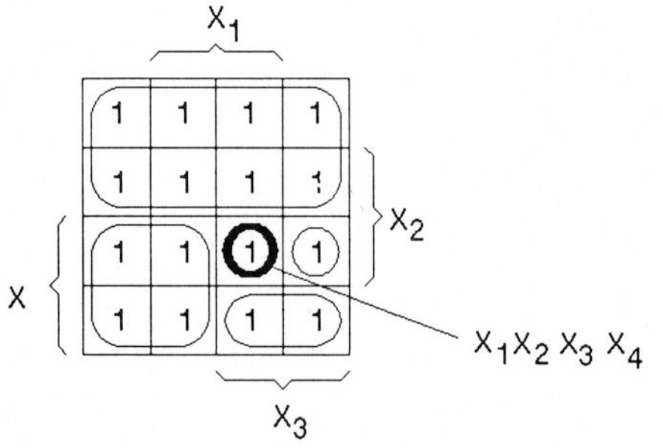

Fig. 4.4.2. Alternative minimal partition of the set of minterms with 4 variables

Notice the non-uniqueness of $\check{1}_m$. Figure 4.4.2 gives an example for 1_4. It shows that even the $\check{M}^*$ of (4.4.13a) are not unique.

We are equipped with the main optimization instrument to model a good "set addition" algorithm. This is an algorithm by Abraham [6] whose main idea will be explained shortly and where a short DDNF of a maxterm is urgently needed.

Abraham Algorithm

The main algorithm discussed now transforms every term $\hat{T}_j$ of a given DNF

$$\varphi = \bigvee_{i=1}^{m} \hat{T}_i \tag{4.4.14}$$

to a DNF $\overset{\diamond}{T}_j$ which is disjoint with $\bigvee_{i=1}^{j-1} \hat{T}_i$ and whose terms $\hat{T}'_l$ are mutually disjoint. Finally, one has the DDNF

$$\varphi = \bigvee_{i=1}^{m} \overset{\diamond}{T}_i = \bigvee_{j=1}^{m'} \hat{T}'_j \,, \quad m' \geqq m \,. \tag{4.4.15}$$

The basic idea of this algorithm can be illustrated by the Venn diagram of Fig. 4.4.3. The three circles correspond to the three terms $\hat{T}_1, \hat{T}_2, \hat{T}_3$ (made up of minterms).

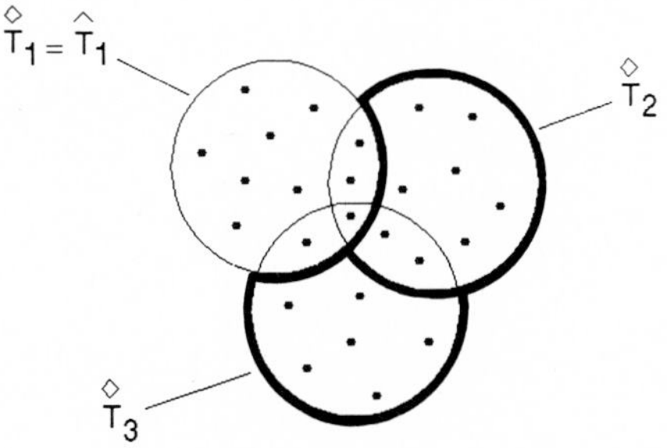

Fig. 4.4.3. Venn diagram for the partition of the set $\hat{M}(\varphi) = M(T_1) \cup M(T_2) \cup M(T_3)$ of minterms which are depicted as points in the plane

The partition of

$$\hat{M}(\varphi) = \hat{M}(\hat{T}_1) \cup \hat{M}(\hat{T}_2) \cup \hat{M}(\hat{T}_3)$$

to the three sets $\hat{M}(\overset{\diamond}{T}_1)$, $\hat{M}(\overset{\diamond}{T}_2)$ and $\hat{M}(\overset{\diamond}{T}_3)$ with

$$\hat{M}(\varphi) = \hat{M}(\overset{\diamond}{T}_1) \cup \hat{M}(\overset{\diamond}{T}_2) \cup \hat{M}(\overset{\diamond}{T}_3)$$

and

$$\hat{M}(\overset{\diamond}{T}_1) \cap \hat{M}(\overset{\diamond}{T}_2) = \hat{M}(\overset{\diamond}{T}_1) \cap \hat{M}(\overset{\diamond}{T}_3) = \hat{M}(\overset{\diamond}{T}_2) \cap \hat{M}(\overset{\diamond}{T}_3) = \phi$$

is done in a straightforward way as follows:

1. Define $\overset{\diamond}{T}_1 = \hat{T}_1$.

2. The minterms of $\hat{M}(\hat{T}_1) \cap \hat{M}(\hat{T}_2)$ are taken away from $\hat{M}(\hat{T}_2)$. The rest of $\hat{M}(\hat{T}_2)$ is $\hat{M}(\overset{\diamond}{T}_2)$ that defines $\overset{\diamond}{T}_2$, which is transformed to a DDNF.
3. The minterms of $\hat{M}(\hat{T}_3) \cap \hat{M}(\hat{T}_1)$ and of $\hat{M}(\hat{T}_3) \cap \hat{M}(\overset{\diamond}{T}_2)$ are taken away from $\hat{M}(\hat{T}_3)$ leaving $\hat{M}(\overset{\diamond}{T}_3)$ that defines $\overset{\diamond}{T}_3$, which is transformed to a DDNF.

Example 4.4.1 Two-out-of-three system function

Let [as in Eq. (4.3.2)]

$$\varphi = X_1 X_2 \vee X_1 X_3 \vee X_2 X_3 \,, \tag{4.4.16}$$

such that

$$\hat{T}_1 = X_1 X_2 \,, \quad \hat{T}_2 = X_1 X_3 \,, \quad \hat{T}_3 = X_2 X_3 \,.$$

Now, the CDNF of $\hat{T}_2$ is

$$\hat{T}_2 = X_1 X_2 X_3 \vee X_1 \bar{X}_2 X_3 \,.$$

Since

$$X_1 X_2 X_3 \in \hat{M}(\hat{T}_1) \tag{4.4.17}$$

we have

$$\overset{\diamond}{T}_2 = X_1 \bar{X}_2 X_3 \,,$$

and, provisionally,

$$\varphi = \hat{T}_1 \vee \overset{\diamond}{T}_2 \vee \ldots = X_1 X_2 \vee X_1 \bar{X}_2 X_3 \vee \ldots$$

Further,

$$\hat{T}_3 = X_1 X_2 X_3 \vee \bar{X}_1 X_2 X_3 \,.$$

Because of Eq. (4.4.17), and, since $\bar{X}_1 X_2 X_3$ is disjoint with $X_1 X_2 \vee X_1 \bar{X}_2 X_3$:

$$\overset{\diamond}{T}_3 = \bar{X}_1 X_2 X_3 \,,$$

and the final result is according to Eq. (4.4.15)

$$\varphi = X_1 X_2 \vee X_1 \bar{X}_2 X_3 \vee \bar{X}_1 X_2 X_3 \,. \text{——}$$

On choosing $\hat{T}_1 = X_2 X_3$ the result is

$$\varphi = X_2 X_3 \vee X_1 X_2 \bar{X}_3 \vee X_1 \bar{X}_2 X_3 \,. \tag{4.4.18a}$$

Note that in general $\hat{T}_i$ can yield (as $\overset{\diamond}{T}_i$) more than one $\hat{T}'_j$!

The general step-by-step procedure to determine the DDNF is:

1. Given φ as in Eq. (4.4.14), choose a proper (short) $\hat{T}_i$ to be $\overset{\diamond}{T}_1$ for Eq. (4.4.15).
2. If φ has been transformed partly to

$$\varphi_{k-1} := \left(\bigvee_{i=1}^{k-1} \overset{\diamond}{T}_i \right) , \tag{4.4.19}$$

take a new $\hat{T}_k$, and take away from $\hat{M}(\hat{T}_k)$ all the minterms contained in φ_{k-1}. The rest is a set of minterms making up a function $\overset{\diamond}{T}_k$ disjoint with φ_{k-1}. (How to find an optimal DDNF for $\overset{\diamond}{T}_k$ is shown below.)

3. If there is no new $\hat{T}_k$ left, the final DDNF φ_m has been found for φ.

Remark. At first sight, one could feel uncomfortable with the task of finding a DNF with disjoint terms for the minterms of a new $\hat{T}_i$ that are not contained in the terms of the intermediate result, Eq. (4.4.19). Fortunately, using a minimal DDNF of $\varphi = 1$, viz. 1_k solves this problem nicely.

Theorem 4.4.5. Let two terms $\hat{T}_1$ and $\hat{T}_2$ be non-disjoint, i.e. $\hat{T}_1 \hat{T}_2 \neq 0$ and let $\tilde{X}_1, \tilde{X}_2, \ldots, \tilde{X}_m$ (if necessary after renumeration) be all those literals which are contained in $\hat{T}_1$ but not in $\hat{T}_2$, then
(a) that part of $\hat{T}_2$ that is disjoint with $\hat{T}_1$ is

$$\hat{T}_2^* := \hat{T}_2(\bar{\tilde{X}}_1 \vee \tilde{X}_1 \bar{\tilde{X}}_2 \vee \tilde{X}_1 \tilde{X}_2 \bar{\tilde{X}}_3 \vee \ldots \vee \tilde{X}_1 \ldots \tilde{X}_{m-1} \bar{\tilde{X}}_m) , \tag{4.4.20}$$

and
(b) the terms $\hat{T}_2 \bar{\tilde{X}}_1, \hat{T}_2 \tilde{X}_1 \bar{\tilde{X}}_2, \hat{T}_2 \tilde{X}_1 \tilde{X}_2 \bar{\tilde{X}}_3, \ldots, \hat{T}_2 \tilde{X}_1 \ldots \tilde{X}_{m-1} \bar{\tilde{X}}_m$ are pairwise disjoint.

Proof. (b) is an immediate consequence of lemma 4.4.2.
(a) According to lemma 4.4.1

$$\hat{T}_2 = \hat{T}_2 \tilde{X}_1 \ldots \tilde{X}_m \vee \hat{T}_2^* .$$

Since $\hat{T}_2$ contains also the literals common to $\hat{T}_1$ and $\hat{T}_2$, the term

$$\hat{T}_2 \tilde{X}_1 \ldots \tilde{X}_m$$

contains $\hat{T}_1$ i.e. its minterms are also minterms of $\hat{T}_1$. Furthermore, each term of $\hat{T}_2^*$ contains a negated variable which appears non-negated in $\hat{T}_1$. Hence $\hat{T}_2^*$ is disjoint with $\hat{T}_1$, q.e.d.

Example 4.4.2 Simplified ARPA net [7], [8]

Figure 4.3.6 shows a very strongly simplified version of the ARPA net. If

$$X_i = \begin{cases} 1 , & \text{if link } i \text{ works properly} \\ 0 , & \text{else} \end{cases}$$

and if it is assumed (for simplicity) that nodes never fail, then the possibility of communication between v_1 and v_2 is given by Eq. (4.3.30) – after permuting terms – as the negated fault-tree function

$$\varphi(X_1, \ldots, X_7) = X_6X_7 \vee X_1X_2X_3 \vee X_1X_4X_7 \vee X_3X_5X_6 \vee X_1X_2X_5X_7 \vee X_1X_3X_4X_5 \vee X_2X_3X_4X_6 \,. \quad (4.4.21)$$

Figure 4.4.4 shows a simple corresponding reliability block diagram.

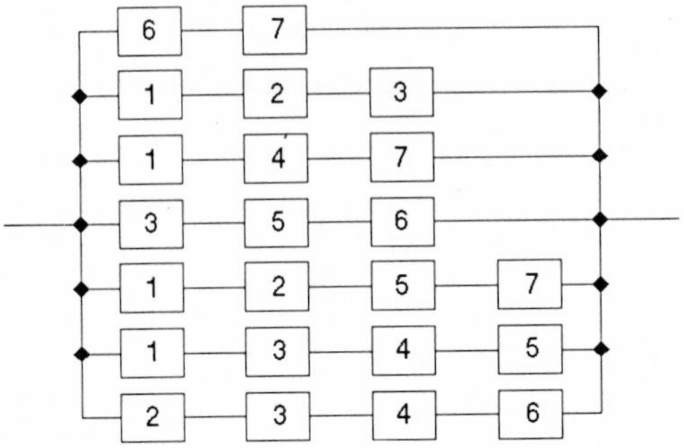

Fig. 4.4.4. Reliability block diagram for ARPA net

It is also a graphical description of the minimal path sets extracted from Fig. 4.3.6.

Now, the detailed procedure (1, 2, 3) preceding theorem 4.4.5:

We define

$$\hat{T}_1 = X_6X_7\,, \quad \hat{T}_2 = X_1X_2X_3\,, \quad \hat{T}_3 = X_1X_4X_7\,, \quad \hat{T}_4 = X_3X_5X_6\,,$$
$$\hat{T}_5 = X_1X_2X_5X_7\,, \quad \hat{T}_6 = X_1X_3X_4X_5\,, \quad \hat{T}_7 = X_2X_3X_4X_6\,. \quad (4.4.22)$$

First of all, since $\overset{\diamond}{T}_1 = \hat{T}_1$,

$$\overset{\diamond}{T}_1 = X_6X_7.$$

From theorem 4.4.5

$$\overset{\diamond}{T}_2 = X_1X_2X_3(\bar{X}_6 \vee X_6\bar{X}_7) = X_1X_2X_3\bar{X}_6 \vee X_1X_2X_3X_6\bar{X}_7 \,.$$

From theorem 4.4.5 with respect to $\hat{T}_1$, provisionally (marked by the upper index)

$$\overset{\diamond}{T}{}_3^{(1)} = X_1X_4X_7\bar{X}_6 \,,$$

with respect to $\hat{T}_2$,

$$\overset{\diamond}{T}_3 = X_1X_4X_7\bar{X}_6(\bar{X}_2 \vee X_2\bar{X}_3) = X_1\bar{X}_2X_4\bar{X}_6X_7 \vee X_1X_2\bar{X}_3X_4\bar{X}_6X_7 \,.$$

Analogously, regarding $\hat{T}_1$,

$$\overset{\diamond}{T}{}_4^{(1)} = X_3X_5X_6\bar{X}_7 \,,$$

and, regarding $\hat{T}_2$,

$$\overset{\diamond}{T}{}_4^{(2)} = X_3X_5X_6\bar{X}_7(\bar{X}_1 \vee X_1\bar{X}_2) = \bar{X}_1X_3X_5X_6\bar{X}_7 \vee X_1\bar{X}_2X_3X_5X_6\bar{X}_7 \ .$$

Finally both the terms of $\overset{\diamond}{T}{}_4^{(2)}$ are disjoint with $\hat{T}_3$ because they contain $\bar{X}_7$. Therefore,

$$\overset{\diamond}{T}_4 = \overset{\diamond}{T}{}_4^{(2)} \ .$$

Further, regarding $\hat{T}_1$,

$$\overset{\diamond}{T}{}_5^{(1)} = X_1X_2X_5X_7\bar{X}_6 \ ,$$

regarding $\hat{T}_2$

$$\overset{\diamond}{T}{}_5^{(2)} = X_1X_2X_5X_7\bar{X}_6\bar{X}_3 \ ,$$

and, regarding $\hat{T}_3$,

$$\overset{\diamond}{T}{}_5^{(3)} = X_1X_2X_5X_7\bar{X}_6\bar{X}_3\bar{X}_4 \ .$$

Regarding $\hat{T}_4$, it is obvious that $\overset{\diamond}{T}{}_5^{(3)}\hat{T}_4 = 0$. Hence

$$\overset{\diamond}{T}_5 = \overset{\diamond}{T}{}_5^{(3)} \ .$$

Further, regarding $\hat{T}_1$,

$$\overset{\diamond}{T}{}_6^{(1)} = X_1X_3X_4X_5(\bar{X}_6 \vee X_6\bar{X}_7) = X_1X_3X_4X_5\bar{X}_6 \vee X_1X_3X_4X_5X_6\bar{X}_7 \ ,$$

and, regarding $\hat{T}_2$,

$$\overset{\diamond}{T}{}_6^{(2)} = X_1X_3X_4X_5\bar{X}_6\bar{X}_2 \vee X_1X_3X_4X_5X_6\bar{X}_7\bar{X}_2 \ .$$

Regarding $\hat{T}_3$, only the second term of $\overset{\diamond}{T}{}_6^{(2)}$ is disjoint with it: Hence

$$\overset{\diamond}{T}{}_6^{(3)} = X_1X_3X_4X_5\bar{X}_6\bar{X}_2\bar{X}_7 \vee X_1X_3X_4X_5X_6\bar{X}_7\bar{X}_2 \ .$$

Regarding $\hat{T}_4$, the first term of $\overset{\diamond}{T}{}_6^{(3)}$ is disjoint with it, and the second term is absorbed by it: Hence

$$\overset{\diamond}{T}{}_6^{(4)} = X_1X_3X_4X_5\bar{X}_6\bar{X}_2\bar{X}_7 \ .$$

Regarding $\hat{T}_5$, this is also disjoint with $\overset{\diamond}{T}{}_6^{(4)}$. Hence,

$$\overset{\diamond}{T}_6 = \overset{\diamond}{T}{}_6^{(4)} \ .$$

Finally, regarding $\hat{T}_1$,

$$\overset{\diamond}{T}{}_7^{(1)} = X_2X_3X_4X_6\bar{X}_7 \ .$$

Regarding $\hat{T}_2$,

$$\overset{\diamond}{T}{}_7^{(2)} = X_2X_3X_4X_6\bar{X}_7\bar{X}_1 \ .$$

Regarding $\hat{T}_3$, $\overset{\diamond}{T}{}_7^{(2)}$ is disjoint with it. Regarding $\hat{T}_4$,

$$\overset{\diamond}{T}{}_7^{(3)} = X_2 X_3 X_4 X_6 \bar{X}_7 \bar{X}_1 \bar{X}_5 \ .$$

Regarding $\hat{T}_5$, $\hat{T}_6$: $\overset{\diamond}{T}{}_7^{(3)}$ is disjoint with both of them. Hence,

$$\overset{\diamond}{T}_7 = \overset{\diamond}{T}{}_7^{(3)} \ .$$

The final overall result is

$$\begin{aligned} \varphi &= \bigvee_{i=1}^{7} \overset{\diamond}{T}_i \\ &= X_6 X_7 \vee X_1 X_2 X_3 \bar{X}_6 \vee X_1 X_2 X_3 X_6 \bar{X}_7 \vee X_1 \bar{X}_2 X_4 \bar{X}_6 X_7 \\ &\quad \vee X_1 X_2 \bar{X}_3 X_4 \bar{X}_6 X_7 \vee \bar{X}_1 X_3 X_5 X_6 \bar{X}_7 \vee X_1 \bar{X}_2 X_3 X_5 X_6 \bar{X}_7 \\ &\quad \vee X_1 X_2 \bar{X}_3 \bar{X}_4 X_5 \bar{X}_6 X_7 \vee X_1 \bar{X}_2 X_3 X_4 X_5 \bar{X}_6 \bar{X}_7 \\ &\quad \vee \bar{X}_1 X_2 X_3 X_4 \bar{X}_5 X_6 \bar{X}_7 \ . \text{——} \end{aligned} \tag{4.4.23}$$

Comparing this with Eq. (4.4.21) shows that the property of disjointness has its price also in the lengths and the number of terms. However, in this case the CDNF would have

$$\underset{\overset{\diamond}{T}_1}{32} + \underbrace{8+4}_{\overset{\diamond}{T}_2} + \underbrace{4+2}_{\overset{\diamond}{T}_3} + \underbrace{4+2}_{\overset{\diamond}{T}_4} + \underset{\overset{\diamond}{T}_5}{1} + \underset{\overset{\diamond}{T}_6}{1} + \underset{\overset{\diamond}{T}_7}{1} = 59 \text{ minterms} \ .$$

(Remember that a term of length m of an n-variable function corresponds to 2^{n-m} minterms.)

4.4.3 The Shannon Decomposition Algorithm

(In [9] this procedure is called a Lagrangian decomposition.) As can be seen from Eq. (4.3.3) or its short-hand version

$$\varphi(\mathbf{X}) = X_i \varphi(X_i = 1) \vee \bar{X}_i \varphi(X_i = 0) \ , \tag{4.4.24}$$

the two terms $X_i\, \varphi(X_i = 1)$ and $\bar{X}_i \varphi(X_i = 0)$ are disjoint. However, neither of them needs be a DNF. Fortunately, here the way to the DNF of disjoint terms is – at least conceptually – much simpler than with the algorithm of §4.4.2. It is only necessary to treat recursively (with one variable, namely X_i, gone) $\varphi(X_i = 1)$ and $\varphi(X_i = 0)$ as $\varphi(\mathbf{X})$ was treated in Eq. (4.4.24). Finally, one will end in all the leaves of this binary tree algorithm with single terms (see Fig. 4.4.5), and by theorem 4.3.2, working one's way up through the tree, one gets a DNF of disjoint terms. The algorithm consists of the following basic steps:

(a) Application of (4.4.24) with an i not used before.
(b) Simplification of $\varphi(X_i = 1)$ and $\varphi(X_i = 0)$ by applying the absorption rule (2.2.14) and the first distributive law (2.2.3) for factoring out common terms.

(The proof of the correctness of this algorithm will be given after two examples.)

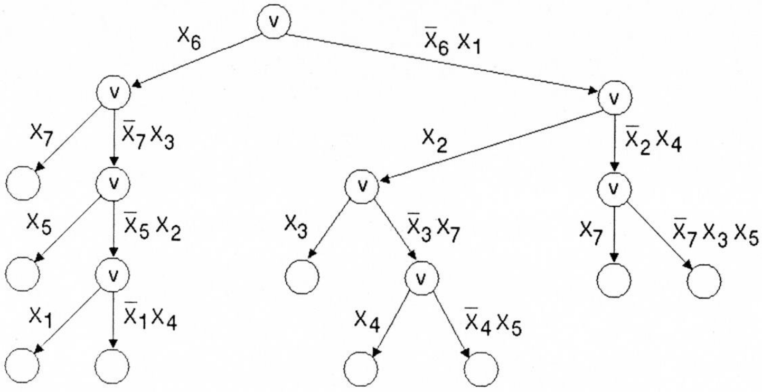

Fig. 4.4.5. Representation of the nested structure of the r.h.s. of Eq. (4.4.33) by a rooted tree

Example 4.4.3 Two-out-of-three system function

As is well known, here

$$\varphi = X_1 X_2 \vee X_1 X_3 \vee X_2 X_3 \ .$$

With $i = 1$ in (4.4.24)

$$\varphi = X_1(X_2 \vee X_3 \vee X_2 X_3) \vee \bar{X}_1 X_2 X_3 \ .$$

By (2.2.14) $X_2 X_3$ is absorbed by X_2 or (and) X_3:

$$\varphi = X_1(X_2 \vee X_3) \vee \bar{X}_1 X_2 X_3 \ .$$

By (2.2.18), finally,

$$\varphi = X_1(X_2 \vee \bar{X}_2 X_3) \vee \bar{X}_1 X_2 X_3 = X_1 X_2 \vee X_1 \bar{X}_2 X_3 \vee \bar{X}_1 X_2 X_3 \ , \quad (4.4.25)$$

which happens to equal Eq. (4.4.18). — —

Let us revisit the simplified ARPA net of example 4.4.2!

Example 4.4.4 Simplified ARPA net

For clarification we repeat Eq. (4.4.21):

$$\varphi(X_1, \ldots, X_7) = X_6 X_7 \vee X_1 X_2 X_3 \vee X_1 X_4 X_7 \vee X_3 X_5 X_6 \vee X_1 X_2 X_5 X_7 \vee X_1 X_3 X_4 X_5 \vee X_2 X_3 X_4 X_6 \ . \quad (4.4.26)$$

With $i = 6$ (4.4.24) yields

$$\varphi = X_6(X_7 \vee X_1X_2X_3 \vee \underline{\underline{X_1X_4X_7}} \vee X_3X_5 \vee \underline{\underline{X_1X_2X_5X_7}}$$
$$\vee \underline{\underline{X_1X_3X_4X_5}} \vee X_2X_3X_4)$$
$$\vee \bar{X}_6(\underline{\underline{X_1}}X_2X_3 \vee \underline{\underline{X_1}}X_4X_7 \vee \underline{\underline{X_1}}X_2X_5X_7 \vee \underline{\underline{X_1}}X_3X_4X_5) \; .$$

According to the basic step *b* of the algorithm (described before example 4.4.3), the underlined terms are absorbed via Eq. (2.2.14) and the underlined X_1's can be "factored" out via Eq. (2.2.3) leading to

$$\begin{aligned}\varphi &= X_6(X_7 \vee X_1X_2X_3 \vee X_3X_5 \vee X_2X_3X_4)\\ &\quad \vee \bar{X}_6X_1(X_2X_3 \vee X_4X_7 \vee X_2X_5X_7 \vee X_3X_4X_5)\\ &=: X_6\varphi_{1,1} \vee \bar{X}_6X_1\varphi_{1,2} \; .\end{aligned} \tag{4.4.27}$$

The functions in the parentheses are called $\varphi_{1,1}$[†] and $\varphi_{1,2}$[†] for which the same type of expansion is performed: With $i = 7$ in Eq. (4.4.24), using Eq. (2.2.11)

$$\begin{aligned}\varphi_{1,1} &= X_7 \vee \bar{X}_7(X_3X_5 \vee X_1X_2X_3 \vee X_2X_3X_4)\\ &= X_7 \vee \bar{X}_7X_3(X_5 \vee X_1X_2 \vee X_2X_4)\\ &=: X_7 \vee \bar{X}_7X_3\varphi_{2,1} \; ,\end{aligned} \tag{4.4.28}$$

where

$$\varphi_{2,1} = X_5 \vee \bar{X}_5X_2(X_1 \vee \bar{X}_1X_4) \; . \tag{4.4.29}$$

Furthermore,

$$\begin{aligned}\varphi_{1,2} &= X_2(X_3 \vee X_4X_7 \vee X_5X_7 \vee \underline{\underline{X_3X_4X_5}}) \vee \bar{X}_2(X_4X_7 \vee X_3X_4X_5)\\ &= X_2(X_3 \vee X_4X_7 \vee X_5X_7) \vee \bar{X}_2X_4(X_7 \vee X_3X_5)\\ &=: X_2\varphi_{2,2} \vee \bar{X}_2X_4\varphi_{2,3} \; ,\end{aligned} \tag{4.4.30}$$

where

$$\varphi_{2,2} = X_3 \vee \bar{X}_3X_7(X_4 \vee \bar{X}_4X_5) \; , \tag{4.4.31}$$

$$\varphi_{2,3} = X_7 \vee \bar{X}_7X_3X_5 \; . \tag{4.4.32}$$

[†] The first index gives the nesting depth of the analysis; the second is just a running number.

Inserting in bottom-up mode yields

$$\begin{aligned}\varphi = & X_6\{X_7 \vee \bar{X}_7 X_3[X_5 \vee \bar{X}_5 X_2(X_1 \vee \bar{X}_1 X_4)]\} \\ & \vee \bar{X}_6 X_1\{X_2[X_3 \vee \bar{X}_3 X_7(X_4 \vee \bar{X}_4 X_5)] \\ & \vee \bar{X}_2 X_4(X_7 \vee \bar{X}_7 X_3 X_5)\} \qquad (4.4.33) \\ = & X_6 X_7 \vee X_6 \bar{X}_7 X_3 X_5 \vee X_6 \bar{X}_7 X_3 \bar{X}_5 X_2 X_1 \vee X_6 \bar{X}_7 X_3 \bar{X}_5 X_2 \bar{X}_1 X_4 \\ & \vee \bar{X}_6 X_1 X_2 X_3 \vee \bar{X}_6 X_1 X_2 \bar{X}_3 X_7 X_4 \vee \bar{X}_6 X_1 X_2 \bar{X}_3 X_7 \bar{X}_4 X_5 \\ & \vee \bar{X}_6 X_1 \bar{X}_2 X_4 X_7 \vee \bar{X}_6 X_1 \bar{X}_2 X_4 \bar{X}_7 X_3 X_5 \ .\end{aligned}$$

Finally, arranging the indices in ascending order,

$$\begin{aligned}\varphi = & X_6 X_7 \vee X_3 X_5 X_6 \bar{X}_7 \vee X_1 X_2 X_3 \bar{X}_5 X_6 \bar{X}_7 \vee \bar{X}_1 X_2 X_3 X_4 \bar{X}_5 X_6 \bar{X}_7 \\ & \vee X_1 X_2 X_3 \bar{X}_6 \vee X_1 X_2 \bar{X}_3 X_4 \bar{X}_6 X_7 \vee X_1 X_2 \bar{X}_3 \bar{X}_4 X_5 \bar{X}_6 X_7 \\ & \vee X_1 \bar{X}_2 X_4 \bar{X}_6 X_7 \vee X_1 \bar{X}_2 X_3 X_4 X_5 \bar{X}_6 \bar{X}_7 \ . \qquad (4.4.34)\end{aligned}$$

Comment. The fact that Eq. (4.4.34) is by one term shorter than Eq. (4.4.23) (nine terms instead of ten) does not imply anything concerning a general comparison of the two algorithms. Equation (4.4.33) can be depicted as a switching circuit graph. However, the special digraph of Fig. 4.4.5 with marked edges, which has some similarity with the decision diagrams of §3.2, is perhaps more instructive. All the inner nodes, including the root, are meant for disjunction and the marks of the edges mean the conjunction of these marks with the function of the node the relevant edge is pointing at. (Clearly, from a graph theoretic point of view it is not very nice, that in Fig. 4.4.5 there are, formally, no leaves. Hence, "dummy" leaves were drawn. However, Fig. 4.4.5 can be transformed to a usual Boolean algebra syntax diagram as introduced by Fig. 3.4.2.)

From the height of the rooted binary tree of Fig. 4.4.5 it is obvious that, despite the seven variables of this example, only four expansion steps were necessary. This shows the degree of optimization achieved here. — —

Proof of Correctness

The correctness of this algorithm, i.e. the fact that it really transforms a DNF to a DDNF, is easily seen as follows. By every expansion step $\hat{M}(\varphi)$ of the given function φ is split in two subsets. Hence, also the final subsets of minterms, belonging to terms of the DDNF, are disjoint such that through theorem 4.4.2 these terms too are disjoint, q.e.d.

4.4.4 A Binary Decision Tree Algorithm

The following algorithm is, like that described in §4.4.3, of remarkable simplicity though rather efficient, as subsequent examples will show. The main idea is to partition (halve) Ω the set of all the 2^n minterms recursively by means of a binary decision tree, where it is always decided if the new halves belong to $\hat{M}(\varphi)$, $\hat{M}(\bar{\varphi})$, or if they are mixtures of minterms some of which are belonging to $\hat{M}(\varphi)$ and others to $\hat{M}(\bar{\varphi})$. In other words, the general argument n-vector

$$\mathbf{X}=(d, d, \ldots, d) \ ,$$

where d stands for don't care, as in the case of not completely specified Boolean functions (§5.3.3), is specified step by step by setting more and more of the d's to 0 or to 1, and then it is checked if

$$\varphi(\mathbf{X})=\begin{cases}0\\1 \ ; \quad X_i = 0, 1, d \ ; \quad i=1, \ldots, n \ .\\d\end{cases} \tag{4.4.35}$$

To formalize this process of changing a Boolean function of n variables to a ternary function of one variable as in Eq. (4.4.35) we define, beyond the obvious rules like $d \vee 0=d$, $d \vee 1=1$, and $dd=d$,

$$\bar{d}=d \ . \tag{4.4.36}$$

Other rules are not needed, even for deeply nested φ, as long as φ is given in the AND-OR-NOT system of operators and as long as no simplifications of the type $Y\bar{Y}=0$ are possible in the given representation of φ, since, formally, for $Y=d$ we get by Eq. (4.4.36) the wrong result

$$Y\bar{Y}=dd \ .$$

The relation between $\mathbf{X}$ and the corresponding term is obvious. For instance, in the case of $n=4$

$$(1, 0, 1, 1) \mathrel{\hat{=}} X_1\bar{X}_2X_3X_4 \ , \qquad (1, d, d, 0) \mathrel{\hat{=}} X_1\bar{X}_4 \ .$$

Note that in applications in the field of reliability technology, where φ is the fault tree function, it is practical to call states characterized by $\varphi=0$ and $\varphi=1$, "good" and "bad" states, respectively. Remember also that minterms correspond to elementary states and that the same is true for specific argument n-tuples (state vectors) $\mathbf{X}$.

Example 4.4.5 Two-out-of-three system with voter

Figure 4.4.6 shows a plausible fault tree of this system.

Figure 4.4.7 shows the decision tree of the above algorithm. Good, bad, and mixed system states $\mathbf{X}$ are found by evaluating

$$X_\varphi=\varphi(\mathbf{X})=X_1X_2 \vee X_2X_3 \vee X_3X_1 \vee X_4 \tag{4.4.37}$$

using Eq. (4.4.35). For instance, $\varphi(d, d, d, 0)=d$, $\varphi(d, d, d, 1)=1$.

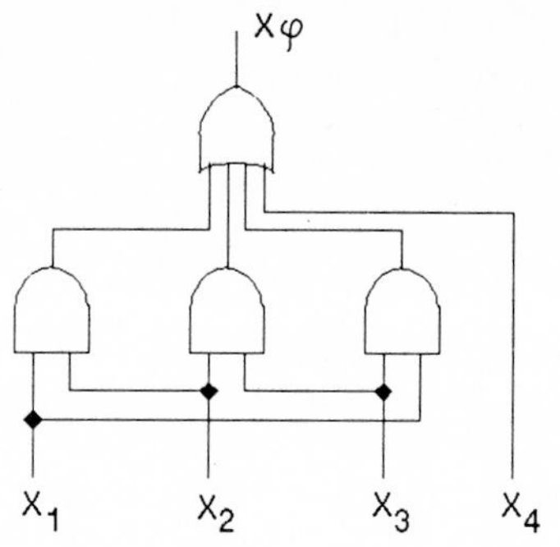

Fig. 4.4.6. Fault tree of two-out-of-three system with voter

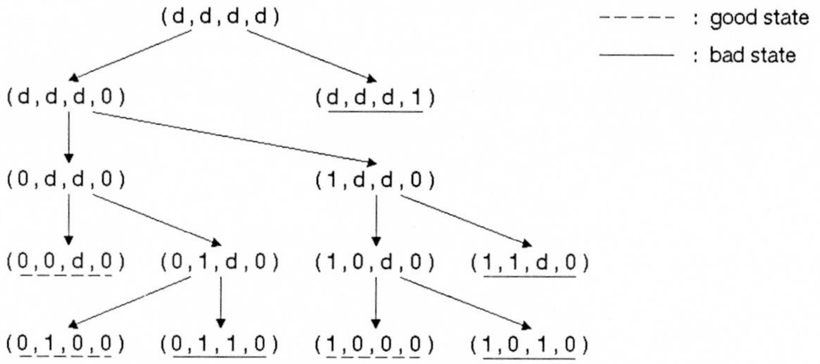

Fig. 4.4.7. Decision tree to find all the good or bad system states, respectively

From the underlined "bad" states of Fig. 4.4.7 one immediately gets the disjoint sum of products form (DDNF)

$$X_\varphi = X_4 \vee X_1 X_2 \bar{X}_4 \vee \bar{X}_1 X_2 X_3 \bar{X}_4 \vee X_1 \bar{X}_2 X_3 \bar{X}_4 \ . \tag{4.4.38}$$

For later use it should be pointed out that, as to counting levels of states in the decision tree of the type of Fig. 4.4.7, if the highest level (the tree's root) has the level

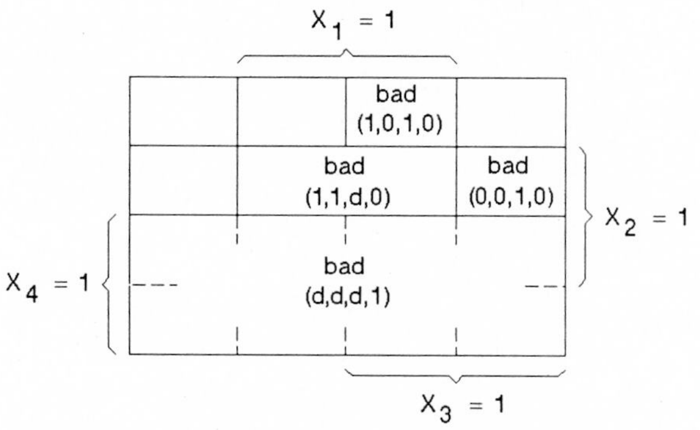

Fig. 4.4.8. Set nesting expressed by the tree of Fig. 4.4.7

No. 0, level m contains system states with 2^{n-m} elementary states each. Hence, Fig. 4.4.7 could also be replaced by a box-nesting as in Fig. 4.4.8.

Alternatively, with the 4-variable Karnaugh map (§3.3), Fig. 4.4.9 shows the sets of minterms implied by Eq. (4.4.38) and by the "bad" sets of Fig. 4.4.7.— —

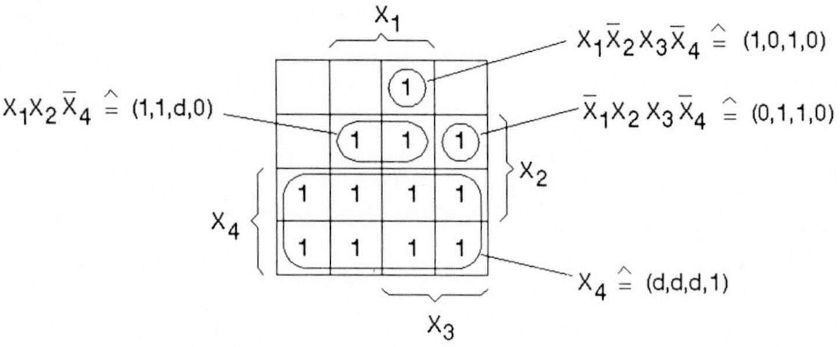

Fig. 4.4.9. Karnaugh map of Eq. (4.4.38)

For deeper insight the "standard" example discussed in most of the relevant literature is re-examined.

Example 4.4.6	Simplified ARPA net

We start with the DNF Eq. (4.4.21, 4.4.26) which is repeated here for convenience:

$$\varphi = X_6X_7 \vee X_1X_2X_3 \vee X_1X_4X_7 \vee X_3X_5X_6 \vee X_1X_2X_5X_7 \vee X_1X_3X_4X_5 \vee X_2X_3X_4X_6 \,. \tag{4.4.39}$$

The relevant sequential decision tree is depicted in Fig. 4.4.10. Whenever a term of Eq. (4.4.39) becomes 1, one has found a good state; whenever all the terms of Eq. (4.4.39) become 0, one has found a bad state. Again it was possible to limit the number of (disjoint) good sets to 9, whose indicator vectors are underlined in Fig. 4.4.10.

The most interesting result of this example is the fact that by some trial and error it was, except for the first two decisions, always possible to avoid a growth in the breadth of the tree. Even though this is partly a peculiarity of this example, it will be wise to remember this fact in a generalized optimized implementation of the sequential decision algorithm presented here.

The underlined nodes in Fig. 4.4.10 yield the DDNF

$$\begin{aligned}\varphi = {} & X_6X_7 \vee X_1X_4\bar{X}_6X_7 \vee X_3X_5X_6\bar{X}_7 \vee X_1X_2X_3\bar{X}_6\bar{X}_7 \vee X_1X_2X_3\bar{X}_4\bar{X}_6X_7 \\ & \vee X_1X_2X_3\bar{X}_5X_6\bar{X}_7 \vee X_1\bar{X}_2X_3X_4X_5\bar{X}_6\bar{X}_7 \vee X_1X_2\bar{X}_3\bar{X}_4X_5\bar{X}_6X_7 \\ & \vee \bar{X}_1X_2X_3X_4\bar{X}_5X_6\bar{X}_7 \,. \text{— —}\end{aligned} \tag{4.4.40}$$

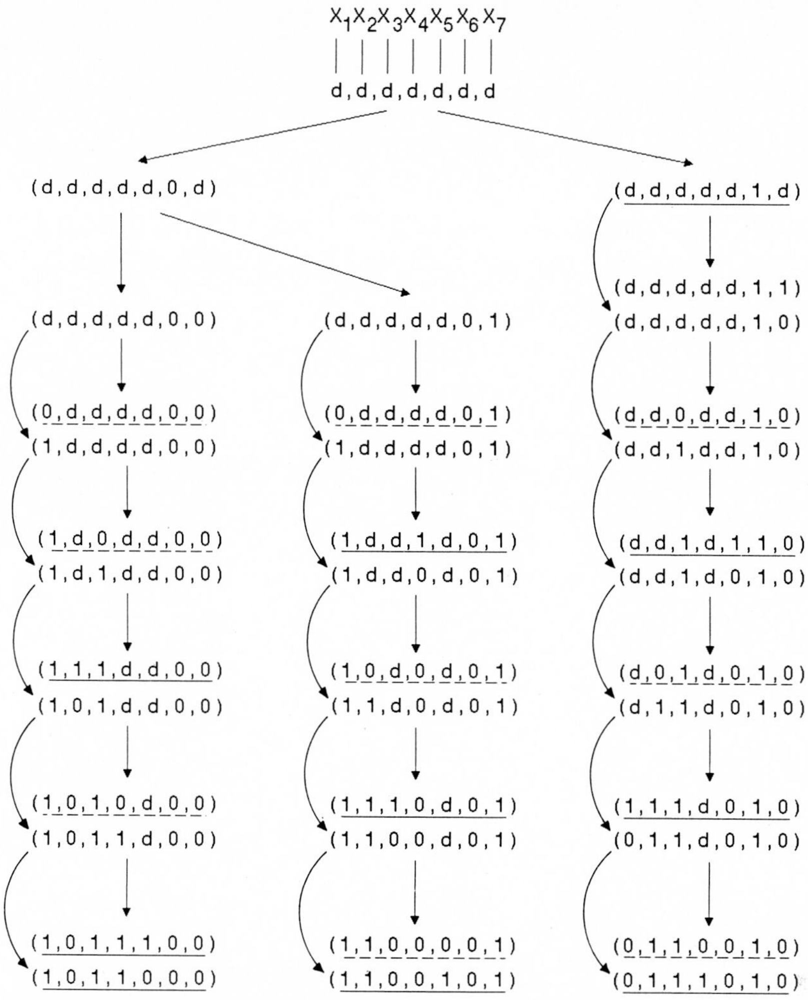

Fig. 4.4.10. Decision tree for states of simplified ARPA net; underlining not as in Fig. 4.4.7. The nine good states correspond to 59 elementary good states

At this point the decision diagrams of §3.2 should be reviewed. Example 4.4.5 is tightly coupled to example 3.2.1. Specifically, there is a one-to-one correspondence between Figs. 3.2.2 and 4.4.7. This offers a nice opportunity to deepen the analysis of §3.2. In the inner nodes of decision diagrams there is not only the rather trivial binary decision for the left or the right son but, moreover, the non-trivial decision for the next d to be replaced by 1 or 0, respectively. This can be done with some kind of optimization in mind, e.g. a short DDNF as above. For any choice of the sequencing of variables to be given fixed values, it is clear how to construct the proper (unique) decision tree. On the other hand, given a decision tree, an algebraic form of the tree's Boolean function φ is found by the following little algorithm:

Algorithm for Decision Tree Evaluation

(1) For every leaf marked by 1, note the sequence of markings of links and incident nodes on the (directed) path from this leaf to the root, yielding a sequence

$$b_1 X_{i_1} b_2 X_{i_1} b_3 X_{i_3} \ldots b_k X_{i_k} , \quad b_j \in \{0, 1\} , \tag{4.4.41}$$

where X_{i_k} is the mark of the root (of the tree) and the b_j and X_{ij} are marks of links and nodes, respectively; see Fig. 4.4.11.

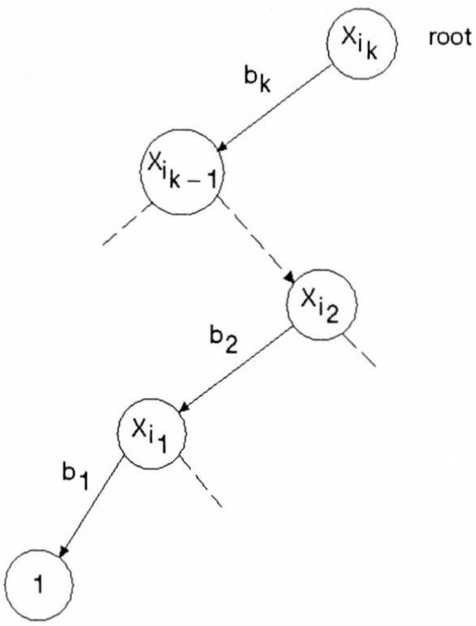

Fig. 4.4.11. Marking paths of a binary decision tree

(2) Replace every pair $b_j X_{i_j}$ by

$$\tilde{X}_{i_j} := \begin{cases} X_{i_j} & \text{if } b_j = 1 , \\ \bar{X}_{i_j} & \text{if } b_j = 0 , \end{cases}$$

whereby Eq. (4.4.41) turns into a term of a DDNF.

Proof. The term

$$\hat{T}_i := \bigwedge_{j=1}^{k} \tilde{X}_{i_j}$$

representing path *i* from the root to a 1-type leaf can be a term of a DNF of φ, since $\hat{T}_i = 1$ implies $X_\varphi = 1$. Every such path *i* differs at least in the node, where it separates from another one *i** in the marking variable X_a of this node. Hence $\hat{T}_i$ will contain $\tilde{X}_a$, and $\hat{T}_{i*}$ will contain $\bar{\tilde{X}}_a$; see Fig. 4.4.12 for illustration. Hence $\hat{T}_i \hat{T}_{i*} = 0$, q.e.d.

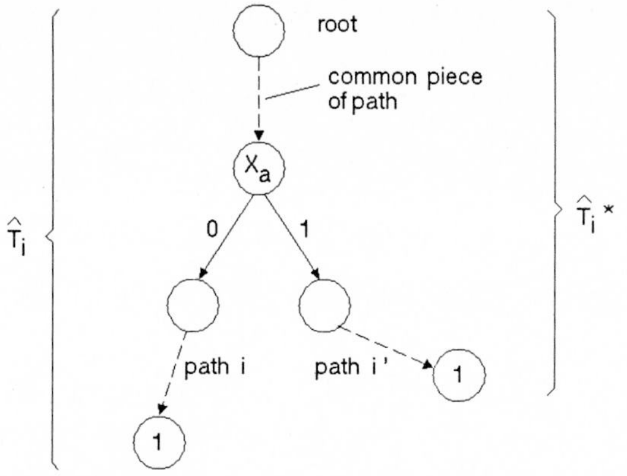

Fig. 4.4.12. On the disjointness of terms of different paths of the decision tree of a Boolean function

4.4.5 Comparison of DDNF Algorithms

The three algorithms of §§4.4.2–4.4.4 are of similar efficiency as could be shown by PASCAL implementations.† Computational complexity depends on the number m of DNF terms and their length, which is at most n for a function $\varphi : B^n \to B$. In almost all cases the number of terms of the resulting DDNF will be much smaller than the number of its minterms. However, the "set addition" algorithm of §4.4.2 differs strongly from the two others in that it does not allow for parallelism of execution; and it can only start with a DNF. The Shannon decomposition of a maxterm produces the optimal DDNF needed in the set addition algorithm of Abraham:

$$\begin{aligned} X_1 \vee X_2 \vee X_3 \vee \ldots \vee X_n &= X_1 \vee \bar{X}_1\{X_2 \vee \bar{X}_2[X_3 \vee \bar{X}_3(\ldots)]\} \\ &= X_1 \vee \bar{X}_1 X_2 \vee \bar{X}_1 \bar{X}_2 X_3 \vee \bar{X}_1 \bar{X}_2 \bar{X}_3 X_4 \\ &\quad \vee \ldots \vee \bar{X}_1 \ldots \bar{X}_{n-1} X_n \,. \end{aligned}$$

Clearly the last r.h.s. follows immediately upon application of the Abraham algorithm on the l.h.s.

The binary tree-type algorithms of §§4.4.3, 4.4.4 differ in that the Shannon algorithm allows for more intermediate optimizations, namely application of absorption rules, etc., whereas the decision tree algorithm needs less memory space for intermediate formulae but, possibly, more CPU time for the frequent evaluation of the complete original φ. However, by a deeper analysis of intermediate optimization steps for the decision tree algorithm the difference in performance between the latter two algorithms can be made smaller.

Last not least the three algorithms of this section offer an almost ideal triple for three-version programming in the context of majority voting with fault tolerant software [50].

† For details see §10.

In §§8 and 9 DDNFs will be used to calculate the probability of a Boolean function being 1 and the mean duration of the value of a binary random process, respectively. In these contexts further comments on the comparison of the above three algorithms will be in order.

4.5 Boolean Functions with Special Properties

Apart from the various representations of Boolean functions, it is of theoretical and practical interest to look for properties that are independent of these representations, e.g. monotonicity or symmetry, even though such properties can be made more apparent by suitably chosen representations. Some of the following points will be very short, i.e. just definitions for the sake of completeness of this text, others will be more amply discussed.

Linearity

Definition 4.5.1. A *linear* Boolean function is one that allows for the representation

$$\varphi(X_1, \ldots, X_n) = c_0 \not\equiv c_1 X_i \not\equiv c_2 X_2 \not\equiv \ldots \not\equiv c_n X_n \; ; \; c_i \in \{0, 1\} \; . \text{— —} \tag{4.5.1}$$

Obviously, an alternative for (4.5.1) is

$$\varphi = c_0 \not\equiv \left(\mathop{\not\equiv}_{i \in I} X_i \right) ; \quad I \subseteq \{1, \ldots, n\} \; . \tag{4.5.2}$$

Linear Boolean functions are good for the algebraic description of certain algebraic coding schemes. In fact, for $c_0 = 0$ Eq. (4.5.2) could describe the determination of a (single) check bit, which could be inverted in case of $c_0 = 1$; see Eq. (4.1.7). Elaborating a little: when encoding an information word

$$\mathbf{i} = (i_0, i_1, \ldots, i_{k-1}) \tag{4.5.3}$$

the associated check word is (with r for redundancy) the linear Boolean vector function $\mathbf{r} = (r_0, r_1, \ldots, r_m)$ defined by

$$\mathbf{r} = \mathbf{i}(\wedge)\mathbf{C} \; , \tag{4.5.4}$$

where $\mathbf{C}$ is the binary check matrix. For instance, in case of a Hamming code, $\mathbf{C}$ can bring about that errors in two bits can be detected or a single bit error can be corrected. The complete codeword $\mathbf{c}$ is the concatenation of $\mathbf{r}$ and $\mathbf{i}$:

$$\mathbf{c} = (\mathbf{r}, \mathbf{i}) \; . \tag{4.5.5}$$

Example 4.5.1 Hamming code for 1 byte

In Eq. (4.5.3) let $k=8$. Then

$$\mathbf{C}=\begin{bmatrix} 1 & 1 & 0 & 0 \\ 1 & 0 & 1 & 0 \\ 1 & 0 & 0 & 1 \\ 0 & 1 & 1 & 0 \\ 0 & 1 & 0 & 1 \\ 0 & 0 & 1 & 1 \\ 1 & 1 & 1 & 0 \\ 1 & 1 & 0 & 1 \end{bmatrix}$$

is a useful check matrix for a Hamming code. Figure 4.5.1 shows the corresponding encoder circuitry; see Eq. (4.5.4). For instance, from Eq. (4.5.4) and the above **C**:

$$r_0 = i_0 \not\equiv i_1 \not\equiv i_2 \not\equiv i_6 \not\equiv i_7 \ . \text{ — —}$$

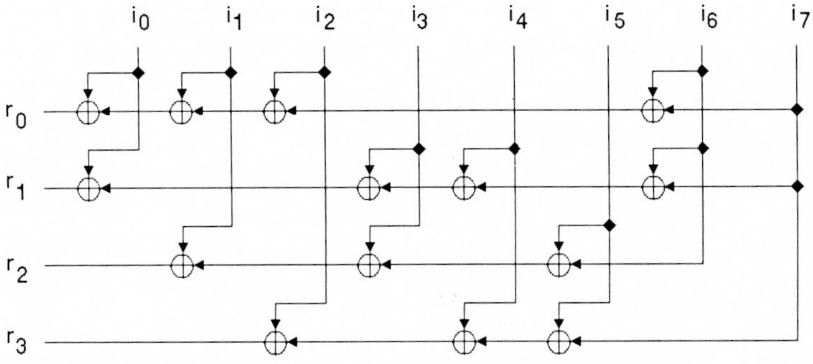

Fig. 4.5.1. Encoder for a (12, 8) Hamming code; see Fig. 3.4.3

In the decoder, first of all, the received binary word

$$\mathbf{c}' = \mathbf{c} \not\equiv \mathbf{e} = (\mathbf{r}', \mathbf{i}') \tag{4.5.6}$$

is checked if it has the proper code word form. This is done by evaluating the so-called *syndrome*

$$\mathbf{s} := \mathbf{r}' \not\equiv \mathbf{r}'' \ , \tag{4.5.7}$$

where

$$\mathbf{r}'' = \mathbf{i}' (\wedge) \mathbf{C} \ . \tag{4.5.8}$$

Hence, an important part of the decoder is another encoder to produce $\mathbf{r}''$ from the received information word $\mathbf{i}'$. If no error $\mathbf{e}$ has occurred (during data transmission or storage in a computer memory), then $\mathbf{s}=\mathbf{0}$. For single or double bit errors (in $\mathbf{i}$)

$\mathbf{r}'' \neq \mathbf{r}'$, if all the lines of $\mathbf{C}$ are different (and contain at least two 1s each). Therefore, in these cases $\mathbf{s} \neq \mathbf{0}$, i.e. single and double bit errors are detected. Moreover, if the single bit error described by

$$e_j = \begin{cases} 0, & j \neq l \\ 1, & j = l \end{cases}$$

yields a special syndrome value $\mathbf{s} = \mathbf{s}^{(l)}$ uniquely, then, on finding $\mathbf{s}^{(l)}$, a simple decoding will identify l and the correct i_l will be found via negation of i'_l.

Example 4.5.2 (Example 4.5.1 contd.)

The decoding mechanism (circuitry) for example 4.5.1 is shown in Fig. 4.5.2. The negation is done using Eq. (4.1.7), i.e.

$$\overline{i'_l} = i'_l \not\equiv 1 \ . \text{——}$$

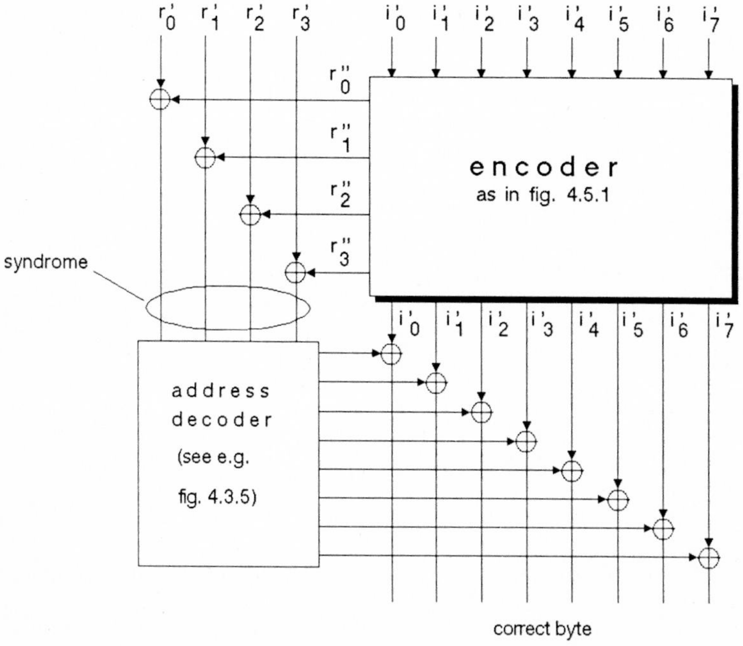

Fig. 4.5.2. Hamming decoder for a coded byte, i.e. a (12, 8) code

Parity Function as "Chess Board" Function

In this context some non-coding properties of the parity function

$$\varphi(X_1, \ldots, X_n) = \mathop{\not\equiv}_{i=1}^{n} X_i \tag{4.5.9}$$

shall be briefly reviewed. First, in the K-maps it has a chess board pattern of 1s; see Fig. 4.5.3. By the process of doubling a K-map it is clear from Eq. (4.1.7) that in the new half the values of X_φ are the negated corresponding values of the old half. Consequently, the CDNF of the parity function is the only possible DNF. As to error detection, it is clear that errors in an even number of bits will not change X_φ, i.e. they are not detectable; whereas errors in an odd number of bits will change X_φ, i.e. they are detectable.

A parity function is linear, and it is also symmetrical:

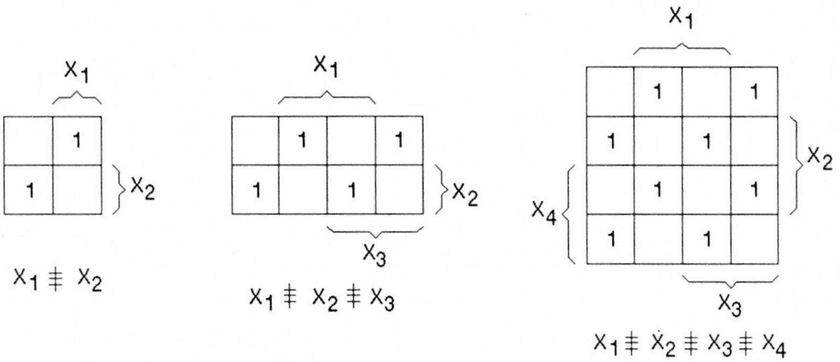

Fig. 4.5.3. The parity function on K-maps

Symmetry

Definition 4.5.2. A *symmetrical* Boolean function is one that allows for any permutation of the variables. — —

Example 4.5.3 (Two-out-of-three majority function)

It is easily checked that

$$\varphi(X_1, X_2, X_3) = X_1 X_2 \vee X_1 X_3 \vee X_2 X_3 \tag{4.5.10}$$

is symmetrical. Exchanging 1 and 2 or 1 and 3 or 2 and 3 does not change φ. — —

Degeneratedness

Definition 4.5.3. A *degenerated* Boolean function of $X_1, \ldots, X_n$ is one that does not depend on all these variables. — —

Note that a given form of a function, e.g. a DNF can become degenerated simply by asserting that it should be interpreted as a function of more variables. Of course, a function can be degenerated even though all its n variables show up in a given representation. For instance, looked upon as a function of X_1 and X_2,

$$\varphi = X_1 X_2 \vee X_1 \bar{X}_2 = X_1$$

is degenerated.

Decomposability

Definition 4.5.4. A *decomposable* Boolean function $\varphi: B^n \to B$ allows for a representation

$$\varphi(\mathbf{X}) = \tilde{\varphi}(\psi'(\mathbf{X}'), \psi''(\mathbf{X}'')), \ \tilde{\varphi}, \psi', \psi'' \text{ Boolean functions,} \tag{4.5.11}$$

where the components of $\mathbf{X}'$ are different from those of $\mathbf{X}''$ but their composed set is the set of the components of $\mathbf{X}$. — —

Example 4.5.4 (A counter-example)

At least at first sight the function

$$\varphi(X_1, X_2, X_3, X_4) = X_1(X_2 \vee \bar{X}_3 X_4) \vee \bar{X}_1 \bar{X}_3 \tag{4.5.12}$$

does not appear to be "nicely" decomposable, since in the DNF

$$\varphi = X_1 X_2 \vee X_1 \bar{X}_3 X_4 \vee \bar{X}_1 X_3 \tag{4.5.13}$$

X_1 or $\bar{X}_1$ are part of every term. — —

Obviously, functions for which one can find a tree-type syntax diagram are decomposable, often in many different ways.

Example 4.5.5 Tree decomposition

Let (see Fig. 4.5.4)

$$\varphi = X_1 \vee X_2 X_3 \vee X_4 X_5 X_6 .$$

Here the function $\tilde{\varphi}$ of Eq. (4.5.11) could be the disjunction and ψ' and ψ'' could be $X_1 \vee X_2 X_3$ and $X_4 X_5 X_6$, respectively. — —

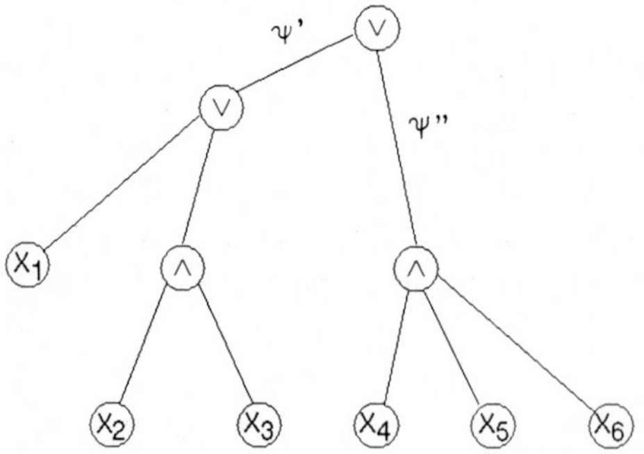

Fig. 4.5.4. Decomposition of a tree

Monotonicity and Coherence

Primarily in reliability theory, a certain type of monotonicity of Boolean functions has some importance. It is called *coherence* and means something like technical soundness in the sense that a coherent system is good (i.e. works according to its specifications) when all its n components are good, and bad whenever all its n components are bad, and can neither become bad when one of its repaired components is put back to service, nor can it become good again by the failure of another component.

In the "language" of Boolean indicator functions, coherence can be expressed by *extremal value consistency*:

$$\varphi(\mathbf{0}) = 0 \ , \quad \mathbf{0} := (0, \ldots, 0) \ ; \quad \varphi(\mathbf{1}) = 1 \ , \quad \mathbf{1} := (1, \ldots, 1) \tag{4.5.14}$$

together with *isotonousness*:

$$\varphi(\mathbf{X}_1) \leqq \varphi(\mathbf{X}_2) \ ; \quad \mathbf{X}_1 \leqq \mathbf{X}_2 \ , \tag{4.5.15}$$

the latter in the sense that for all $i \in \{1, \ldots, n\}$

$$X_{1,i} \leqq X_{2,i} \ .$$

Coherent functions are easily identified by the following theorem.

Theorem 4.5.1. DNFs with only normal variables are coherent.

Proof. From

$$\varphi(\mathbf{X}) = \bigvee_{i=1}^{m} \hat{T}_i \ , \quad \hat{T}_i = \bigwedge_{j=1}^{n_i} X_{l_{i,j}}$$

it follows that the subsequent trivial expansion is possible for every component X_k of $\mathbf{X}$:

$$\varphi = X_k \varphi' \vee \varphi'' \ ; \quad \varphi', \varphi'' \text{ independent of } X_k \text{ and Boolean.} \tag{4.5.16}$$

Hence, changes of X_k from 0 to 1 will not change φ (if $\varphi' = 0$ or $\varphi'' = 1$) or they will change φ from 0 to 1 (if $\varphi' = 1$ and $\varphi'' = 0$), but never from 1 to 0.

Further, $\varphi(\mathbf{0}) = 0$, because for $\mathbf{X} = \mathbf{0}$ all $\hat{T}_i = 0$. Also $\varphi(\mathbf{1}) = 1$, because for $\mathbf{X} = \mathbf{1}$ all $\hat{T}_i = 1$, q.e.d.

If in at least one $\hat{T}_i$ at least one literal $\tilde{X}_k = \bar{X}_k$, then, in general, on applying Eq. (2.2.3) one finds the form (representation)

$$\varphi = X_k \varphi' \vee \bar{X}_k \varphi'' \vee \varphi''' \tag{4.5.17}$$

with φ', φ'', φ''' independent of X_k and Boolean. For $\mathbf{X}$ such that $\varphi' = \varphi''' = 0$ and $\varphi'' = 1$, a change of X_k from 0 to 1 would mean a change of φ from 1 to 0. This seems to indicate that a stronger statement than theorem 4.5.1 is true, saying that negated variables in a DNF would mean non-coherence. However, this would not be true, as the following trivial counter example shows. By Eq. (2.2.18), with φ'' independent from X_k, the function

$$\varphi = X_k \vee \bar{X}_k \varphi'' = X_k \vee \varphi''$$

is coherent if φ'' is.

Example 4.5.6 Exclusive OR as an example of non-coherence

One of the simplest non-coherent functions is [see Eq. (2.1.1)]

$$\varphi = X_1 \not\equiv X_2 = X_1 \bar{X}_2 \vee \bar{X}_1 X_2 \ . \tag{4.5.18}$$

(An illustration is the company or institute run by two capable directors who hate each other, which runs smoothly only when one and only one is present.)

Clearly, here

$$\varphi(1, 0) \geqq \varphi(1, 1) \ ,$$

violating Eq. (4.5.15). — —

Corollary 4.5.1. Any Boolean function φ that can be represented using only conjunction and disjunction (but not negation) is coherent.

Proof. Such a φ can be easily transformed to a DNF without negated variables. The rest is evident by theorem 4.5.1, q.e.d.

For future use we add the following definition of unate-ness [45].

Definition 4.5.5. A Boolean function representation containing only normal or only negated variables is a *unate* one. A function with at least one unate DNF is a *unate function.* — —

The trivial example function

$$\varphi = X_1 \vee X_2 = X_1 \vee \bar{X}_1 X_2$$

is unate. However, the last representation is not a unate one.

Corollary 4.5.2. A unate DNF with normal variables is coherent.

Proof. Theorem 4.5.1.

Of course, not all unate functions are coherent. Counter example: $\varphi(X_1, X_2) = \bar{X}_1 X_2$, since $\varphi(1, 1) = 0$.

Monotonicity also has an interesting consequence in the field of path sets and cut sets (§ 4.3.2):

Theorem 4.5.2. A set of edges of a graph is either a cut set or the set of all the other edges of this graph is a path set.

We prove the following equivalent version.

Theorem 4.5.3. For any monotonously non-decreasing $\varphi(\mathbf{X})$, where $\mathbf{X}=(\mathbf{X}_a, \mathbf{X}_b)$, either

$$\varphi(\mathbf{X}_a, \mathbf{0})=0 \quad \text{or} \quad \varphi(\mathbf{1}, \mathbf{X}_b)=1 \ .$$

Proof. $\varphi(\mathbf{X}_a, \mathbf{0}) \neq 0$ is equivalent to

$$\varphi(\mathbf{X}_a, \mathbf{0})=1 \ ,$$

and, in case of non-decreasing monotonicity, since $\mathbf{1} \geqq \mathbf{X}_a, \mathbf{X}_b \geqq \mathbf{0}$,

$$\varphi(\mathbf{1}, \mathbf{X}_b)=1 \ ,$$

q.e.d.

In order to comprehend the equivalence of the last two theorems, notice that $X_i=0$ and $X_i=1$ can be interpreted as "link i is bad" and "link i is good", respectively, so that the $\mathbf{X}_c$ of $\varphi(\mathbf{X}_c)=0$ and $\varphi(\mathbf{X}_c)=1$ indicate a cut set and a path set, respectively.

Similarity

Comparing $\varphi(\mathbf{X}_1)$ with $\varphi(\mathbf{X}_2)$ leads the way to comparing $\varphi_1(\mathbf{X})$ with $\varphi_2(\mathbf{X})$. If almost always, i.e. for almost all $\mathbf{X}$, $\varphi_1(\mathbf{X})=\varphi_2(\mathbf{X})$, it makes sense to call φ_1 and φ_2 similar (Boolean) functions. A good measure for the degree of similarity is the number of common minterms, i.e. the number of common lines in the truth table. (Note that φ_1 and φ_2 should be functions of the same vector variable $\mathbf{X}$.)

The concept of similarity is of some practical interest in digital circuits diagnostics. Generally faults, in a (combinational) digital electronics circuit with the nominal output function φ, which produce a similar output function φ_α, are difficult to detect, since the number of test patterns for which $\varphi_\alpha(\mathbf{X}) \neq \varphi(\mathbf{X})$, is small. On the other hand, fortunately, such faults tend to produce only few errors.

4.6 Recursive Definition of Boolean Functions

In a number of situations, especially in combinational switching circuits synthesis, it is practical to implement hardware according to

$$\varphi_n(X_1, \ldots, X_{kn}) = \psi[\varphi_{n-1}(X_1, \ldots, X_{kn-k}), X_{kn-k+1}, \ldots, X_{kn}] \ , \qquad (4.6.1)$$

i.e. φ_n depends on $X_{\varphi_{n-1}}$ and k further variables. A trivial example with $k=1$ is the serial calculation of a parity bit according to

$$\mathop{\not\equiv}_{i=1}^{n} X_i = \left(\mathop{\not\equiv}_{i=1}^{n-1} X_i \right) \not\equiv X_n \ .$$

Here $\psi=\psi(a, b)$ is simply the EXOR operation. An interesting realization of the parity function, which shows recursion in a non-trivial way, is the binary decision

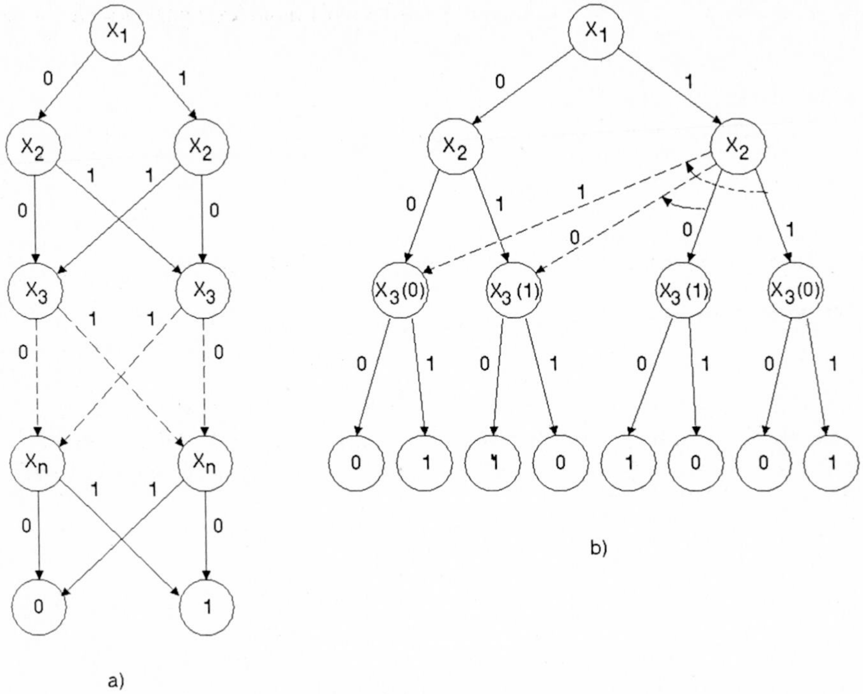

Fig. 4.6.1a, b. Binary decision diagrams for the parity function

diagram of Fig. 4.6.1(a) whose construction is obvious from Fig. 4.6.1(b). (Remember in this context exercise 3.6!)

In order to see that at every step of adding another variable the typical rotation of the right-hand subtree, consisting of the right-hand pair of leaves and their father, is possible in the X_3-nodes, the old function values are added in parentheses. They form from left to right the quadruple 0110.

Now, let us return to Eq. (4.6.1) and look at an example where $k=2$.

Example 4.6.1 Ripple-carry adder

Let $s=a+b$, where

$$a=(a_{m-1}\ \ldots\ a_1 a_0)\ ,$$

$$b=(b_{m-1}\ \ldots\ b_1 b_0)\ ,$$

$$s=(s_m s_{m-1}\ \ldots\ s_1 s_0)$$

are binary addends and their sum, respectively. Further, let the carry bit c_i be added to a_i and b_i. Now, by the elementary way binary addition works,

$$s_i=a_i \not\equiv b_i \not\equiv c_i\ ,\qquad c_0=0\ , \tag{4.6.2}$$

$$c_{i+1}=a_i b_i \vee a_i c_i \vee b_i c_i\ . \tag{4.6.3}$$

(Notice that s_i is the parity of a_i, b_i, and c_i, and that the value of c_{i+1} is that of the majority of a_i, b_i, and c_i.)

Figure 4.6.2 shows a switching network graph for Eq. (4.6.2, 4.6.3) which are the equations of the 1-bit full adder circuit.

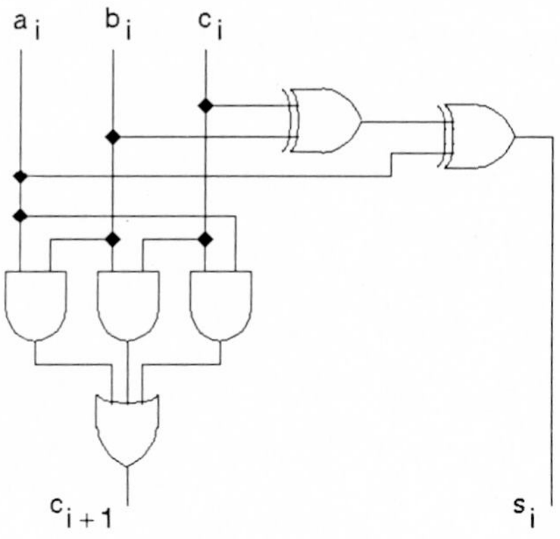

Fig. 4.6.2. Full adder bit slice logic diagram

From Eq. (4.6.3)

$$c_{i+1} = c_i(a_i \vee b_i) \vee a_i b_i \ ,$$

or, in notation nearer to that of Eq. (4.6.1),

$$\varphi(a_0, b_0, \ldots, a_i, b_i) = [\varphi_{i-1}(a_0, b_0, \ldots, a_{i-1}, b_{i-1})](a_i \vee b_i) \vee a_i b_i \ ;$$
$$\varphi_1 = 0 \ . \qquad (4.6.4)$$

Here

$$\psi = \psi(a, b, c) = a(b \vee c) \vee bc \ .$$

Figure 4.6.3 shows the complete m-bit adder.

Note how unpractically long and involved any non-recursive form of φ_i could become for $i > 2$ or 3. Yet, for increased speed of operation, i.e. to have only few

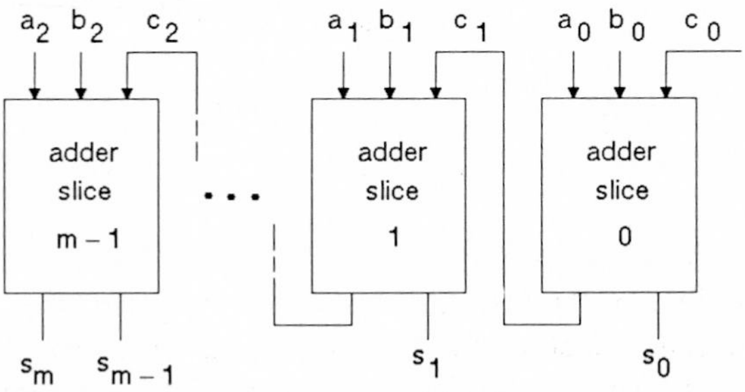

Fig. 4.6.3. Complete m-bit adder. (Slice 0 can be simplified.)

(minimally 2) gate delays a non-recursive, so-called carry-look-ahead type of adder may be a preferred implementation of Eqs. (4.6.2, 4.6.3).

The description of binary tree-type switching functions is similar to Eq. (4.6.1):

$$X_{\varphi_{n,i}} = \varphi_{n,i}(X_{\varphi_{n-1,1}}, X_{\varphi_{n-1,2}}) \; ; \quad i = 1, 2 \; , \tag{4.6.5}$$

where the values 1 and 2 of i describe left and right-hand sub-trees respectively. A trivial example is a binary tree for the fast calculation of a parity bit, viz.

$$\not\equiv_{i=1}^{2n} X_i = \left(\not\equiv_{i=1}^{n} X_i \right) \not\equiv \left(\not\equiv_{i=n+1}^{2n} X_i \right) . \tag{4.6.6}$$

A less trivial example is a network to determine the number of 1s in a $2n$-bits word using half adders only, which can be used for so-called Berger codes.

Example 4.6.2 Berger-encoder

Figure 4.6.4 shows the case $n = 2$. There are interwoven binary trees for the 1s, and the 2s. And for more inputs there would be more trees for more 2^n's, where only the 1^n's tree is complete.

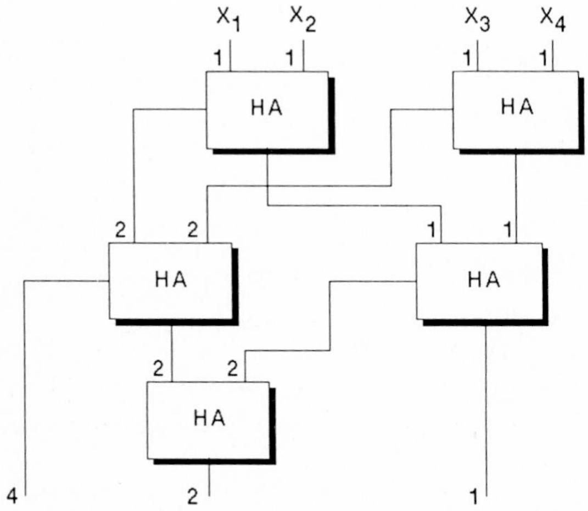

Fig. 4.6.4. Berger-encoding network. (See Fig. 3.4.4 for details.)

4.7 Hazards

Until now, electronic implementations of Boolean functions were discussed irrespective of fundamental laws of physics, such as the limitation of the velocity of signal propagation through wires and through switching devices, namely gates; see Fig. 1.2. The phenomena produced by these speed limitations and lacks of strict synchronism are traditionally called *hazards*. (No doubt hazards can produce various problems, even dangers in electronics circuitry.)

Theoretically, when switching from $\mathbf{X}_i$ to $\mathbf{X}_j$ the value of $\varphi(\mathbf{X})$ will change or not, depending on the given φ, its form, and its physical implementation. Typically, in electronic switching circuits the change from $\mathbf{X}_i$ to $\mathbf{X}_j$ will take place in several deterministic or random steps; usually, with only one component of **X** changing at a time†, even though the whole change may be over in a small fraction of a microsecond. Now, two situations are possible:

(1) So-called *static* hazard:
If $\varphi(\mathbf{X}_j)=\varphi(\mathbf{X}_i)$, $\mathbf{X}_i$ may change to some $\mathbf{X}_k$ with $\varphi(\mathbf{X}_k)\neq\varphi(\mathbf{X}_j)$ before ending up with $\mathbf{X}_j$. This phenomenon is called a static hazard, since in switching circuits "spikes" of the type shown in Fig. 4.7.1 can produce problems, e.g. undesired "noise".

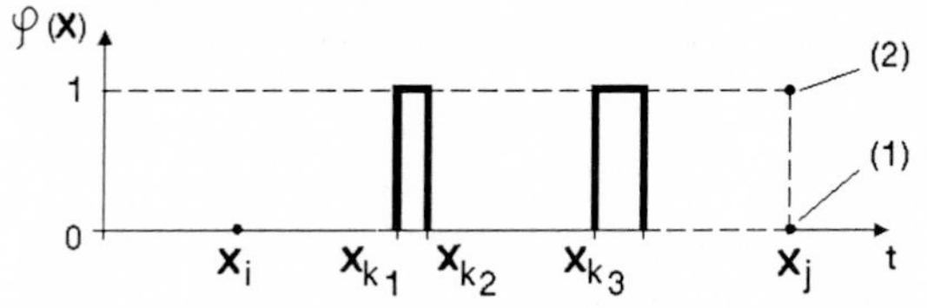

Fig. 4.7.1. Hazards of a switching function. At five points in time the associated values of **X** are noted

(2) So-called *dynamic* hazard:
If $\varphi(\mathbf{X}_j)\neq\varphi(\mathbf{X}_i)$, $\mathbf{X}_i$ may change to some $\mathbf{X}_{k_1}$ with $\varphi(\mathbf{X}_{k_1})\neq\varphi(\mathbf{X}_i)$ and then to some $\mathbf{X}_{k_2}$ with $\varphi(\mathbf{X}_{k_2})=\varphi(\mathbf{X}_i)$.
This may also be undesirable.

Comment. The distinction between static and dynamic hazards is not the only one made in this field. It makes sense to differentiate also between hazards due to a special technological representation of φ or to φ "itself", i.e. irrespective of the chosen representation. However, here this distinction will not be pursued at any depth.

Besides the unit hypercubes also Karnaugh maps can visualize the above phenomena. In Fig. 4.7.2 on the way from any $\mathbf{X}_i$ with $\varphi(\mathbf{X}_i)=1$ to $\mathbf{X}_j$ with $\varphi(\mathbf{X}_j)=1$ along edges of the square of Fig. 4.7.2(a) or along the paths of Fig. 4.7.2(b) $\mathbf{X}_k$s with $(\mathbf{X}_k)=0$ are inevitably found.

If the set of little 1-squares in the Karnaugh map is connected as in Fig. 4.7.3, there is some hope that paths from any 1-square to any other can be confined to the 1-set. In the case of a DNF of φ this is possible by the addition of superfluous terms‡ to a minimal form. Clearly, paths within the 1-squares of a single term cannot produce a hazard, since the literals that are negated on such a path are not part of this term. Hence, the crossing of borders of terms must be avoided, unless this

† i.e. moving on the edges of the hypercube of the type of Fig. 3.1.3 on a shortest path from $\mathbf{X}_i$ to $\mathbf{X}_j$

‡ And corresponding gates

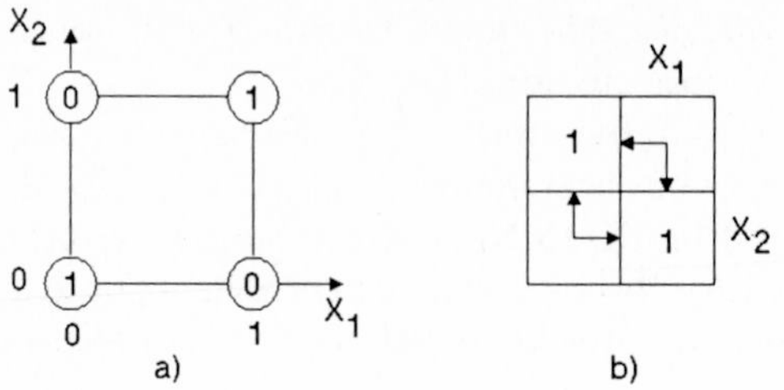

Fig. 4.7.2

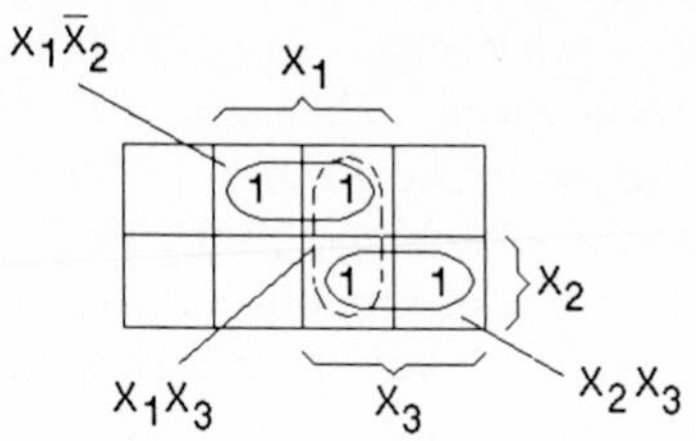

Fig. 4.7.3

Fig. 4.7.2a, b. Example function with hazards

Fig. 4.7.3. A Boolean function whose implementation could be made free of (type (1)-) hazards

crossing occurs, at the same time, inside another term. For instance, the 1-step position of the path indicated by the arrow in Fig. 4.7.3 means crossing the boundaries of $X_1\bar{X}_2$ and X_2X_3, but still remaining in X_1X_3. Hence, in

$$\varphi = X_1\bar{X}_2 \vee X_2X_3 \tag{4.7.1}$$

a change of X_2 might mean a (very) short change of φ to 0, since, for a (very) short period of time, both signals $\bar{X}_2(t)$ and $X_2(t)$ could equal 0. However, with the same

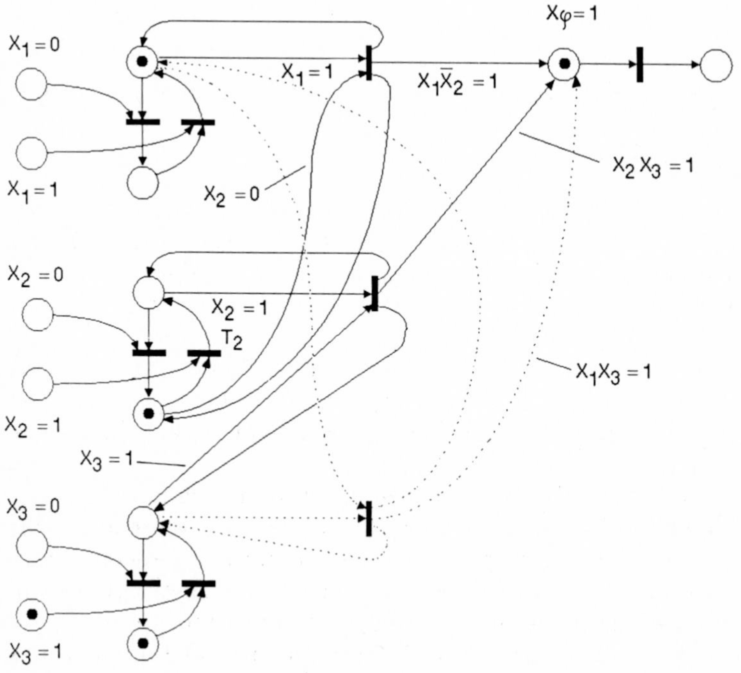

Fig. 4.7.4. Petri net for Eqs. (4.7.6, 4.7.7). The places at the far left are used to change the input vector $\mathbf{X}=(X_1, X_2, X_3)$. A change to $X_i=j; j\in\{0, 1\}$, is initialized by putting a token in the indicated place

φ changed to the form

$$\varphi = X_1\bar{X}_2 \vee X_2X_3 \vee X_1X_3 \tag{4.7.2}$$

this could not happen for $X_1 = X_3 = 1$.

Details of the above can be modelled by the Petri net of Fig. 4.7.4; see also §3.10. Tokens in places mean that the annotated events happen. It is assumed throughout that it suffices that such events happen with a sufficiently high frequency. Figure 4.7.4 shows the moment of switching from $\mathbf{X} = (X_1, X_2, X_3) = (1, 0, 0)$ to $\mathbf{X} = (1, 0, 1)$. If the next input is $\mathbf{X} = (1, 1, 1)$, then a delay in the transition T_2 will mean that no token comes to the place for $X_\varphi = 1$, which is interpreted as $X_\varphi = 0$. Adding the DNF term X_1X_3 which corresponds to the dotted part of Fig. 4.7.4, this cannot happen, since then there is a token in the place for $X_3 = 1$.

A good overview with many interesting results is given in [44]. A more theoretical discussion can be found in [9].

Comment. From a practical computer engineering point of view the theory of hazards is not that important, since almost all circuits are clocked. The really important problem is to avoid excessive delays by non-DNF designs. However, PLAs (programmed logic arrays) and ROMs (read only memories) guarantee a temporally balanced design.

Exercises

4.1

Prove Eq. (4.3.23), i.e. (in short)

$$\varphi = [X_i \vee \varphi(X_i = 0)]\,[\bar{X}_i \vee \varphi(X_i = 1)] \tag{E4.1}$$

using Shannon's inversion rule and Shannon's decomposition theorem (4.3.3)!

4.2

Show the correctness of Eq. (4.3.31) via the algorithm following Eq. (4.3.31) if Eq. (4.3.30) is given!

Hint. If that algorithm is used without care, even in this simple example an excessive amount of work may result. Therefore, extensive use should be made of

$$X_iX_i = X_i\ , \qquad (X_i \vee X_j)(X_i \vee X_k) = X_i \vee X_jX_k\ , \qquad X_i \vee X_iX_j = X_i\ .$$

Furthermore, in the first "multiplication" step combine the first, the second and the last pair of terms of Eq. (4.3.30) with each other.

4.3

Find a DDNF of

$$\varphi = X_1X_4 \vee \bar{X}_1\bar{X}_4 \vee \bar{X}_2X_5 \vee X_1X_3X_4 \vee X_2X_3\bar{X}_4 \vee X_1X_2X_3X_5 \tag{E4.2}$$

(a) by the set addition method of Abraham,
(b) by the Shannon decomposition method,
(c) by the decision tree method.

4.4

Take the two-out-of-three majority function

$$\varphi = X_1 X_2 \vee X_1 X_3 \vee X_2 X_3 ,$$

which is coherent and change it by a minimal change of the CDNF of φ to two non-coherent functions for which alternatively

(a) $\varphi(\mathbf{0}) = 1$,
(b) $\varphi(\mathbf{1}) = 0$.
(c) Generalize the results for coherent functions of n variables!

4.5

Prove that a DNF of the type given in Eq. (4.4.14) can be transformed to

$$\varphi = \hat{T}_1 \vee \bar{\hat{T}}_1 \hat{T}_2 \vee \bar{\hat{T}}_1 \bar{\hat{T}}_2 \hat{T}_3 \vee \ldots \vee \bar{\hat{T}}_1 \ldots \bar{\hat{T}}_{m-1} \hat{T}_m ,$$

and that all the terms of this DNF (of "literals" $\hat{T}_i$ and $\bar{\hat{T}}_j$) are pairwise disjoint.

4.6

How big is the number N_n of different non-zero conjunction terms of n Boolean variables?

4.7

Prove lemma 4.3.2 by looking at truth tables only.

4.8

Discuss the decomposition

$$\varphi = X_i \varphi'(X_i = 1) \vee \bar{X}_i \varphi'(X_i = 0) \vee \varphi_i , \tag{E4.3}$$

where φ_i is independent of X_i. How can φ_i be found?

4.9

A demultiplexer is the inverse of a multiplexer.
(a) Draw the switching diagram of a tree-type demultiplexer for three variables. Use two-input AND gates and inverters only.!
(b) Redraw the above demultiplexer using 2-outputs demultiplexers only.
Hint. The desired circuits have 8 outputs, one for each minterm.

5 Minimal Disjunctive Normal Forms

The optimization of Boolean functions is a wide field. One of the few regions where results have achieved a definite appraisal is the minimization of DNFs. Primarily developed for switching circuit synthesis, it has become of some importance also in fault tree analysis (in reliability theory).

In this context it is interesting to note that choosing between a CDNF and a CCNF can depend on the number of minterms of $\varphi: B^n \rightarrow B$. If it is smaller than 2^{n-1} the CDNF needs less n-input gates, otherwise the CCNF needs less gates.

Traditionally, in switching circuiits synthesis (partly even today for VLSI design) the main goal was to find a digital switching network of a given logical quality with as few and "simple" as possible gates, where often simple meant NOR gates with few inputs and few gates to be driven by any gate output (fan out limitation). Hence, roughly speaking, short Boolean expressions – possibly for coupled functions of the same variables – are (or were) of interest.

A highly interesting approach towards a general theory of the "inherent" complexity of Boolean functions is presented in [43]. Since [43] is a German paper, I quote from its English abstract: "The paper proposes a model and a measure of the inherent complexity of switching functions. The concept of inherent complexity is based on the structure of the function in Boolean n-space, in order, and symmetry. ... Inherent complexity is regarded as the cause of representational complexity, and it is shown how the cost of implementing or computing a switching function can be related to the inherent complexity measure."

5.1 General Considerations

In the context of optimization it is sometimes required to decide if a term $\hat{T}_i$ implies a Boolean function φ or not (of the same set of indicator variables); in short:

$$\hat{T}_i = 1 \overset{?}{\Rightarrow} \varphi = 1 \ . \tag{5.1.1}$$

(It is, of course, understood that φ is not given as a DNF comprising $\hat{T}_i$.) This problem is easily solved by observing that

$$\hat{T}_i := \bigwedge_{j=1}^{n_i} \tilde{X}_{l_{i,j}} = 1 \tag{5.1.2}$$

iff

$$\tilde{X}_{l_i,j}=1 \ , \quad j=1,2,\ldots,n_i \ . \tag{5.1.3}$$

Hence, inserting Eq. (5.1.3) in $\varphi(X_1,\ldots,X_n)$ has to give the constant function 1 if Eq. (5.1.1) is true.

Example 5.1.1 Test for implication

As is visible from the K-map of Fig. 5.1.1(a), the term X_1X_3 implies

$$\varphi(X_1,X_2,X_3)=X_1X_2 \vee \bar{X}_2X_3 \ . \tag{5.1.4}$$

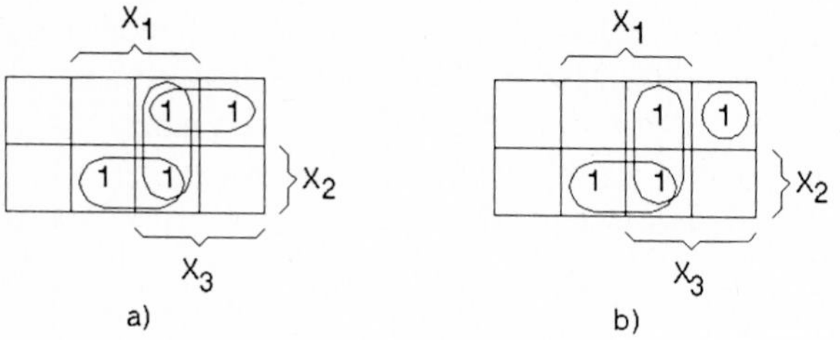

Fig. 5.1.1a, b. Minimizing a three-variable function

In fact,

$$\varphi(1,X_2,1)=X_2 \vee \bar{X}_2=1 \ .$$

In other words: In

$$\varphi(X_1,X_2,X_3)=X_1X_2 \vee X_1X_3 \vee \bar{X}_2X_3 \tag{5.1.5}$$

the term X_1X_3 is superfluous. — —

In 1952 W. Quine [10] coined the term of an "irredundant" DNF. It is what remains of a given DNF after deletion of all superfluous terms and – in the remaining terms – of all superfluous literals.

Example 5.1.2 Test for a superfluous literal

As can be readily seen from Fig. 5.1.1(b) in the last term of

$$\varphi=X_1X_2 \vee X_1X_3 \vee \bar{X}_1\bar{X}_2X_3$$

the literal $\bar{X}_1$ is superfluous. In fact, the test of $\bar{X}_1\bar{X}_2X_3$ without $\bar{X}_1$, i.e.

$$\bar{X}_2X_3=1 \overset{?}{\Rightarrow} \varphi=1,$$

results in

$$\varphi = X_1 \vee \bar{X}_1 = 1. \text{ — —}$$

Since a non-superfluous term of φ, consisting of non-superfluous literals only, is a prime implicant, the following is obvious.

Theorem 5.1.1. Any shortest DNF must consist of prime implicants. — —

In this context the following theorem is of some interest, since it says that in certain cases a (the) shortest DNF can be found without knowing all the prime implicants of that function.

To simplify the proof of that theorem, we first present a small lemma, which is obvious from the K-maps.

Lemma 5.1.1. A DNF assumes the value 1, iff at least one term assumes the value 1.

Proof. (a) $\hat{T}_i = 1 \Rightarrow \varphi = 1$ is trivial.

(b) If $\varphi(\mathbf{X}_i) = 1$, this $\mathbf{X}_i$ corresponds to a minterm $\hat{M}_j$, which must "belong" to at least one term $\hat{T}_i$ of φ and $\hat{M}_j = 1 \Rightarrow \hat{T}_i = 1$, q.e.d.

Theorem 5.1.2. With unate functions, i.e. starting with an unate DNF, the shortest DNF is unique.

Proof. Let us assume† that all literals are normal (non negated), such that a shortest DNF of φ is

$$\varphi = \bigvee_{i=1}^{m} \hat{T}_i \ , \quad \hat{T}_i = \bigwedge_{i=1}^{n_i} X_{l_{i,j}} \ ; \quad l_{ij} \in \{1, \ldots, n\} \ ,$$

where no term is part of another and is without superfluous letters. If φ were not unique, then at least one φ' existed as

$$\varphi' = \varphi = \bigvee_{i=1}^{m'} \hat{T}'_i$$

with at least one $\hat{T}'_k \notin \{\hat{T}_1, \ldots, \hat{T}_m\}$. Trivially,

$$\hat{T}'_k = 1 \Rightarrow \varphi = 1 \ .$$

Now, by lemma 5.1.1

$$\hat{T}'_k = 1 \Rightarrow \text{at least one } \hat{T}_i = 1 \ , \quad i \in \{1, \ldots, m\} \ .$$

This means (for normal variables terms) that $\hat{T}'_k$ contains all these $\hat{T}_i$ or it equals one of them, i.e. $\hat{T}'_k$ is not a shortest term, contradicting the above assumption of the

† This is possible without loss of generality, since one can, in this case, introduce new variables none of which are, formally, negated.

minimality of φ', or

$$\hat{T}'_k \in \{\hat{T}_1, \ldots, \hat{T}_m\} ,$$

which contradicts the above assumption of the non-uniqueness of the shortest DNF, q.e.d.

5.2 Finding All Prime Implicants of a Boolean Function

As will be shown in an example, it is in general indispensible to find all prime implicants (PIs) of (a Boolean function) φ to achieve a minimal DNF of φ. Therefore, three widely different procedures are presented here to find all PIs of a given φ.

Definition 5.2.1. A prime implicant (PI), which is the only one implied by at least one minterm (of the given Boolean function) is a *core* PI. In K-map language a core PI is one which covers at least one minterm as the only PI.

Example 5.2.1 (From [10])

As can be shown algebraically in detail or from Fig. 5.2.1

$$\varphi = X_1\bar{X}_2 \vee \bar{X}_1 X_2 \vee X_2\bar{X}_3 \vee \bar{X}_2 X_3 \tag{5.2.1}$$

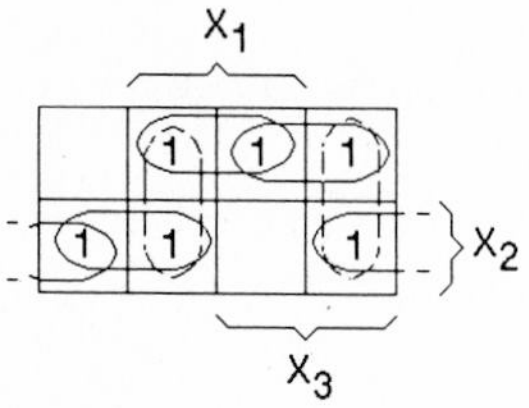

Fig. 5.2.1. K-map of an irredundant DNF

is an irredundant DNF, since all PIs are core PIs. Still, both

$$\varphi = X_1\bar{X}_2 \vee \underline{\underline{\bar{X}_1 X_3}} \vee X_2\bar{X}_3 \tag{5.2.2}$$

and

$$\varphi = \bar{X}_1 X_2 \vee \underline{\underline{X_1\bar{X}_3}} \vee \bar{X}_2 X_3 \tag{5.2.3}$$

are shorter by using a further PI (dotted in Fig. 5.2.1) not contained in Eq. (5.2.1). Note, that prime implicants for minimization are superfluous if an unate DNF is known, in which case, after applying the absorption rule, Eq. (2.2.14), all the terms are PIs.

In order to get a reference solution for a standard example in §§5.2.1, 2, 3 we present this example, whose solution is based on the K-map methodology, also in this section.

Example 5.2.2 (From [11])

$$\varphi(X_1, \ldots, X_5) = X_1 X_4 \vee \bar{X}_1 \bar{X}_4 \vee \bar{X}_2 X_5 \vee X_1 X_3 X_4 \vee X_2 X_3 \bar{X}_4 \vee X_1 X_2 X_3 X_5 \ . \quad (5.2.4)$$

The minterms of φ are represented in Fig. 5.2.2. In Fig. 5.2.3 the prime implicants (PIs) are shown as groups of minterms.

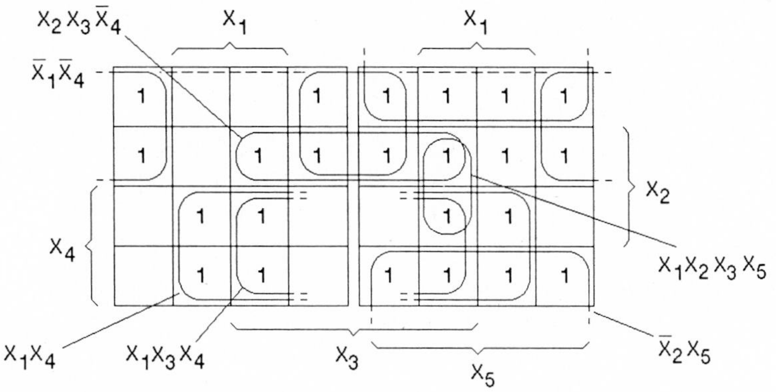

Fig. 5.2.2. K-map with terms of Eq. (5.2.4)

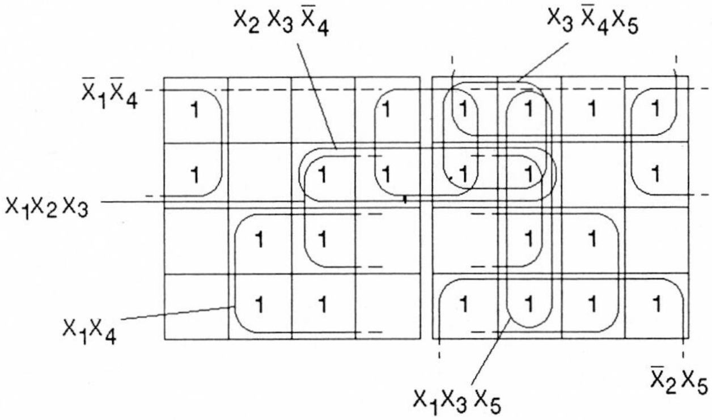

Fig. 5.2.3. All the PIs of φ of Eq. (5.2.4)

5.2.1 The Quine-McCluskey Procedure

(The main idea is to be found in [10].)

The Quine–McCluskey procedure has the advantage of great clarity, and it is easy to program even on assembly language level. However, its starting point is the

CDNF and this can have up to $2^n - 1$ terms which can be – even for a moderate number n of system components – a formidable number, both with respect to computation time and to computer memory demand.

The procedure: Given φ: $B^n \to B$.

(1) Find all pairs of minterms differing in only one literal and use for each pair Eq. (2.2.16) to find a term $\hat{T}^{(1)}$ with $n-1$ literals:

$$\hat{T}^{(1)} X_i \vee \hat{T}^{(1)} \bar{X}_i = \hat{T}^{(1)}; \quad \hat{T}^{(1)} X_i, \hat{T}^{(1)} \bar{X}_i \in \hat{M}(\varphi) . \tag{5.2.5}$$

Mark the terms $\hat{T}^{(1)} X_i$ and $\hat{T}^{(1)} \bar{X}_i$ (e.g. by a little hook)!

(2) Treat the (shorter) terms of the $\hat{T}^{(1)}$ type in the same way as the minterms, i.e. try to find $\hat{T}^{(2)}$-type terms (with $n-2$ literals) from $\hat{T}^{(1)}$-type pairs $\hat{T}^{(2)} X_j$ and $\hat{T}^{(2)} \bar{X}_j$ according to the absorption (merging rule)

$$\hat{T}^{(2)} X_j \vee \hat{T}^{(2)} \bar{X}_j = \hat{T}^{(2)} . \tag{5.2.6}$$

Mark the terms $\hat{T}_2 X_j$ and $\hat{T}_2 \bar{X}_j$!

(3) Continue by increasing the index of the $\hat{T}^{(i)}$ by 1 at every step.

(4) When the procedure has come to a stop – at the latest after finding $\hat{T}^{(n-1)}$'s which are literals – the *non*-marked terms are all the PIs.

Proof. The non-marked terms are implicants of φ, and since it is not possible to shorten them any more (remember $\hat{T}^{(i+1)}$ is one literal shorter than $\hat{T}^{(i)}$) they must be PIs, q.e.d.

The first example is almost trivial.

Example 5.2.3 Two-out-of-three function

See Table 5.2.1 for details!

Table 5.2.1. Prime implicants of the two-out-of-three majority function

minterms	$\hat{T}^{(1)}$-terms	
$X_1 X_2 X_3$	$X_1 X_2$	PIs
$X_1 X_2 \bar{X}_3$	$X_1 X_3$	PIs
$X_1 \bar{X}_2 X_3$	$X_2 X_3$	PIs
$\bar{X}_1 X_2 X_3$		

The next example looks "harmless", yet it proves to be at the brink of what can be done by "paper and pencil".

Example 5.2.4 (Example 5.2.2 redone)

$$\varphi(X_1, \ldots, X_5) = X_1X_4 \vee \bar{X}_1\bar{X}_4 \vee \bar{X}_2X_5 \vee X_1X_3X_4 \vee X_2X_3\bar{X}_4 \vee X_1X_2X_3X_5 \,. \tag{5.2.7}$$

For finding all the minterms of φ, we use the "truth table", Table 5.2.2.

Table 5.2.2. Truth table of φ of Eq. (5.2.7)

i of $\hat{M}_i$	$X_5X_4X_3X_2X_1$	φ	i of $\hat{M}_i$	$X_5X_4X_3X_2X_1$	φ
0	0 0 0 0 0	1	16	1 0 0 0 0	1
1	0 0 0 0 1	0	17	1 0 0 0 1	1
2	0 0 0 1 0	1	18	1 0 0 1 0	1
3	0 0 0 1 1	0	19	1 0 0 1 1	0
4	0 0 1 0 0	1	20	1 0 1 0 0	1
5	0 0 1 0 1	0	21	1 0 1 0 1	1
6	0 0 1 1 0	1	22	1 0 1 1 0	1
7	0 0 1 1 1	1	23	1 0 1 1 1	1
8	0 1 0 0 0	0	24	1 1 0 0 0	1
9	0 1 0 0 1	1	25	1 1 0 0 1	1
10	0 1 0 1 0	0	26	1 1 0 1 0	0
11	0 1 0 1 1	1	27	1 1 0 1 1	1
12	0 1 1 0 0	0	28	1 1 1 0 0	1
13	0 1 1 0 1	1	29	1 1 1 0 1	1
14	0 1 1 1 0	0	30	1 1 1 1 0	0
15	0 1 1 1 1	1	31	1 1 1 1 1	1

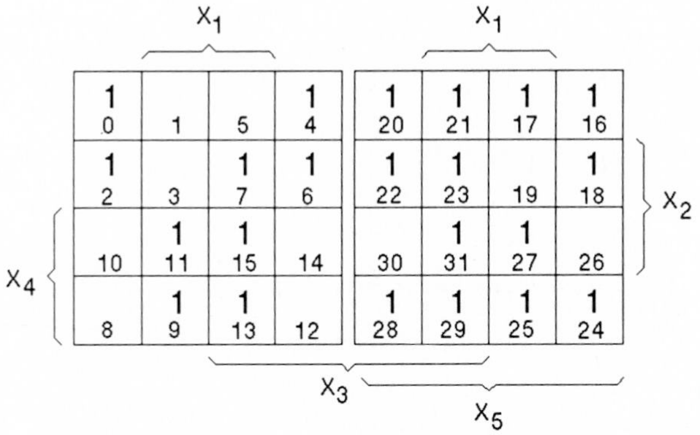

Fig. 5.2.4. Minterms of Eq. (5.2.7) with their standard numbering

The result is in accordance with the K-map of Fig. 5.2.4. (Zeroes are not noted.) As to the "standard" minterm numbering, consult Fig. 3.3.3.

Definition 5.2.2. As in coding theory the number of 1s in a binary n-tuple is called its *weight*.

Now, for easier notation, instead of the minterms of φ, the argument vectors $\mathbf{X}$ with $\varphi(\mathbf{X})=1$ are gathered in the left-most column of a further table with increasing weight; see Table 5.2.3. Then the elimination of variables according to Eq. (5.2.5) begins. The eliminated variables are denoted by "–" (just to avoid an empty space). Terms $\hat{T}^{(i)}$ of Eq. (5.2.5) are marked by a tick. This process is repeated as long as possible.

Table 5.2.3. Table for the merging process yielding the PIs of φ of Eq. (5.2.7). Duplicates of terms are put in parentheses

i of $\hat{M}_i$	$X_5X_4X_3X_2X_1$		i,j of $\hat{M}_i, \hat{M}_j$	$X_5X_4X_3X_2X_1$		i,j of $\hat{M}_i, \hat{M}_j$	$X_5X_4X_3X_2X_1$	
0	0 0 0 0 0	✓	0, 2	0 0 0 – 0	✓	7, 15	0 – 1 1 1	✓
			0, 4	0 0 – 0 0	✓	7, 23	– 0 1 1 1	✓
2	0 0 0 1 0	✓	0, 16	– 0 0 0 0	✓	11, 15	0 1 – 1 1	✓
4	0 0 1 0 0	✓				11, 27	– 1 0 1 1	✓
16	1 0 0 0 0	✓	2, 6	0 0 – 1 0	✓	13, 15	0 1 1 – 1	✓
			2, 18	– 0 0 1 0	✓	13, 29	– 1 1 0 1	✓
6	0 0 1 1 0	✓	4, 6	0 0 1 – 0	✓	21, 23	1 0 1 – 1	✓
9	0 1 0 0 1	✓	4, 20	– 0 1 0 0	✓	21, 29	1 – 1 0 1	✓
17	1 0 0 0 1	✓	16, 17	1 0 0 0 –	✓	22, 23	1 0 1 1 –	✓
18	1 0 0 1 0	✓	16, 18	1 0 0 – 0	✓	25, 27	1 1 0 – 1	✓
20	1 0 1 0 0	✓	16, 20	1 0 – 0 0	✓	25, 29	1 1 – 0 1	✓
24	1 1 0 0 0	✓	16, 24	1 – 0 0 0	✓	28, 29	1 1 1 0 –	✓
7	0 0 1 1 1	✓	6, 7	0 0 1 1 –	✓	15, 31	– 1 1 1 1	✓
11	0 1 0 1 1	✓	6, 22	– 0 1 1 0	✓	23, 31	1 – 1 1 1	✓
13	0 1 1 0 1	✓	9, 11	0 1 0 – 1	✓	27, 31	1 1 – 1 1	✓
21	1 0 1 0 1	✓	9, 13	0 1 – 0 1	✓	29, 31	1 1 1 – 1	✓
22	1 0 1 1 0	✓	9, 25	– 1 0 0 1	✓			
25	1 1 0 0 1	✓	17, 21	1 0 – 0 1	✓			
28	1 1 1 0 0	✓	17, 25	1 – 0 0 1	✓			
			18, 22	1 0 – 1 –	✓			
15	0 1 1 1 1	✓	20, 21	1 0 1 0 –	✓			
23	1 0 1 1 1	✓	20, 22	1 0 1 – 0	✓			
27	1 1 0 1 1	✓	20, 28	1 – 1 0 0	✓			
29	1 1 1 0 1	✓	24, 25	1 1 0 0 –	✓			
			24, 28	1 1 – 0 0	✓			
31	1 1 1 1 1	✓						

Table 5.2.3 (continued)

i,j,k,l	$X_5X_4X_3X_2X_1$		i,j,k,l	$X_5X_4X_3X_2X_1$		PI
0, 2, 4, 6	0 0 – – 0	✓	6, 7, 22, 23	– 0 1 1 –		$X_2X_3\bar{X}_4$
0, 2, 16, 18	– 0 0 – 0	✓	6, 22, 7, 23	(– 0 1 1 –)		
0, 4, 2, 6	(0 0 – – 0)		9, 11, 13, 15	0 1 – – 1	✓	
0, 4, 16, 20	– 0 – 0 0	✓	9, 11, 25, 27	– 1 0 – 1	✓	
0, 16, 2, 18	(– 0 0 – 0)		9, 13, 11, 15	(0 1 – – 1)		
0, 16, 4, 20	(– 0 – 0 0)		9, 13, 25, 29	(– 1 – 0 1	✓	
			9, 25, 11, 27	(– 1 0 – 1)		

Table 5.2.3 (continued)

i, j, k, l	$X_5X_4X_3X_2X_1$		i, j, k, l	$X_5X_4X_3X_2X_1$		PI
2, 6, 18, 22	– 0 – 1 0	✓	9, 25, 13, 29	(– 1 – 0 1)		
2, 18, 6, 22	(– 0 – 1 0)		17, 21, 25, 29	1 – – 0 1	✓	
4, 6, 20, 22	– 0 1 – 0	✓	17, 25, 21, 29	(1 – – 0 1)		
4, 20, 6, 22	(– 0 1 – 0)		20, 21, 22, 23	1 0 1 – –		$X_3\hat{X}_4X_5$
16, 17, 20, 21	1 0 – 0 –	✓	20, 21, 28, 29	1 – 1 0 –	✓	
16, 17, 24, 25	1 – 0 0 –	✓	20, 22, 21, 23	(1 0 1 – –)		
16, 18, 20, 22	1 0 – – 0	✓	20, 28, 21, 29	(1 – 1 0 –)		
16, 20, 17, 21	(1 0 – 0 –)		24, 25, 28, 29	1 1 – 0 –	✓	
16, 20, 18, 22	(1 0 – – 0)		24, 28, 25, 29	(1 1 – 0 –)		
16, 20, 24, 28	1 – – 0 0	✓				
16, 24, 17, 25	(1 – 0 0 –)		7, 15, 23, 31	– – 1 1 1		$X_1X_2X_3$
16, 24, 20, 28	(1 – – 0 0)		7, 23, 15, 31	(– – 1 1 1)		
			11, 15, 27, 31	– 1 – 1 1	✓	
			11, 27, 15, 31	(– 1 – 1 1)		
			13, 15, 29, 31	– 1 1 – 1	✓	
			13, 29, 15, 31	(– 1 1 – 1)		
			21, 23, 29, 31	1 – 1 – 1		$X_1X_3X_5$
			21, 29, 23, 31	(1 – 1 – 1)		
			25, 27, 29, 31	1 1 – – 1	✓	
			25, 29, 27, 31	(1 1 – – 1)		

indices of 8 minterms	$X_5X_4X_3X_2X_1$	PI
0, 2, 4, 6, 16, 18, 20, 22	– 0 – – 0	$\bar{X}_1\bar{X}_4$
0, 2, 16, 18, 4, 6, 20, 22	(– 0 – – 0)	
0, 4, 16, 20, 2, 6, 18, 22	(– 0 – – 0)	
16, 17, 20, 21, 24, 25, 28, 29	1 – – 0 –	$\bar{X}_2X_5$
16, 17, 24, 25, 20, 21, 28, 29	(1 – – 0 –)	
16, 20, 24, 28, 17, 21, 25, 29	(1 – – 0 –)	
9, 11, 13, 15, 25, 27, 29, 31	– 1 – – 1	X_1X_4
9, 11, 25, 27, 13, 15, 29, 31	(– 1 – – 1)	
9, 13, 25, 29, 11, 15, 27, 31	(– 1 – – 1)	

Note, in the table how very useful the grouping of terms according to equal weight is: This way only pairs with neighbouring weights need be compared for possible merging. Finally, all the terms which were never used in a merging process of Eq. (5.2.5) type are the PIs of φ, for they imply φ and are the shortest terms which do so. The PIs found check completely with those of Fig. 5.2.3.

5.2.2 The Consensus Procedure of Quine

In the following non-bottom up (from minterms) procedure to find all the prime implicants of a Boolean function the concept of the *consensus* of two terms plays a central role.

Definition 5.2.3. Let $\hat{T}_i$ and $\hat{T}_j$ be non-disjoint terms without X_k and $\bar{X}_k$. Let $\hat{T}_i X_k$ and $\hat{T}_j \bar{X}_k$ be terms of φ, i.e. $\hat{T}_i \hat{T}_j \neq 0$ and

$$\hat{M}(\hat{T}_i X_k),\ \hat{M}(\hat{T}_j \bar{X}_k) \subset \hat{M}(\varphi) \ . \tag{5.2.8}$$

Then $\hat{T}_i \hat{T}_j$ is called the *consensus* of $\hat{T}_i X_k$ and $\hat{T}_j \bar{X}_k$, i.e.

$$C(\hat{T}_i X_k,\ \hat{T}_j \bar{X}_k) := \hat{T}_i \hat{T}_j \ . \text{——} \tag{5.2.9}$$

The extreme usefulness of the consensus is evident from the following theorem.

Theorem 5.2.1. The consensus of two terms of φ is another term that implies φ.

Proof. From Eq. (2.2.16)

$$\hat{T}_i \hat{T}_j = \hat{T}_i \hat{T}_j X_k \vee \hat{T}_i \hat{T}_j \bar{X}_k \ .$$

From theorem 4.2.1

$$\hat{M}(\hat{T}_i \hat{T}_j X_k) \subset \hat{M}(\hat{T}_i X_k),\ \hat{M}(\hat{T}_i \hat{T}_j \bar{X}_k) \subset \hat{M}(\hat{T}_j \bar{X}_k) \ .$$

Because of Eq. (5.2.8)

$$\hat{M}(\hat{T}_i \hat{T}_j X_k) \subset \hat{M}(\varphi) \ , \qquad \hat{M}(\hat{T}_i \hat{T}_j \bar{X}_k) \subset \hat{M}(\varphi) \ .$$

Hence, since

$$\hat{M}(\hat{T}_i \hat{T}_j X_k) \cup \hat{M}(\hat{T}_i \hat{T}_j \bar{X}_k) = \hat{M}(\hat{T}_i \hat{T}_j),\ [\hat{M}(\hat{T}_i \hat{T}_j X_k) \cup \hat{M}(\hat{T}_i \hat{T}_j \bar{X}_k)] \subset \hat{M}(\varphi) \ ,$$

we have found that

$$\hat{M}(\hat{T}_i \hat{T}_j) \subset \hat{M}(\varphi) \Leftrightarrow \hat{T}_i \hat{T}_j \text{ implies } \varphi \ ,$$

q.e.d.

The following example can help to deepen the concept of consensus by means of a K-map.

Example 5.2.5 K-map for consensus

Let $\varphi = X_1 X_2 X_3 \vee X_4 \bar{X}_3 \vee \ldots,$

$$\hat{T}_i X_k \triangleq X_1 X_2 X_3 \Rightarrow \hat{T}_i = X_1 X_2 \ , X_k = X_3 \ , \hat{T}_j \bar{X}_k \triangleq X_4 \bar{X}_3 \Rightarrow \hat{T}_j = X_4 \ .$$

Here

$$C(\hat{T}_i X_k,\ \hat{T}_j \bar{X}_k) = \hat{T}_i \hat{T}_j = X_1 X_2 X_4 \ .$$

Figure 5.2.5 shows that $\hat{M}(\hat{T}_i \hat{T}_j) \subset \hat{M}(\varphi)$.

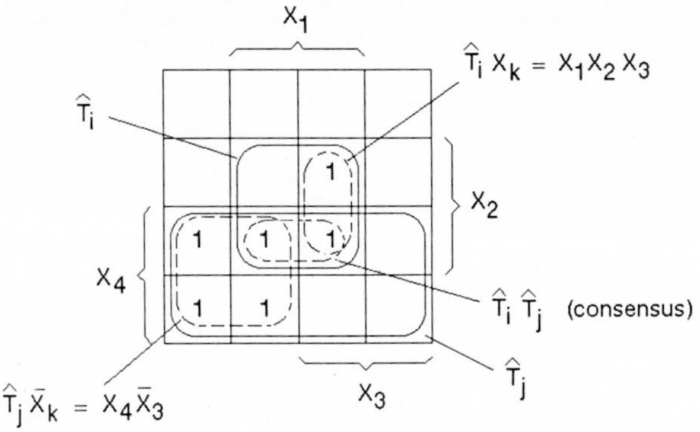

Fig. 5.2.5. K-map of consensus of $X_1X_2X_3$ and $X_4\bar{X}_3$

Theorem 5.2.2. (credited 1955 by Quine [11] to Samson & Mills [12]). Given φ as a DNF, after a finite cyclic sequence of

(I) absorption according to Eq. (2.2.14), i.e.

$$\hat{T}_i \vee \hat{T}_i\hat{T}_j = \hat{T}_i \tag{5.2.10}$$

and simplification according to Eq. (2.2.18), i.e.

$$\tilde{X}_i \vee \bar{\tilde{X}}_i\hat{T}_j = \tilde{X}_i \vee \hat{T}_j\ ; \tag{5.2.10a}$$

(II) adding (in a disjunctive way) the consensus of two terms of φ to φ, if the consensus is not of the form $\hat{T}_i\hat{T}_j$ with $\hat{T}_i$ in φ†,

namely, as soon as neither (I) nor (II) are applicable, φ is a DNF with all its PIs. (The proof shall be preceded by an example.)

Example 5.2.6 (Example 5.2.2 redone).

Let (in our notation) φ be that of Eq. (5.2.4), viz.

$$\varphi(X_1, \ldots, X_5) = X_1X_4 \vee \bar{X}_1\bar{X}_4 \vee \bar{X}_2X_5 \vee X_1X_3X_4 \vee X_2X_3\bar{X}_4 \vee X_1X_2X_3X_5\ .$$

By (I) $X_1X_3X_4$ is absorbed by X_1X_4.
By (II) we add (OR φ with) $C(X_1X_4, X_2X_3\bar{X}_4) = X_1X_2X_3$.
By (I) $X_1X_2X_3X_5$ is absorbed by $X_1X_2X_3$. Now

$$\varphi = X_1X_4 \vee \bar{X}_1\bar{X}_4 \vee \bar{X}_2X_5 \vee X_2X_3\bar{X}_4 \vee X_1X_2X_3\ .$$

† Otherwise the algorithm would not terminate since by Eq. (5.2.10) $\hat{T}_i\hat{T}_j$ would be absorbed in step (I).

By (II) we OR φ with

$$C(\bar{X}_2X_5, X_2X_3\bar{X}_4)=X_3\bar{X}_4X_5$$

and with

$$C(\bar{X}_2X_5, X_1X_2X_3)=X_1X_3X_5 \ .$$

Now

$$\varphi=X_1X_4 \vee \bar{X}_1\bar{X}_4 \vee \bar{X}_2X_5 \vee X_2X_3\bar{X}_4 \vee X_1X_2X_3 \vee X_3\bar{X}_4X_5 \vee X_1X_3X_5 \ , \tag{5.2.11}$$

which turns out to be the final result. It checks with the K-map of Fig. 5.2.3. — —

Proof of Theorem 5.2.2 (from [11], see also [2]):

Let us assume the procedures (I) and (II) have come to an end but at least one PI (be it $\hat{T}$) is not a term of the final DNF φ_f. Let $\hat{T}^*$ be the longest term, i.e. the one with the greatest number of literals with the following properties:

(a) $\hat{T}^*$ has at least the literals of $\hat{T}$ (it "contains" $\hat{T}$).
(Note: By theorem 4.2.1 $\hat{T}^*$ implies φ since $\hat{T}$ implies φ.)

(b) $\hat{T}^*$ does not contain any $\hat{T}_i$ of φ_f.
(Note: $\hat{T}_i$, even though perhaps not a PI, implies φ_f, but so does $\hat{T}$ with an irredundant set of literals.)

(c) $\hat{T}^*$ contains only literals or negated literals of φ_f.

$\hat{T}$ implies φ_f and therefore cannot have other literals. Yet at least one $\tilde{X}_k$ of φ_f will not be part of $\hat{T}^*$. Otherwise, (b) would be violated: $\hat{T}^*$ would contain at least one $\hat{T}_i$, since with all the literals of φ_f in $\hat{T}^*$, $\hat{T}^*$ would be a minterm of φ_f. Now $X_k\hat{T}^*$ and $\bar{X}_k\hat{T}^*$ obviously satisfy (a) and (c)†. By the above assumption of $\hat{T}^*$ being the longest term to satisfy (a), (b) and (c), clearly $X_k\hat{T}^*$ and $\bar{X}_kT^*$ do not satisfy (b)†. This means, that $\tilde{X}_k\hat{T}^*$ contains a term of φ_f (containing $\tilde{X}_k$). Both those terms of φ_f cannot just be X_k and $\bar{X}_k$, because this would give the trivial result $\varphi=1$. Only two non-trivial alternatives exist: If φ_f had among its terms the pairs

$$X_k, \bar{X}_k\hat{T}_i \text{ or } \bar{X}_k, X_k\hat{T}_j$$

then procedure (I) would be applicable. If it had the pair

$$X_k\hat{T}_i, \bar{X}_k\hat{T}_j$$

then procedure (II) would be applicable. In any case – contrary to the initial assumption – the existence of a PI not included in the terms of φ_f would mean that the process (I), (II), (I), . . . had not yet terminated.

The proof that no further terms (but only PIs) are in a correct φ_f is now easy: To any term $\hat{T}_k$ that implies φ there belongs at least one shortest (with certain literals

† With $\hat{T}^*$ replaced by $X_k\hat{T}^*$ and $\bar{X}_k\hat{T}^*$, respectively.

deleted) which is a PI of φ. With every PI of φ present in φ_f, $\hat{T}_k$ would have been deleted by the absorption process of (I), q.e.d.

5.2.3 The Double Negation Procedure of Nelson

An easy to handle (and remember) way to find all the PIs of a given Boolean function $\varphi(X_1, \ldots, X_n)$ was published in 1954 by Nelson [13].

Theorem 5.2.3. Let a Boolean function φ be given as a CNF. Then, on changing it to a DNF (by Eq. (E2.1)) and absorbing shorter terms (by Eq. (2.2.14)) this DNF is the conjunction of all the PIs of φ. (Note that in this text any DNF is supposed to be one, where neither $\tilde{X}_i\bar{\tilde{X}}_i = 0$ nor $\tilde{X}_i\tilde{X}_i = \tilde{X}_i$ can be applied any more.)

Before turning to the proof, we discuss an illustrating example.

Example 5.2.7 (Example 4.3.4 contd.)

Let by (4.3.22),

$$\begin{aligned}\varphi &= X_1X_2 \vee \bar{X}_2X_3X_4 \vee X_3\bar{X}_4 \\ &= (X_1 \vee X_3)(X_1 \vee \bar{X}_2 \vee \bar{X}_4)(X_2 \vee X_3)\ .\end{aligned}$$

"Factoring" out gives

$$\begin{aligned}\varphi &= (X_1 \vee \bar{X}_2X_3 \vee X_3\bar{X}_4)(X_2 \vee X_3) \\ &= X_1X_2 \vee X_1X_3 \vee \bar{X}_2X_3 \vee X_3\bar{X}_4\ .\end{aligned} \tag{5.2.12}$$

From Fig. 5.2.6 it is obvious that Eq. (5.2.12) contains all the PIs of φ and no further terms.

Before proving theorem 5.2.3 we prove several helpful lemmas.

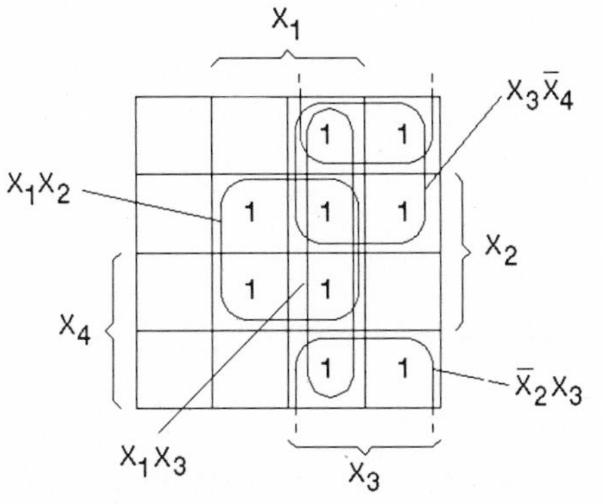

Fig. 5.2.6. Verification of Eq. (5.2.12)

A well-known standard procedure to produce a short DNF (not necessarily the shortest) from a given CNF† is the following one:

Definition 5.2.4 (standard compression of a CNF to a short DNF). The application of the distributive law, Eq. (2.2.3), followed by

(a) the orthogonality rule $A\bar{A}=0$,
(b) the idempotence rule $AA=A$,
(c) the absorption rule $A \vee AB=A$; $A,B\in\{0,1\}$

is called the standard simplification rule (SSR). — —

Note that in the SSR in practice the distributive law, Eq. (2.2.3), is used in the extended form

$$\left(\bigvee_{i=1}^{k}\tilde{X}_{l_{1,i}}\right)\left(\bigvee_{j=1}^{m}\tilde{X}_{l_{2,j}}\right)=\bigvee_{i=1}^{k}\bigvee_{j=1}^{m}\tilde{X}_{l_{1,i}}\tilde{X}_{l_{2,j}};\, l_{1,i},\, l_{2,j}\in\{1,\ldots,n\}\,. \tag{5.2.13}$$

Lemma 5.2.1. The application of the SSR on a given CNF has a unique result, i.e. a unique (short) DNF except for permutations of terms and literals.

Proof. If the successive literals of the given CNF are given different names, even though they are the same, then the resulting raw DNF after successive applications of Eq. (5.2.13) is unique because of the formal equality of Eq. (2.2.3) with the distributive law of elementary algebra and the uniqueness of a sum of terms found from multiplying several such terms with each other. Therefore, the SSR applied to the raw DNF will yield the same short DNF irrespective of the details of the application of the SSR, q.e.d.

Lemma 5.2.2. The expansion of a disjunction term $\check{T}_i$ to a CCNF according to

$$\begin{aligned}\check{T}_i&=(\check{T}_i \vee X_j)(\check{T}_i \vee \bar{X}_j)\\&=(\check{T}_i \vee X_j \vee X_k)(\check{T}_i \vee X_j \vee \bar{X}_k)(\check{T}_i \vee \bar{X}_j \vee X_k)(\check{T}_i \vee \bar{X}_j \vee \bar{X}_k)\\&=\ldots=\bigwedge_{j=1}^{2^m}(\check{T}_i \vee \check{M}'_j)\,,\end{aligned} \tag{5.2.14}$$

where $\check{M}'_j$ is maxterm j of the m variables not appearing in $\check{T}_i$, is the inverse of the SSR in case the raw DNF for $\check{T}_i$ is found by successive applications of the distributive law, Eq. (5.2.13).

Proof. When working one's way back from the last r.h.s. of Eq. (5.2.14) to the l.h.s., constant use is made of

$$(\check{T}_k \vee X_l)(\check{T}_k \vee \bar{X}_l)=\check{T}_k\check{T}_k \vee \check{T}_k\bar{X}_l \vee X_l\check{T}_k \vee X_l\bar{X}_l \tag{5.2.15}$$

† Note that no term $\check{T}_i$ of a CNF must contain $X_j \vee \bar{X}_j$. Such terms are ruled out (deleted) by definition.

for different $\check{T}_k$ which shrink to $\check{T}_i$. Now, by definition 5.2.4

$$(a)\ X_l\bar{X}_l = 0\ , \quad (b)\ \check{T}_k\check{T}_k = \check{T}_k\ , \quad (c)\ \check{T}_k \vee \tilde{X}\check{T}_k = \check{T}_k\ .$$

Hence, using the SSR:

$$(\check{T}_k \vee X_l)(\check{T}_k \vee \bar{X}_l) = \check{T}_k\ , \tag{5.2.15a}$$

which means that the SSR can reduce the last r.h.s. of Eq. (5.2.14) to its l.h.s., q.e.d.

Lemma 5.2.3. Given a Boolean function φ and a CNF of φ, the (short) DNF of φ found via the SSR is the same for this CNF and the CCNF of φ.

Proof. By Eq. (5.2.13) one first expands the given CNF to the CCNF of φ, which is then reduced to a short DNF, which is unique in the sense of lemma 5.2.1. Since this reduction can, by lemma 5.2.2, yield the given CNF the short DNF of this CNF and its CCNF must be the same except for permutations of terms and/or literals, q.e.d.

Lemma 5.2.4. Two different CNFs of a given φ yield the same shortest DNF on applying the SSR to both of them.

Proof. Application of lemma 5.2.3 to both DNFs.

Lemma 5.2.5. The application of the SSR produces a DNF of PIs of φ only.

Proof. The simplification rules of the SSR (of definition 5.2.4) are all that is needed to find the shortest non-superfluous terms except for the merging rule

$$\hat{T}_i X_j \vee \hat{T}_i \bar{X}_j = \hat{T}_i$$

used extensively in the Quine–McCluskey procedure of §5.2.1. However, in the framework of the SSR the pair $\hat{T}X_j$ and $\hat{T}_i\bar{X}_j$ would have been produced via the expansion of

$$\hat{T}_i(X_j \vee \bar{X}_j \vee \ldots)\ ,$$

which is ruled out by the definition 4.2.7 of a disjunction "term", q.e.d.

Lemma 5.2.6. If $\tilde{\varphi} = \varphi \vee \hat{T}$, where $\hat{T}$ is a PI of φ, is transformed to a CNF and then transformed to a short DNF via the SSR, then $\hat{T}$ is a term of the latter DNF.

Proof. Let $\hat{T} = \tilde{X}_{i_1}\tilde{X}_{i_2} \ldots \tilde{X}_{i_k}$. Using the conjunctive decomposition of Eq. (4.3.23), φ can be decomposed such that on applying the distribution law used in the SSR (see definition 5.2.4) one term of the raw DNF is $\hat{T}$ and by its primeness $\hat{T}$ cannot be changed by the rules (a), (b), and (c) of the SSR, so that it remains in the final short DNF of that procedure. In fact, by Eq. (4.3.23), on decomposing the expressions in the relevant brackets successively:

$$\begin{aligned}\varphi &= [\tilde{X}_{i_1} \vee \varphi(\tilde{X}_{i_1}=0)][\bar{\tilde{X}}_{i_1} \vee \varphi(\tilde{X}_{i_1}=1)]\\ &= [\tilde{X}_{i_1} \vee \tilde{X}_{i_2} \vee \varphi(\tilde{X}_{i_1}=0, \tilde{X}_{i_2}=0)]\,[\tilde{X}_{i_1} \vee \bar{\tilde{X}}_{i_2}\\ &\quad \vee \varphi(\tilde{X}_{i_1}=0, \tilde{X}_{i_2}=1)]\\ &\quad \wedge [\bar{\tilde{X}}_{i_1} \vee \tilde{X}_{i_2} \vee \varphi(X_{i_1}=1, X_{i_2}=0)][\bar{\tilde{X}}_{i_1} \vee \bar{\tilde{X}}_{i_2} \vee \varphi(X_{i_1}=1, X_{i_2}=1)]\\ &= \ldots = \ldots [\bar{\tilde{X}}_{i_1} \vee \bar{\tilde{X}}_{i_2} \vee \ldots \vee \bar{\tilde{X}}_{i_k}\\ &\quad \vee \varphi(\tilde{X}_{i_1}=1, \tilde{X}_{i_2}=1, \ldots, \tilde{X}_{i_k}=1)]\ . \end{aligned} \tag{5.2.16}$$

Obviously, in the final decomposed form each bracket contains either at least one literal of $\hat{T}$ or the constant

$$\varphi(\tilde{X}_{i_1}=1, \tilde{X}_{i_2}=1, \ldots, \tilde{X}_{i_k}=1)=1\ . \tag{5.2.17}$$

Equation (5.2.17) is true because $\hat{T}$ implies φ. Hence, on applying the SSR to φ one can find at least one term, which becomes $\hat{T}$ after the application of the idempotence rule, q.e.d.

Now, every thing is prepared for the proof of theorem 5.2.3: Theorem 5.2.3 can be rephrased as follows: "The application of the SSR produces a DNF consisting of all PIs of φ".

By Lemma 5.2.5 the above DNF consists of PIs only. Assume that the PI $\hat{T}$ were not part of the short DNF φ_s produced by the SSR (applied to any CNF of φ). Then $\varphi_s \vee \hat{T}$ could be transformed to a CNF of φ and reduced by the SSR to φ_s. This would contradict lemma 5.2.6 since by that lemma $\hat{T}$ should show up in φ_s. Hence $\hat{T}$ must have been a term of φ_s, q.e.d.

To show the remarkable performance of the Nelson algorithm a non-trivial example is redone.

Example 5.2.8 (Example 5.2.2 redone)

From Eq. (5.2.4), on applying Eq. (4.3.23) repeatedly:

$$\begin{aligned}\varphi &= X_1X_4 \vee \bar{X}_1\bar{X}_4 \vee \bar{X}_2X_5 \vee X_1X_3X_4 \vee X_2X_3\bar{X}_4 \vee X_1X_2X_3X_5\\ &= (X_1 \vee \bar{X}_4 \vee \bar{X}_2X_5)(\bar{X}_1 \vee X_4 \vee \bar{X}_2X_5 \vee X_2X_3\bar{X}_4 \vee X_2X_3X_5)\\ &= (X_2 \vee X_1 \vee \bar{X}_4 \vee X_5)(\bar{X}_2 \vee X_1 \vee \bar{X}_4)(X_2 \vee \bar{X}_1 \vee X_4 \vee X_5)\\ &\quad \wedge (\bar{X}_2 \vee \bar{X}_1 \vee X_4 \vee X_3\bar{X}_4 \vee X_3X_5)\ . \end{aligned} \tag{5.2.18}$$

In order to avoid excessive algebraic manipulations, the last term is simplified as follows:

$$\bar{X}_2 \vee \bar{X}_1 \vee X_4 \vee X_3\bar{X}_4 \vee X_3X_5 = \bar{X}_2 \vee \bar{X}_1 \vee X_4 \vee X_3\ .$$

Hence, after the reordering of literals:

$$\varphi = (X_1 \vee X_2 \vee \bar{X}_4 \vee X_5)(X_1 \vee \bar{X}_2 \vee \bar{X}_4)$$
$$\wedge (\bar{X}_1 \vee X_2 \vee X_4 \vee X_5)(\bar{X}_1 \vee \bar{X}_2 \vee X_3 \vee X_4) \; . \tag{5.2.19}$$

Now, the raw DNF of the SSR is produced, first combining the first and the second pairs of parentheses, respectively, and simplifying according to the SSR:

$$\varphi = (X_1 \vee \bar{X}_4 \vee \bar{X}_2 X_5)(\bar{X}_1 \vee X_2 X_3 \vee X_4 \vee \bar{X}_2 X_5 \vee X_3 X_5)$$
$$= X_1 X_2 X_3 \vee X_1 X_4 \vee \bar{X}_2 X_5 \vee X_1 X_3 X_5$$
$$\vee \bar{X}_1 \bar{X}_4 \vee X_2 X_3 \bar{X}_4 \vee X_3 \bar{X}_4 X_5 \; ; \tag{5.2.20}$$

which equals Eq. (5.2.11) but for trivial permutations.

5.3 Minimization

Minimality within the class of DNFs of a given Boolean function φ means, first of all, irredundancy in the sense that no term and, inside terms, no literal is superfluous. Furthermore, among the irredundant DNFs of φ, one can find a minimal DNF as the one with the least number of terms or with the least total number of literals, etc.

The type of minimization discussed here belongs to the more general problem of covering a given set (of elements of any kind) by an optimal choice from a given set of subsets. Since this optimization is a typical application of Boolean analysis to operations research it is discussed in an extra sub-section, and the optimization of DNFs will be regarded as an example thereafter.

5.3.1. Optimal Choice of Subsets

In applied engineering and operations research it happens frequently that a given basic set

$$s_e = \{e_1, \ldots, e_m\}$$

is to be covered optimally by a subset S_o of the set

$$S = \{s_1, \ldots, s_n\} \; ; \qquad s_i \subset s_e \; ; \qquad i \in \{1, \ldots, n\}$$

of subsets of s_e. Figure 5.3.1 shows a Venn diagram for illustration, where the elements e_j are points in the plane. The problem is, of course, trivial in case all the s_i are core sets in the sense that each s_i contains at least one e_j singularly. In that case $S_o = S$. Returning to the example of Fig. 5.3.1 notice that the s_i cover s_e in a redundant way in that either s_2 or s_3 are superfluous. Obviously, in this little example the practical question is simply: Should s_2 be deleted or, rather, s_3. Of

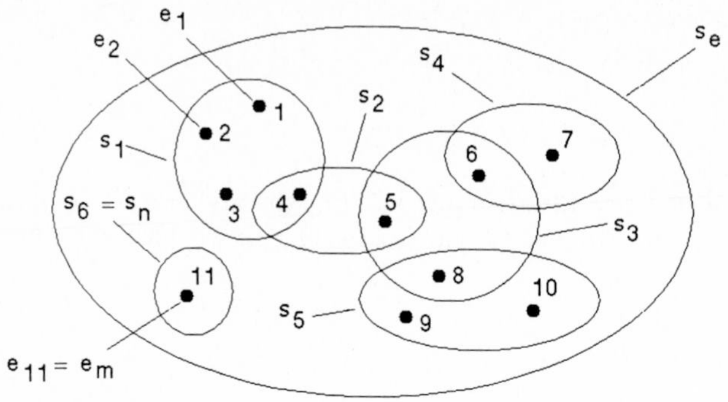

Fig. 5.3.1. Subsets s_i of a set s_e of elements e_j

course, the answer will depend on the optimality criterion used. Yet the first step usually is to find all irredundant (i.e. in this sense minimal) covers $S_{o,i}$ of s_e.

Let us formalize things a little bit: Candidates of S_o must have the property that e_1 through e_m are elements of at least one of the s_i of S_o:

$$S_o = \{s_{l_1}, \ldots, s_{l_k}\} \ ; \quad e_i \in \bigcup_{j=1}^{k} s_{l_j} =: s_o \ ; \quad l_j \in \{1, \ldots, n\} \ ; \quad s_o \subseteq s_e \ . \tag{5.3.1}$$

Of course only $s_o = s_e$ yields a proper S_o. Now, with

$$Y_j := \begin{cases} 1, & \text{if} \quad e_j \in s_o \\ 0, & \text{else,} \end{cases} \tag{5.3.2}$$

and

$$Y_{j,i} := \begin{cases} 1, & \text{if} \quad e_j \in s_{l_i} \\ 0, & \text{else,} \end{cases} \tag{5.3.3}$$

an adequate Boolean functions' formulation corresponding to (5.3.1) is

$$\psi = \bigwedge_{j=1}^{m} Y_j = \bigwedge_{j=1}^{m} \bigvee_{i=1}^{k} Y_{j,i} = \begin{cases} 1, & \text{if} \quad S_o \text{ covers } s_e \\ 0, & \text{else.} \end{cases} \tag{5.3.4}$$

Often $Y_{j,i}$ can be replaced by

$$X_i = \begin{cases} 1, & \text{if a given } e_j \in s_i \\ 0, & \text{else,} \end{cases} \tag{5.3.5}$$

resulting in

$$\psi = \bigwedge_{j=1}^{m} \bigvee_{i \in I_j} X_i \ , \tag{5.3.6}$$

where I_j is the index set of all the s_i containing e_j.

Obviously, here every term of a DNF of ψ corresponds to a candidate S_o. In simple cases, where only the cardinality of S_o must be small any shortest term of ψ would yield an optimal S_o.

Example 5.3.1 Minimal covers for the points of Fig. 5.3.1

A short investigation of Fig. 5.3.1 yields the obvious results:

$$S_{o,1} = \{s_1, s_2, s_4, s_5, s_6, s_7\} ,$$

and

$$S_{o,2} = \{s_1, s_3, s_4, s_5, s_6, s_7\} ,$$

i.e. either s_2 or s_3 are superfluous. Now, let us derive this result formally via Eq. (5.3.4): By Eq. (5.3.6) and Fig. 5.3.1

$$\psi = \bigwedge_{j=1}^{11} Y_j ,$$

where

$$Y_1 = X_1 , \quad Y_2 = X_1 , \quad Y_3 = X_1 , \quad Y_4 = X_1 \vee X_2 ,$$

$$Y_5 = X_2 \vee X_3 , \quad Y_6 = X_3 \vee X_4 , \quad Y_7 = X_4 , \quad Y_8 = X_3 \vee X_5 ,$$

$$Y_9 = X_5 , \quad Y_{10} = X_5 , \quad Y_{11} = X_6 .$$

By elementary simplification rules of §2.2

$$\begin{aligned} \psi &= X_1(X_1 \vee X_2)(X_2 \vee X_3)(X_3 \vee X_4)X_4(X_3 \vee X_5)X_5X_6 \\ &= X_1(X_2 \vee X_3)X_4X_5X_6 \\ &= X_1X_2X_4X_5X_6 \vee X_1X_3X_4X_5X_6 . \end{aligned} \tag{5.3.7}$$

Clearly, the first term corresponds to $S_{o,1}$, the second to $S_{o,2}$. ——

Comment. Eq. (5.3.6) is one of the relatively few examples where a practical problem first leads to a CNF instead of a DNF.

5.3.2 Shortest Disjunctive Normal Forms

Now the general ideas of section 5.3.1 are applied to the sets of prime implicants of section 5.2 to determine (a) shortest DNF(s). Since the general concepts are clear, it seems best to continue with a well-known example.

Example 5.3.2 Optimization of the DNF of Fig. 5.2.3

Given all prime implicants (PIs), we first look for irredundant DNFs. For this purpose one must know which PIs cover (contain, are canonical DNFs with) which minterms. In Table 5.3.1 crosses mark the minterms of the relevant PIs.

Table 5.3.1. Minterms of the prime implicants of Eq. (5.2.4) from Table 5.2.3

Standard index (No.) of minterms

PI	**0**,	**2**,	**4**	6,	7,	**9**	**11**,	**13**,	15	16,	**17**,	**18**	20,	21,	22	23,	**24**,	25	**27**,	**28**,	29	31
X_1X_4†						(x)	(x)	(x)	(x)									(x)	(x)		(x)	(x)
$\bar{X}_1\bar{X}_4$†	(x)	(x)	(x)	(x)						(x)		(x)	(x)		(x)							
$\bar{X}_2X_5$†										(x)	(x)		(x)	(x)			(x)	(x)		(x)	(x)	
$X_1X_2X_3$					x				x							x						x
$X_1X_3X_5$														x		x					x	x
$X_2X_3\bar{X}_4$				x	x										x	x						
$X_3\bar{X}_4X_5$													x	x	x	x						

† Necessary (core) prime implicant

One should then look for indispensible or "core" PIs, namely those being the only ones to cover one or several minterms. In Table 5.3.1 minterms contained in only one PI are given in fat print, and the crosses of core PIs are marked by small circles. From Table 5.3.1 obviously $X_1X_4, \bar{X}_1\bar{X}_4$ and $\bar{X}_2X_5$ are core PIs of the φ of (5.2.4). Since, generally, the core PIs contain also other minterms, generally, the number of minterms to be "covered" by further (non-core) PIs is reduced. Table 5.3.2 shows this with reference to φ of Eq. (5.2.4).

Now, there are at least two choices for the set of further PIs necessary to cover those minterms not covered by core PIs. In the example of Table 5.3.2 one further PI will do; either $X_1X_2X_3$ or $X_2X_3\bar{X}_4$. The other two PIs are superfluous anyway. The final result (minimal DNF) is

$$\varphi = X_1X_4 \vee \bar{X}_1\bar{X}_4 \vee \bar{X}_2X_5 \vee \begin{cases} X_1X_2X_3 \\ X_2X_3\bar{X}_4 \end{cases} . \quad (5.3.8)$$

In general, irredundant DNFs of PIs are found via the method of §5.3.1: Let PI i cover minterm $\hat{M}_j; j \in I_i$, and let $Y_{j,i}$ be the indicator variable of this fact; $i \in I_{NCP}$ (with NCP for non core PIs). Then, to cover $\hat{M}_j$, the irredundant form of φ must contain at least one PI covering this $\hat{M}_j$. Now, with I_{NCM} (with NCM for non core minterms) the index set of the minterms not covered by the core PIs and Y_j the

Table 5.3.2. Minterms that are covered only in non-core PIs

No.	PI	$\hat{M}_7$	$\hat{M}_{23}$
1	$X_1X_2X_3$	x	x
2	$X_1X_3X_5$		x
3	$X_2X_3\bar{X}_4$	x	x
4	$X_3\bar{X}_4X_5$		x

indicator variable for the fact that $\hat{M}_j$ is covered by the irredundant φ, we have for covering all $\hat{M}_j$'s of I_{NCM} the Boolean function

$$\psi := \bigwedge_{j\in I_{NCM}} Y_j = \bigwedge_{j\in I_{NCM}} \bigvee_{i\in I_{NCP}} Y_{j,i} \; ; \quad Y_{j,i} = \begin{cases} 1, & \text{if } \hat{M}_j \text{ is a minterm of PI}i \\ 0, & \text{else,} \end{cases} \tag{5.3.9}$$

Obviously, in the final result the first index is irrelevant. Rather one can start a further DNF minimization with $X_i := Y_{j,i}$.

Example 5.3.3 (Example 5.3.2 contd.)

From Table 5.3.2

$$\psi = Y_7 Y_{23} = (Y_{1,7} \vee Y_{3,7})(Y_{1,23} \vee Y_{2,23} \vee Y_{3,23} \vee Y_{4,23}) \tag{5.3.10}$$

With $Y_{j,i} \leftarrow X_i$, as indicated above, we find the plausible result

$$\psi = (X_1 \vee X_3)(X_1 \vee X_2 \vee X_3 \vee X_4) = X_1 \vee X_3 \; , \tag{5.3.11}$$

which means that PI 1 or PI 3 have to be added to the core PIs. — —

From irredundant forms there can be a short or a long way to ultimate optimum forms, depending on the criterion of optimality. Often the overall length of the DNF (counting literals and ORs) is to be minimized. For more advanced results see [51].

Application to Testing for Unateness

The set of minimal DNFs found from the set of all the prime implicants of φ gives rise to a simple test for unateness of φ (see §4.5).

Theorem 5.3.1. A Boolean function φ is unate, iff at least one of its minimal DNFs contains each variable of φ either only normal or only negated.

Proof. The if-part is clear from the definition 4.5.5 of unateness. The only-if part follows from the process of trying to transform a given DNF to contain only X_i or only $\bar{X}_i$, if any. In this process it is desirable to have as few terms as possible, each being as short as possible. In fact, if a given term $\hat{T}_j$ is one of the fewer which contain $\tilde{X}_i$ it would be best if $\hat{T}_j$ would either be superfluous or if it could be merged with another term containing $\bar{\tilde{X}}_i$. Hence, unateness can be judged from looking exclusively at all the shortest DNFs of φ, q.e.d.

For example, φ of Eq. (5.3.8) is not unate.

5.3.3 Incompletely Specified Boolean Functions

In practice it happens frequently that a Boolean function is incompletely specified; i.e. for certain $\mathbf{X} = \mathbf{X}_i$ practitioners don't care whether $\varphi(\mathbf{X}_i) = 0$ or 1. This is difficult to describe algebraically, but is simple in tables, e.g. in the K-map. Clearly, the "don't cares" (often denoted by "d" in the K-map) can be used for optimization.

Example 5.3.4 BCD-code

Often, decimal digits are coded as four-bit binary numbers, i.e. in BCD (binary coded decimals) code. Since, in the error-free case of the possible 16 4-bit words $\mathbf{X}_i$ only ten appear and are to be decoded, one doesn't care what would happen with the rest. The K-map of Fig. 5.3.2 shows the Boolean function for the upper horizontal bar of a seven-segment representation of the decimals (see Fig. 5.3.2b). Setting all the ds to 1, the shortest DNF of the function switching on the upper horizontal bar is

$$\varphi_a = X_2 \vee X_4 \vee X_1 X_3 \vee \bar{X}_1 \bar{X}_3 \ . \tag{5.3.12}$$

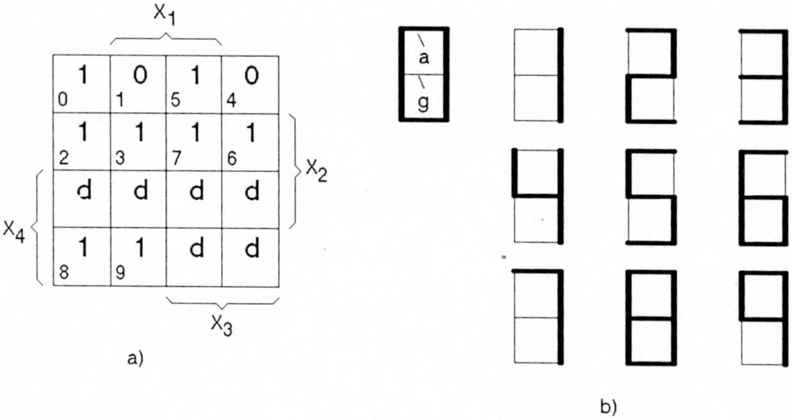

Fig. 5.3.2a, b. Encoding a seven-segment unit for decimal digits. The bar a is used in the digits 0, 2, 3, 5, 6, 7, 8, 9

With all the ds as 0s we get the different function

$$\tilde{\varphi}_a = X_2 \bar{X}_4 \vee X_1 X_3 \bar{X}_4 \vee \bar{X}_1 \bar{X}_3 \bar{X}_4 \vee \bar{X}_2 \bar{X}_3 X_4 \ . \tag{5.3.13}$$

Notice that, mathematically, an unspecified function is a set of functions consisting of 2^m functions in the case of m don't care variables, since for each don't care literal $\tilde{X}_i$ one may either use X_i or $\bar{X}_i$.

See §4.4.4 for applications of "d" other than for the classical minimization.

Exercises

5.1

In many cases one finds, for a given φ, a DDNF which is as long as the shortest DNF, but it can be longer too. Give a three-variable example for both situations.

5.2

Given a function φ by means of the K-map of Fig. E5.1, find all its PIs via the Quine–McCluskey algorithm.

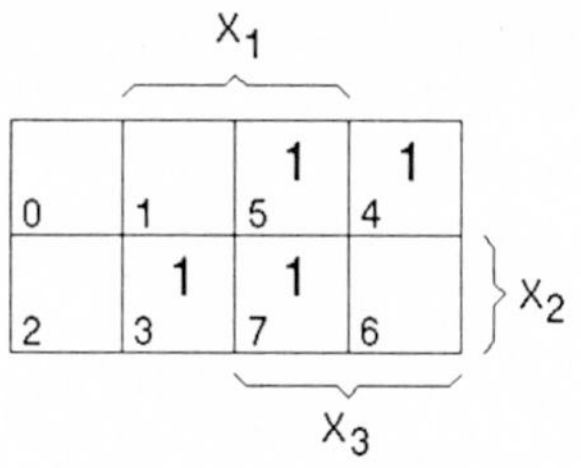

Fig. E5.1. K-map of Fig. 5.1.1 revisited

5.3

(a) Let φ of Fig. E5.1 be given algebraically as its CDNF. Use the consensus method of Quine to find all the PIs of φ.

(b) In which case is the consensus of $\hat{T}_i X_k$ and $\hat{T}_j \bar{X}_k$ simply the disjunction (OR) of both terms?

5.4

Find all the PIs of (S5-1) by Nelson's double negation procedure.

5.5

Given the complete set of PIs of φ in Table E5.1, which is (are) the minimal DNF(s)?

Table E5.1 PIs and their minterms for a given φ. (Minterms need not be identified.)

PI								
$\hat{T}_1$	x			x		x		
$\hat{T}_2$		$\otimes$	$\otimes$			$\otimes$		
$\hat{T}_3$	x				x		x	
$\hat{T}_4$	$\otimes$			$\otimes$	$\otimes$		$\otimes$	$\otimes$

5.6

Do the same as in Example 5.3.4 for the central horizontal bar, i.e. for φ_g.

5.7

Which DNF has maximum shortest length, and how long is it, if all the literals and the OR symbols are counted as characters?

Hint: Shortest length means length after minimization according to §5.3.

6 Boolean Difference Calculus

Questions concerning the sensitivity of a Boolean function with respect to certain changes of argument values are not only of theoretical interest [9].

One of the simpler problems is the logical description of a change of the value of $\varphi(X_1, \ldots, X_n)$ on a change of X_k. It is obvious that this problem is strongly connected with digital diagnostics in that proper test inputs of a network should produce different outputs, etc.

Here the basic theory necessary to understand digital diagnostics will be presented with some examples. As an introduction, a special polynomial-type representation of Boolean functions will be treated, which was announced in §4. It allows for another canonical representation, viz. one with EXOR and AND only.

6.1 Exclusive-Disjunction Form Without Negated Variables

The complete system $\{\not\equiv, \wedge\}$ has been found to be of considerable interest, mainly in digital diagnostics. Among other features it allows for a canonical polynomial type form of a Boolean function. Let us show the completeness of $\not\equiv$ and $\wedge$ first.

Theorem 6.1.1. The pair of Boolean operators $\not\equiv$ (EXOR) and $\wedge$ (AND) can replace the operator triple $\wedge$ (AND), $\vee$ (OR) and $\bar{}$ (NOT).

Proof.
Alternative 1 (truth table approach): Details are obvious from Table 6.1.1. First,

$$X_i \vee X_j = X_i \not\equiv X_j \not\equiv X_i X_j \ . \tag{6.1.1}$$

Also, as is obvious from the small boxes in Table 6.1.1,

$$\bar{X}_i = X_i \not\equiv 1 \ .$$

Alternative 2 (algebraic approach):
From Eq. (2.1.1)

$$X_i \not\equiv X_j = X_i \bar{X}_j \vee \bar{X}_i X_j = X_j \not\equiv X_i \ . \tag{6.1.2}$$

After very simple manipulations

$$\overline{X_i \not\equiv X_j} = X_i \equiv X_j = X_i X_j \vee \bar{X}_i \bar{X}_j \ . \tag{6.1.3}$$

Table 6.1.1. Truth table of the function $X_i \not\equiv X_j \not\equiv X_i X_j$[†]

X_i X_j	$X_i \not\equiv X_j$	$X_i X_j$	Aim $\overbrace{X_i \vee X_j}$	$X_i \not\equiv X_j \not\equiv X_i X_j$
0 0	0	0	0	0
0 1	1	0	1	1
1 0	1	0	1	1
1 1	0	1	1	1

[†] Here, as in similar future expressions, conjunction has priority over EXOR.

Consequently, with $\overline{X_i X_j} = \bar{X}_i \vee \bar{X}_j$ by de Morgan's law, and using Eqs. (2.2.7) and (2.2.13) several times,

$$\begin{aligned}(X_i \not\equiv X_j) \not\equiv X_i X_j &= (X_i \not\equiv X_j)\overline{X_i X_j} \vee \overline{(X_i \not\equiv X_j)} X_i X_j \\ &= (X_i \bar{X}_j \vee \bar{X}_i X_j)(\bar{X}_i \vee \bar{X}_j) \vee (X_i X_j \vee \bar{X}_i \bar{X}_j) X_i X_j \\ &= X_i \bar{X}_j \vee \bar{X}_i X_j \vee X_i X_j \vee X_i X_j = X_i(X_j \vee \bar{X}_j) \vee X_j(X_i \vee \bar{X}_i) \\ &= X_i \vee X_j \ . \end{aligned} \tag{6.1.3a}$$

Finally, it is easily verified, e.g. from Eq. (6.1.2), that

$$\bar{X}_i = X_i \not\equiv 1 \ . \tag{6.1.4}$$

Since conjunction exists any way, the system $\{\wedge, \not\equiv\}$ is "complete", q.e.d.

The systematic construction of functions in the system $\{\wedge, \not\equiv\}$ (and preferably also $^{-}$) is substantially eased by the following lemmas.

Lemma 6.1.1. The disjunction of two disjoint Boolean functions equals their antivalence. In other words: If $\varphi_1 \varphi_2 = 0$, then

$$\varphi_1 \vee \varphi_2 = \varphi_1 \not\equiv \varphi_2 \ . \tag{6.1.5}$$

Proof. As in Eq. (6.1.3)

$$\overline{\varphi_1 \not\equiv \varphi_2} = \varphi_1 \varphi_2 \vee \bar{\varphi}_1 \bar{\varphi}_2 \ .$$

Now, with $\varphi_1 \varphi_2 = 0$, de Morgan's law gives at once Eq. (6.1.5), q.e.d.

Applying this to the Shannon decomposition, Eq. (4.3.3), we immediately have the following.

Corollary 6.1.1.

$$\varphi = X_i \varphi(X_i = 1) \not\equiv \bar{X}_i \varphi(X_i = 0) \ . \tag{6.1.6}$$

Lemma 6.1.2.

$$X_i \equiv X_j = \overline{X_i \not\equiv X_j} = X_i \not\equiv \bar{X}_j = \bar{X}_i \not\equiv X_j \ . \tag{6.1.7}$$

Proof. When replacing in Eq. (6.1.2), X_j by $\bar{X}_j$, or X_i by $\bar{X}_i$ Eq. (6.1.3) is found, q.e.d.

Lemma 6.1.3. Common terms of EXORed operands can be "factored" out:

$$X_i(X_j \not\equiv X_k) = X_i X_j \not\equiv X_i X_k \ . \tag{6.1.8}$$

Proof. The r.h.s. of Eq. (6.1.8) is by Eq. (6.1.2) equal to

$$\varphi_r := X_i X_j \overline{X_i X_k} \vee \overline{X_i X_j} X_i X_k \ .$$

By the usual elementary laws of §2.2

$$\varphi_r = X_i X_j(\bar{X}_i \vee \bar{X}_k) \vee (\bar{X}_i \vee \bar{X}_j) X_i X_k = X_i X_j \bar{X}_k \vee \bar{X}_j X_i X_k$$

$$= X_i(X_j \bar{X}_k \vee \bar{X}_j X_k) = X_i(X_j \not\equiv X_k) = \varphi_l \ ,$$

the l.h.s. of (6.1.8), q.e.d.

Corollary 6.1.2.

$$(X_i \not\equiv X_j)(X_k \not\equiv X_l) = X_i X_k \not\equiv X_i X_l \not\equiv X_j X_k \not\equiv X_j X_l \ . \tag{6.1.9}$$

Proof. By Eq. (6.1.8), applied twice, first with X_i replaced by $X_i \not\equiv X_j$ and X_j by X_l

$$(X_i \not\equiv X_j)(X_k \not\equiv X_l) = [(X_i \not\equiv X_j)X_k] \not\equiv [(X_i \not\equiv X_j)X_l]$$

$$= (X_i X_k \not\equiv X_j X_k) \not\equiv (X_i X_l \not\equiv X_j X_l) \ ,$$

q.e.d. (The parentheses are superfluous because of the next lemma.)

A further lemma concerns the associativeness of the EXOR operation:

Lemma 6.1.4. (Associative law of antivalence)

$$(\varphi_1 \not\equiv \varphi_2) \not\equiv \varphi_3 = \varphi_1 \not\equiv (\varphi_2 \not\equiv \varphi_3) \ . \tag{6.1.10}$$

Proof. This is proved readily by transforming both sides by means of Eqs. (6.1.2), (6.1.3). A more elegant proof looks at the two cases $\varphi_3 = 0, 1$:

$\varphi_3 = 0$: Because of $X \not\equiv 0 = X$ the l.h.s. and r.h.s. of (6.1.10) become

$$\varphi_l = \varphi_1 \not\equiv \varphi_2 \ , \qquad \varphi_r = \varphi_1 \not\equiv (\varphi_2) = \varphi_1 \not\equiv \varphi_2 \ .$$

$\varphi_3 = 1$: Because of $X \not\equiv 1 = \bar{X}$,

$$\varphi_l = \varphi_1 \equiv \varphi_2 \ , \qquad \varphi_r = \varphi_1 \not\equiv \bar{\varphi}_2 \ .$$

But by lemma 6.1.2 $\varphi_r = \varphi_l$, q.e.d.

The above lemmas are very useful for the following.

Theorem 6.1.2 (Reed–Muller representation). Every Boolean function can be transformed to an exclusive-disjunctive form with normal (non complemented) variables only, i.e. $\varphi: B^n \to B$ can be transformed to

$$\varphi = c \not\equiv \overset{m}{\underset{i=1}{\not\equiv}} \hat{T}_i\,, \quad \hat{T}_i = \bigwedge_{j=1}^{n_i} X_{l_j}\,; \quad n_i, l \in \{1, \ldots, n\}; c \in \{0, 1\}; c = \varphi(\mathbf{0})\,. \tag{6.1.11}$$

Proof. Starting with a DDNF of the type of Eq. (4.4.10) then, by lemma 6.1.1

$$\varphi = \underset{i}{\not\equiv} \hat{T}'_i\,, \quad \hat{T}'_i = \bigwedge_j \tilde{X}_{l_{i,j}}\,. \tag{6.1.12}$$

By Eq. (6.1.4) in all the $\hat{T}'_i$ the $\bar{X}_k$ are replaced by $X_k \not\equiv 1$. Then, by corollary 6.1.2, conjunctions of negated variables are transformed according to

$$(X_k \not\equiv 1)(X_l \not\equiv 1) = X_k X_l \not\equiv X_k \not\equiv X_l \not\equiv 1\,. \tag{6.1.13}$$

The result is (in each $\hat{T}'_i$) multiplied via Eq. (6.1.8) with the conjunction of the non-negated variables. The result is Eq. (6.1.11), q.e.d.

Example 6.1.1	Two-out-of-three function

From Eq. (4.4.18a), which is a DDNF,

$$\varphi = X_1 X_2 \bar{X}_3 \vee X_1 \bar{X}_2 X_3 \vee X_2 X_3\,. \tag{6.1.14}$$

Now, according to the procedure in the proof of theorem 6.1.2,

$$\begin{aligned}\varphi &= X_1 X_2 (X_3 \not\equiv 1) \not\equiv X_1 (X_2 \not\equiv 1) X_3 \not\equiv X_2 X_3 \\ &= \underline{\underline{X_1 X_2 X_3}} \not\equiv X_1 X_2 \not\equiv \underline{\underline{X_1 X_2 X_3}} \not\equiv X_1 X_3 \not\equiv X_2 X_3\,.\end{aligned}$$

Further, by Eq. (6.1.2) for any Boolean ψ,

$$\psi \equiv \psi = 0\,. \tag{6.1.15}$$

Hence, finally,

$$\varphi = X_1 X_2 \not\equiv X_1 X_3 \not\equiv X_2 X_3 \tag{6.1.16}$$

has the form Eq. (6.1.11). — —

An (at least theoretically) important feature of a shortest form of Eq. (6.1.11) is its uniqueness, i.e. it is another canonical form.

Theorem 6.1.3. A form

$$\varphi = \overset{m}{\underset{i=1}{\not\equiv}} \hat{T}_i\,, \quad \hat{T}_i = \bigwedge_{j=1}^{n_i} X_{l_{i,j}}\,, \tag{6.1.17}$$

where no $\hat{T}_i$s can be absorbed by others, is unique.

Proof. If φ' and φ'' were different forms of φ of the type of Eq. (6.1.17), then by Eq. (6.1.15)

$$0=\varphi' \not\equiv \varphi''= \not\equiv_{i \in I_d} \hat{T}_i , \tag{6.1.18}$$

where I_d is the set of the indices of all the terms by which φ' and φ'' differ. Now, start the following procedure (Gedanken experiment) to show that $I_d=\phi$.

(I) If any variables can be factored out of all the terms of the r.h.s. of Eq. (6.1.18), do so, and set these variables to 1.

(II) Set a variable X_j contained in at least one of the $\hat{T}_i$ to 0. This reduces the number of terms of Eq. (6.1.18). Continue with this procedure until only one $\hat{T}_i$ remains (with some or all of its variables set to 1).

(III) Set the remaining variables of this $\hat{T}_i$ (if any) to 1.

The result would be for some $\mathbf{X}=\mathbf{X}_0$, specified step by step,

$$\varphi'(\mathbf{X}_0) \not\equiv \varphi''(\mathbf{X}_0)=1$$

in contradiction to $\varphi'=\varphi''$. Hence φ of the form Eq. (6.1.18) cannot be non-unique. q.e.d.

Corollary 6.1.2. Boolean functions of an exclusive-disjunction form with non-negated variables are not constants.

Proof.

(a) Trivially $\varphi(\mathbf{0})= \not\equiv_{i} 0=0$; $\mathbf{0}:=(0, \ldots, 0)$.

(b) By the "procedure" in the proof of theorem 6.1.3 it is possible to find at least one $\mathbf{X}_0$ with $\varphi(\mathbf{X}_0)=1$, q.e.d.

In digital diagnostics, where certain tests are to show whether or not by a fault α a Boolean switching network function φ has been transformed to φ_α, one can call

$$\psi_\alpha := \varphi \not\equiv \varphi_\alpha \tag{6.1.19}$$

the test generating function (TGF), since arguments $\mathbf{X}=\mathbf{X}_i$, for which

$$\psi_\alpha(\mathbf{X}_i)=1 ,$$

are test vectors or test patterns with

$$\varphi_\alpha(\mathbf{X}_i) \neq \varphi(\mathbf{X}_i) .$$

For two simultaneous faults α_1 and α_2 the TGF is

$$\psi_{\alpha_1 \wedge \alpha_2} := \psi_{\alpha_1} \psi_{\alpha_2} . \tag{6.1.20}$$

If one of two faults is to be found, an appropriate TGF is

$$\psi_{\alpha_1 \vee \alpha_2} := \psi_{\alpha_1} \vee \psi_{\alpha_2} = \psi_{\alpha_1} \not\equiv \psi_{\alpha_2} \not\equiv \psi_{\alpha_1} \psi_{\alpha_2} . \tag{6.1.21}$$

6.2 Concepts of Boolean Differences

Definition 6.2.1. The *Boolean difference* of a Boolean function φ with respect to one of its variables X_i is the Boolean function

$$\Delta_{X_i}\varphi(\mathbf{X})^{\dagger} := \varphi(X_1,\ldots,X_n) \not\equiv \varphi(X_1,\ldots,X_{i-1},\bar{X}_i,X_{i+1},\ldots,X_n)\ .$$

In short:

$$\Delta_{X_i}\varphi(\mathbf{X}) = \varphi(\mathbf{X}) \not\equiv \varphi(X_i \leftarrow \bar{X}_i)\ . \tag{6.2.1}$$

That $\Delta_{X_i}\varphi$ does not depend on X_i is clear from the following.

Lemma 6.2.1.

$$\Delta_{X_i}\varphi(\mathbf{X}) = \varphi(X_i=1) \not\equiv \varphi(X_i=0) = \Delta_{\bar{X}_i}\varphi(\mathbf{X})\ . \tag{6.2.2}$$

Proof. By the commutativeness of the antivalence (EXOR) operation, Eq. (6.2.1) makes it obvious that

$$\Delta_{X_i}\varphi(\mathbf{X})/_{X_i=1} = \Delta_{X_i}\varphi(\mathbf{X})/_{X_i=0} = \varphi(X_i=1) \not\equiv \varphi(X_i=0)\ ,$$

q.e.d.

Obviously, if φ is independent of X_i the Boolean difference with respect to X_i vanishes (is 0) because of Eq. (6.1.15).

Equation (6.1.15) offers an interesting possibility to "eliminate" any of the terms of the r.h.s. of Eq. (6.2.1): EXORing both sides of Eq. (6.2.1) with $\varphi(X_i \leftarrow \bar{X}_i)$ yields

$$\varphi = \Delta_{X_i}\varphi \not\equiv \varphi(X_i \leftarrow \bar{X}_i) \tag{6.2.3}$$

and

$$\varphi(X_i \leftarrow \bar{X}_i) = \Delta_{X_i}\varphi \not\equiv \varphi\ . \tag{6.2.4}$$

Equation (6.2.4) means: φ with X_i changed to $\bar{X}_i$ is found by EXORing φ with its Boolean difference with respect to X_i.

Multiple Boolean differences can be defined in several ways. The first, one will think of, is the recursive sequencing of two differences:

$$\begin{aligned}\Delta^2_{X_i,X_j}\varphi &:= \Delta_{X_j}(\Delta_{X_i}\varphi) = \Delta_{X_j}[\varphi(X_i=0) \not\equiv \varphi(X_i=1)]\\ &= [\varphi(X_i=0,\, X_j=0) \not\equiv \varphi(X_i=1,\, X_j=0)] \not\equiv [\varphi(X_i=0,\, X_j=1)\\ &\quad \not\equiv \varphi(X_i=1,\, X_j=1)]\ ,\end{aligned}$$

where the brackets are superfluous.

Hence, the sequencing of single Boolean differences can be done in any order:

$$\Delta^2_{X_i,X_j}\varphi = \Delta^2_{X_j,X_i}\varphi\ . \tag{6.2.5}$$

† Other variants of notation are e.g. $\Delta\varphi(\mathbf{X})/\Delta X_i$ or, simply, φ_{X_i}.

Some ad-hoc notational simplification: Let

$$\varphi[a_i, a_j] := \varphi(X_i \leftarrow a_i, X_j \leftarrow a_j) \; ; \qquad \varphi(\mathbf{X}) = \varphi[X_i, X_j] \; .$$

Definition 6.2.2. The *joint Boolean difference* with respect to the variables X_i and X_j is

$$\begin{aligned} \Delta_{(X_i, X_j)}\varphi &:= \varphi \not\equiv \varphi(X, \ldots, X_{i-1}, \bar{X}_i, \ldots, X_{j-1}, \bar{X}_j, \ldots, X_n) \\ &=: \varphi \not\equiv \varphi[\bar{X}_i, \bar{X}_j] \; . \text{ — —} \end{aligned} \tag{6.2.6}$$

Let us change this to a form similar to Eq. (6.2.2):

By Eq. (6.1.6)

$$\begin{aligned} \varphi &= X_i[X_j\varphi(X_i=1, X_j=1) \not\equiv \bar{X}_j\varphi(X_i=1, X_j=0)] \\ &\not\equiv \bar{X}_i[X_j\varphi(X_i=0, X_j=1) \not\equiv \bar{X}_j\varphi(X_i=0, X_j=0)] \\ &= X_i(X_j\varphi[1,1] \not\equiv \bar{X}_j\varphi[1,0]) \not\equiv \bar{X}_i(X_j\varphi[0,1] \not\equiv \bar{X}_j\varphi[0,0]) \; , \end{aligned}$$

and, replacing X_i and X_j by $\bar{X}_i$ and $\bar{X}_j$, respectively,

$$\varphi[\bar{X}_i, \bar{X}_j] = \bar{X}_i(\bar{X}_j\varphi[1,1] \not\equiv X_j\varphi[1,0]) \not\equiv X_i(\bar{X}_j\varphi[0,1] \not\equiv X_j\varphi[0,0]) \; .$$

Hence, by Eqs. (6.2.6), (2.1.1) and (S1.1)

$$\Delta_{(X_i, X_j)}\varphi = (X_i \not\equiv X_j)(\varphi[0,1] \not\equiv \varphi[1,0]) \not\equiv \overline{(X_i \not\equiv X_j)}(\varphi[0,0] \not\equiv \varphi[1,1]) \; . \tag{6.2.7}$$

Notice that the joint Boolean difference in general depends on X_i and X_j.

It should be mentioned here that the exclusive-disjunction form with non-negated variables can be noted by means of special values of (single and multiple) Boolean differences; for details see [9].

From Eq. (6.1.19) the (diagnostics) test generation function (TGF) is

$$\psi_\alpha = \varphi \not\equiv \varphi_\alpha \; . \tag{6.2.8}$$

Lemma 6.2.2. For α being the fault $X_i = 0$ (X_i stuck at 0) of a combinational switching network S with output $X_\varphi = \varphi(\mathbf{X})$

$$\psi_\alpha = X_i \Delta_{X_i}\varphi \; . \tag{6.2.9}$$

Proof. By Eqs. (6.2.8), (6.1.6) and (6.1.4)

$$\begin{aligned} \psi_\alpha &= \varphi \not\equiv \varphi(X_i=0) = X_i\varphi(X_i=1) \not\equiv \bar{X}_i\varphi(X_i=0) \not\equiv \varphi(X_i=0) \\ &= X_i\varphi(X_i=1) \not\equiv X_i\varphi(X_i=0) = X_i[\varphi(X_i=1) \not\equiv \varphi(X_i=0)] \; , \end{aligned}$$

which equals Eq. (6.2.9) by Eq. (6.2.2), q.e.d.

Corollary 6.2.1. For $\alpha \mathrel{\hat{=}} X_i = 1$

$$\psi_\alpha = \bar{X}_i \Delta_{X_i}\varphi \; . \tag{6.2.10}$$

Proof. From

$$\psi_\alpha = \varphi \not\equiv \varphi(X_i = 1)$$

as in the proof of lemma 6.2.2 the desired result is found, q.e.d.

With the diverse Boolean differences pictures can help us understand the concepts. If for example

$$\Delta_{X_1}\varphi(X_1, X_2, X_3) = \psi(X_2, X_3) \ ,$$

this means that moving on those edges of the unit cube that are parallel to the X_1-axis, φ changes according to ψ.

Example 6.2.1 Two-out-of-three-majority function

Let

$$\varphi(X_1, X_2, X_3) = X_1 X_2 \vee X_1 X_3 \vee X_2 X_3 \ .$$

Then, by Lemma 6.2.1 and (2.2.14)

$$\Delta_{X_1}\varphi = (X_2 \vee X_3) \not\equiv X_2 X_3 \ .$$

By Eq. (6.1.1) and (6.1.15)

$$\Delta_{X_1}\varphi = X_2 \not\equiv X_3 \ . \tag{6.2.11}$$

Hence, in Fig. 6.2.1, on edges parallel with the X_1-axis X_φ changes wherever $X_2 \neq X_3$; see the thick bars in Fig. 6.2.1.

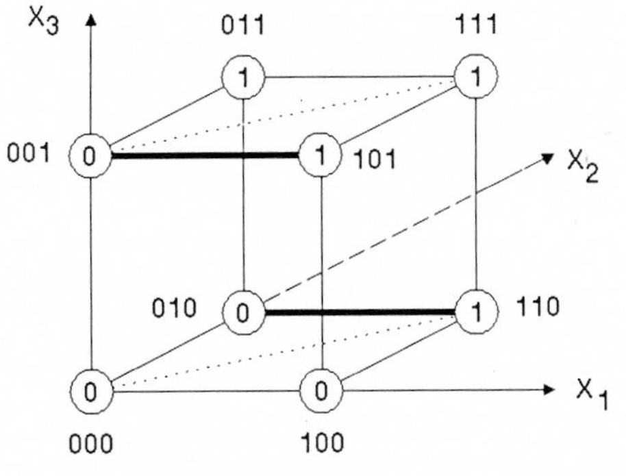

Fig. 6.2.1. Understanding the Boolean differences in B^3

If, for example,

$$\Delta_{(X_1, X_2)}\varphi(X_1, X_2, X_3) = \tilde{\psi}(X_1, X_2 X_3) \ ,$$

then, when moving on surface diagonals parallel with the X_1–X_2-plane, φ changes according to $\tilde{\psi}$.

Example 6.2.2 (Example 6.2.1 contd.)

By Eq. (6.2.6) and (6.1.16) for the φ of example 6.2.1

$$\Delta_{(X_1,X_2)}\varphi = X_1X_2 \not\equiv X_1X_3 \not\equiv X_2X_3 \not\equiv \bar{X}_1\bar{X}_2 \not\equiv \bar{X}_1X_3 \not\equiv \bar{X}_2X_3 \ .$$

Since

$$X_iX_j \not\equiv X_i\bar{X}_j = X_i(X_j \not\equiv \bar{X}_j) = X_i$$

and

$$X_1X_2 \not\equiv \bar{X}_1\bar{X}_2 = X_1X_2(X_1 \vee X_2) \vee (\bar{X}_1 \vee \bar{X}_2)\bar{X}_1\bar{X}_2$$
$$= X_1X_2 \vee \bar{X}_1\bar{X}_2 = X_1 \equiv X_2 = X_1 \not\equiv X_2 \not\equiv 1 \ ,$$

we have

$$\Delta_{(X_1,X_2)}\varphi = X_1 \not\equiv X_2 \not\equiv 1 \not\equiv X_3 \not\equiv X_3 = X_1 \equiv X_2 \ . \tag{6.2.12}$$

Hence, in Fig. 6.2.1 we expect X_φ to change on diagonals in the lower and the upper horizontal plane; see the dotted lines in Fig. 6.2.1.

6.3 Basic Rules of Boolean Difference Calculus

The application of the Boolean difference of $\varphi := \varphi(\mathbf{X})$ is strongly supported by the following rules, which we prove as little lemmas.

Lemma 6.3.1. The negated function has the same Boolean difference as the original one:

$$\Delta_{X_i}\bar{\varphi} = \Delta_{X_i}\varphi \ . \tag{6.3.1}$$

Proof. By Eqs. (6.1.2), (6.1.4) and (6.1.15), applied to Eq. (6.2.2)

$$\Delta_{X_i}\bar{\varphi} := \overline{\varphi(X_i = 1)} \not\equiv \overline{\varphi(X_i = 0)} = [\varphi(X_i = 1) \not\equiv 1] \not\equiv [\varphi(X_i = 0) \not\equiv 1]$$
$$= \varphi(X_i = 1) \not\equiv \varphi(X_i = 0) = \Delta_{X_i}\varphi \ ,$$

q.e.d.

Lemma 6.3.2. The Boolean difference and the exclusive-disjunction are interchangeable in order:

$$\Delta_{X_i}(\varphi_1 \not\equiv \varphi_2) = \Delta_{X_i}\varphi_1 \not\equiv \Delta_{X_i}\varphi_2 \ . \tag{6.3.2}$$

Proof. By Eqs. (6.1.2), (6.2.2), and (6.1.10)

$$\begin{aligned}\Delta_{X_i}(\varphi_1 \not\equiv \varphi_2) &= [\varphi_1(X_i = 0) \not\equiv \varphi_2(X_i = 0)] \not\equiv [\varphi_1(X_i = 1) \not\equiv \varphi_2(X_i = 1)] \\ &= [\varphi_1(X_i = 0) \not\equiv \varphi_1(X_i = 1)] \not\equiv [\varphi_2(X_i = 0) \not\equiv \varphi_2(X_i = 1)]\end{aligned}$$

equals the r.h.s. of Eq. (6.3.2), q.e.d.

Lemma 6.3.3.

$$\Delta_{X_i}(\varphi_1 \varphi_2) = \varphi_1 \Delta_{X_i}\varphi_2 \not\equiv \varphi_2 \Delta_{X_i}\varphi_1 \not\equiv (\Delta_{X_i}\varphi_1)(\Delta_{X_i}\varphi_2) \; . \tag{6.3.3}$$

Proof. By Eqs. (6.2.1), (6.2.4), (6.1.9), (6.1.15)

$$\begin{aligned}\Delta_{X_i}(\varphi_1 \varphi_2) &:= \varphi_1 \varphi_2 \not\equiv \varphi_1(X_i \leftarrow \bar{X}_i)\varphi_2(X_i \leftarrow \bar{X}_i) \\ &= \varphi_1 \varphi_2 \not\equiv (\varphi_1 \not\equiv \Delta_{X_i}\varphi_1)(\varphi_2 \not\equiv \Delta_{X_i}\varphi_2)\end{aligned}$$

equals the r.h.s. of Eq. (6.3.3), q.e.d.

Lemma 6.3.4.

$$\Delta_{X_i}(\varphi_1 \vee \varphi_2) = \bar{\varphi}_1 \Delta_{X_i}\varphi_2 \not\equiv \bar{\varphi}_2 \Delta_{X_i}\varphi_1 \not\equiv (\Delta_{X_i}\varphi_1)(\Delta_{X_i}\varphi_2) \; . \tag{6.3.4}$$

Proof. By (6.1.3a), (6.1.4), (6.3.2), (6.3.3)

$$\begin{aligned}\Delta_{X_i}(\varphi_1 \vee \varphi_2) &= \Delta_{X_i}(\varphi_1 \not\equiv \varphi_2 \not\equiv \varphi_1\varphi_2) = \Delta_{X_i}\varphi_1 \not\equiv \Delta_{X_i}\varphi_2 \not\equiv \Delta_{X_i}(\varphi_1\varphi_2) \\ &= \Delta_{X_i}\varphi_1 \not\equiv \Delta_{X_i}\varphi_2 \not\equiv \varphi_1\Delta_{X_i}\varphi_2 \not\equiv \varphi_2\Delta_{X_i}\varphi_1 \not\equiv (\Delta_{X_i}\varphi_1)(\Delta_{X_i}\varphi_2)\end{aligned}$$

equals the r.h.s. of (6.3.4), q.e.d.

Lemma 6.3.5. (Chain rule of boolean difference calculus).
If $\varphi(X_1, \ldots, X_n)$ is decomposable (according to definition (4.5.11)) as

$$\varphi(X_1, \ldots, X_n) = \tilde{\varphi}(\psi(X_1, \ldots, X_m), X_{m+1}, \ldots, X_n) \; ,$$

then for $i \in \{1, \ldots, m\}$, with ψ as a variable of $\tilde{\varphi}$,

$$\Delta_{X_i}\varphi = (\Delta_\psi \tilde{\varphi})(\Delta_{X_i}\psi) \; . \tag{6.3.5}$$

Proof. By Eqs. (6.1.4), (6.1.6)

$$\begin{aligned}\varphi &= \psi\tilde{\varphi}(1, X_{m+1}, \ldots, X_n) \not\equiv (\psi \not\equiv 1)\tilde{\varphi}(0, X_{m+1}, \ldots, X_n) \\ &= \psi[\tilde{\varphi}(\psi = 1) \not\equiv \tilde{\varphi}(\psi = 0)] \not\equiv \tilde{\varphi}(\psi = 0) \\ &= \psi\Delta_\psi\tilde{\varphi} \not\equiv \tilde{\varphi}(\psi = 0) \; ,\end{aligned}$$

where only the first ψ depends on X_i, $i \in \{1, \ldots, m\}$.

By Eq. (6.2.2) and (6.1.15) for general φ

$$\Delta_{X_i}(\varphi \vee c) = \Delta_{X_i}\varphi; \, c \in \{0, 1\} \; . \tag{6.3.6}$$

Hence, by

$$\Delta_{X_i} c\varphi = c\Delta_{X_i}\varphi;\, c \in \{0, 1\} \ , \tag{6.3.7}$$

finally

$$\Delta_{X_i}\varphi = (\Delta_{X_i}\psi)(\Delta_{\psi}\tilde{\varphi}) \ ,$$

q.e.d.

As a non-trivial application of Boolean vectors, let us look at the Boolean difference. Let $\varphi = \mathbf{A}\,\mathbf{B}^{\mathrm{T}}$ with the components of $\mathbf{A}$ being all the minterms of $\{X_s\} \subset \{X_1, \ldots, X_n\}$.

If $X_i \notin X_s$, then by Eq. (6.2.2) with $\underline{\varphi}_1 \equiv \mathbf{A}$, $\underline{\varphi}_2 \equiv \mathbf{B}$

$$\Delta_{X_i}\varphi = \underline{\varphi}_1 \underline{\varphi}_2^{\mathrm{T}}(X_i = 1) \not\equiv \underline{\varphi}_1 \underline{\varphi}_2^{\mathrm{T}}(X_i = 0) = \underline{\varphi}_1(\Delta_{X_i}\underline{\varphi}_2)^{\mathrm{T}} \ ,$$

where $\Delta_{X_i}\underline{\varphi}_2$ is meant component-wise, i.e.

$$\Delta_{X_i}\underline{\varphi}_2 := (\Delta_{X_i}\varphi_{2,1}, \Delta_{X_i}\varphi_{2,2}, \ldots, \Delta_{X_i}\varphi_{2,n}) \ .$$

This approach can be advantageous in cases where the different $\Delta_{X_i}\varphi_{2,j}$ are easy to derive and where $\Delta_{X_i}\varphi$ is needed for several values of i.

6.4 Diagnosing Permanent Faults in Switching Networks

In this section a simple example will show how some of the above results can be applied to the localization (diagnosis) of permanent faults in electronic circuits that implement a given Boolean function. For instance, Fig. 3.8.2 shows two equivalent realizations of that part of a binary adder that "computes" the carry to the next-higher bit. Figure 6.4.1 shows the network for the computation of the result of a 1-bit binary adder. (For comparison see Fig. 4.6.2.)

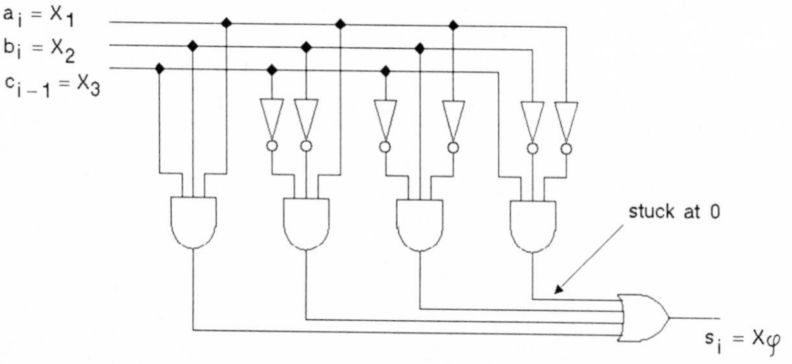

Fig. 6.4.1. One-bit binary adder without carry production; a_i, b_i: addends, c_{i-1} carry from next-lower bit, s_i: sum bit

Now, let us suppose that for some reason the uppermost input of the OR-gate of Fig. 6.4.1 is stuck at 0. This fault α changes the original (correct) switching function

$$\varphi = X_1 X_2 X_3 \vee X_1 \bar{X}_2 \bar{X}_3 \vee \bar{X}_1 X_2 \bar{X}_3 \vee \bar{X}_1 \bar{X}_2 X_3 \tag{6.4.1}$$

to

$$\varphi_\alpha = X_1 X_2 X_3 \vee X_1 \bar{X}_2 \bar{X}_3 \vee \bar{X}_1 X_2 \bar{X}_3 \ . \tag{6.4.2}$$

How can this be found out by observing only input and output signals but no interior point, since the latter are usually not accessible in an LSI circuit chip?

By Eq. (6.2.8) the test generation function is here

$$\psi_\alpha = \varphi \not\equiv \varphi_\alpha = \bar{X}_1 \bar{X}_2 X_3 \ . \tag{6.4.3}$$

Proof. φ and φ_α are CDNFs. Hence, by lemma 4.3.2 the ORs in Eqs. (6.4.1) and (6.4.2) can be replaced by EXORs. Finally, by Eq. (6.1.15) we get Eq. (6.4.3), q.e.d.

The practical meaning of Eq. (6.4.3) is that any input triple, which yields $\varphi_\alpha = 1$, is a test pattern in the sense that the value $s_i = \varphi_\alpha$ of the output would be different from the correct value φ. Specifically, since here ψ_α is a minterm, there exists only one test pattern, namely

$$X_1 = 0 \ , \quad X_2 = 0 \ , \quad X_3 = 1 \ ; \quad \text{i.e.} \quad \mathbf{X}_t = (0, 0, 1) \ .$$

This result is easily interpreted as follows. In the fault-free case the input vector $\mathbf{X}_t$ would make the faulty signal value to be 1, and the correct or the wrong value of this signal can propagate to the circuit output, since the other inputs of the OR-gate of Fig. 6.4.1 are all 0 for $\mathbf{X} = \mathbf{X}_t$.

Exercises

6.1

Prove

(a) lemma 6.1.2,

(b) lemma 6.1.4

under the aspect that the EXOR operation is also the (1-bit-) addition modulo 2.

6.2

Show, by using Eq. (6.1.3), as an alternative of example 6.1.1, that

$$X_1 X_2 \vee X_1 X_3 \vee X_2 X_3 = X_1 X_2 \not\equiv X_1 X_3 \not\equiv X_2 X_3 \ .$$

6.3

Let

$$\varphi = X_1 X_2 \vee X_3 =: \psi \vee X_3 =: \tilde{\varphi}(\psi, X_3) \ .$$

Compare the elementary derivation of $\Delta_{x_1}\varphi$ with that according to the chain rule, Eq. (6.3.5).

6.4
In Fig. E6.1 let the output of the left-hand AND gate be stuck at 0. Calculate a test vector for this fault, and discuss its plausibility.

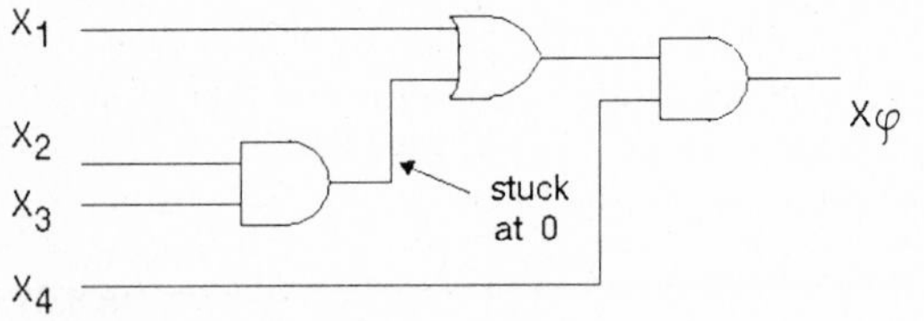

Fig. E6.1. Simple combinational circuit with internal fault

7 Boolean Functions Without Boolean Operators

The motivation for working without Boolean operators can mostly be found in the field of Boolean stochastics which will be treated in §§8 and 9. In the present chapter stochastic aspects will be precluded. Even today practical workers in many fields who apply Boolean functions are easily frustrated by the very idea of working without Boolean operators. The following remarks will show that this concept is not quite so awkward as one may suspect initially.

At least in earlier parts of his work George Boole did not use what we now call Boolean operators. In his famous book *The Laws of Thought* (Macmillan 1854), he writes on p. 57 (Dover edition of 1958): "Rule: Express simple names or qualities by the symbols x, y, z etc., their contraries by $1-x$, $1-y$, $1-z$ etc.; classes of things defined by common names or qualities, by connecting the corresponding symbols as in multiplication; collections of things consisting of portions different from each other by the sign $+$. In particular, let the expression, 'Either xs or ys' be expressed by

$$x(1-y)+y(1-x) \ ,$$

when the classes denoted by x and y are exclusive, by

$$x+y(1-x)$$

when they are not exclusive. . . . "

In essence he would say nowadays

$$\begin{aligned} \text{NOT } x &\hat{=} 1-x \ , \\ x \text{ AND } y &\hat{=} xy \ , \\ x \text{ OR } y &\hat{=} x+y(1-x) = x+y-xy \ , \\ x \text{ EXOR } y &\hat{=} x(1-y)+y(1-x) = x+y-2xy \ , \end{aligned}$$

where the coefficient 2 is non-trivial.

From the above one will not be surprised to find in [9, p. 103] the following: "Obtain the expression of . . . in terms of real sums by using the following well-known identities:

$$a \not\equiv b = a+b-2ab \ , \quad a \vee b = a+b-ab \ , \quad \bar{a} = 1-a \ ."$$

7.1 Fundamental Concepts and Consequences

As will be shown in §7.2, it can become rather cumbersome to calculate the probability of random events of the type $X_\varphi = 1$ if $X_\varphi = \varphi(X_1, \ldots, X_n)$ with n not too small and/or φ of a rather complicated nature (e.g. many terms in the shortest DNF). In such cases [15] contains the hint of calculating $E\{X_\varphi\}$ instead of $P\{X_\varphi = 1\}$. Now, it is not clear how to do that if φ contains typical Boolean operators as $\wedge$ and $\vee$ or others. In [16] and [17] it is shown how to replace Boolean operators by $+$ (plus), $-$ (minus) and concatenation (multiplication). It can be done easily more or less along the lines given by Boole himself.

For Boolean indicator variables X_i and X_j clearly

$$X_i \wedge X_j = X_i X_j \ , \tag{7.1.1}$$

(meaning the product now when omitting an operator); further

$$X_i \vee X_j = X_i + X_j - X_i X_j = 1 - (1 - X_i)(1 - X_j) \tag{7.1.2}$$

and

$$\bar{X}_i = 1 - X_i \ . \tag{7.1.3}$$

(Notice the different meaning of $X_i \bar{X}_i = 0$ compared to that of Eq. (2.2.7).)

The idempotence law

$$X_i^2 = X_i \tag{7.1.4}$$

is trivial, since, trivially, $0^2 = 0$ and $1^2 = 1$.

Another often used version of Eq. (7.1.2) is

$$X_i \vee X_j = X_i + \bar{X}_i X_j \ . \tag{7.1.5}$$

Equation (7.1.1) is readily extended to

$$\bigwedge_i X_i = \prod_i X_i \ . \tag{7.1.1a}$$

The corresponding extension of Eq. (7.1.2) is

$$\bigvee_i X_i = 1 - \prod_i (1 - X_i) = 1 - \prod_i \bar{X}_i \ . \tag{7.1.2a}$$

This is easily proved by induction; but also the observation that on both sides setting one X_i to 1 suffices to change the value from 0 to 1 is a proof. It should be pointed out here that in [16] the axioms of Boolean algebra are shown to be fulfilled. For example, the distributive law

$$X_i \wedge (X_j \vee X_k) = (X_i \wedge X_j) \vee (X_i \wedge X_k)$$

is "proved" as follows:

$$\text{l.h.s.: } X_i \wedge (X_j \vee X_k) = X_i(X_j + X_k - X_j X_k) = X_i X_j + X_i X_k - X_i X_j X_k \ ,$$

$$\text{r.h.s.: } (X_i \wedge X_j) \vee (X_i \wedge X_k) = X_i X_j + X_i X_k - X_i X_j X_k{}^{\dagger} \ .$$

Also, other rules (laws) are easily proved, e.g. the absorption rule

$$X_i \vee (X_i \wedge X_j) = X_i + X_i X_j - X_i X_j = X_i$$

and de Morgan's law

$$\overline{X_i \vee X_j} = 1 - X_i - X_j + X_i X_j = (1 - X_i)(1 - X_j) = \bar{X}_i \bar{X}_j \ .$$

Also, as already mentioned in the introduction of §7, by Eq. (7.1.2)

$$X_i \not\equiv X_j = X_i \bar{X}_j \vee \bar{X}_i X_j = X_i(1 - X_j) + (1 - X_i) X_j = X_i + X_j - 2X_i X_j \ . \quad (7.1.6)$$

(Formally, the last term can be interpreted as an abbreviation for $-X_i X_j - X_i X_j$.) An interesting alternative for Eq. (7.1.6) is

$$X_i \not\equiv X_j = (X_i - X_j)^2 \ , \quad (7.1.7)$$

where the power 2 must not be omitted, since $X_i - X_j$ is not Boolean.

It should be mentioned that also the inverse process of replacing Boolean operators by addition, subtraction, and multiplication is possible in several systematic ways. The most trivial one is to produce the truth table, which leads the way to the CDNF. A better method, producing a shorter DNF than the CDNF, is the application of the Shannon decomposition algorithm of section 4.4.3.

Example 7.1.1 Two-out-of-three majority

Let

$$\varphi = X_1 X_2 + X_1 X_3 + X_2 X_3 - 2X_1 X_2 X_3 \ . \quad (7.1.8)$$

This is a Boolean function, since $\varphi \in \{0, 1\}$. More need not be known here. By Eq. (4.4.24) applied twice

$$\begin{aligned}\varphi &= X_1(X_2 + X_3 + X_2 X_3 - 2X_2 X_3) \vee \bar{X}_1 X_2 X_3 \\ &= X_1[X_2(1 + X_3 + X_3 - 2X_3) \vee \bar{X}_2 X_3] \vee \bar{X}_1 X_2 X_3 \\ &= X_1 X_2 \vee X_1 \bar{X}_2 X_3 \vee \bar{X}_1 X_2 X_3 \ ,\end{aligned} \quad (7.1.9)$$

which is, by Eq. (4.4.18), a DNF of the two-out-of-three majority function. — —

The Shannon decomposition can also be used to check if a given multilinear polynomial (with integer coefficients) is Boolean. At some point in the trans-

† In the last term Eq. (7.1.4) was applied.

formation process the non-Boolean character, if any, of the given polynomial will become apparent.

Example 7.1.2 Non-Boolean polynomial

Let

$$\psi = X_1 X_2 + X_1 X_3 + X_2 X_3 - X_1 X_2 X_3 \ . \tag{7.1.10}$$

(For $X_1 = X_2 = X_3 = 1$: $\psi = 2$. Hence the non-Boolean property of ψ is obvious.) Nevertheless, proceeding as in example 7.1.1 yields

$$\begin{aligned}\psi &= X_1(X_2 + X_3 + X_2 X_3 - X_2 X_3) \vee \bar{X}_1 X_2 X_3 \\ &= X_1(X_2 + X_3) \vee \bar{X}_1 X_2 X_3 \ .\end{aligned} \tag{7.1.11}$$

Since $X_2 + X_3$ is non-Boolean, ψ is non-Boolean. — —

Remark. It is certainly not everybody's taste to mix two algebras in one expression. Yet, with enough brackets this works quite well, as Eq. (7.1.9) shows. No doubt, expressions like

$$X_1 + X_2 \vee X_3$$

should be avoided; but expressions like

$$(X_1 + X_2) \vee X_3 \ , \ X_1 + (X_2 \vee X_3)$$

can make sense, if both of them are Boolean. The problem with them is that, in general, neither $X_1 + X_2$ nor $X_1 + X_4$, where $X_4 := X_2 \vee X_3$, are Boolean.

Discussion. No doubt Eqs. (7.1.1), (7.1.2) could be replaced by

$$X_i \vee X_j = \max\{X_i, X_j\}, \ X_i \wedge X_j = \min\{X_i, X_j\} \tag{7.1.12}$$

as in lattice theory [48]. However, it appears that for the purposes of this text Eqs. (7.1.1), (7.1.2) are the better choice.

In closing it should be emphasized that all the formulae of §7.1 are true for any type of Boolean quantities and not only for variables. Since such quantities – let us call them terms T_i, T_j, etc. – can be disjoint, it is possible that

$$T_i \vee T_j = T_i + T_j$$

or

$$T_i \not\equiv T_j = T_i + T_j \ .$$

But, generally,

$$T_i \vee T_j = T_i + T_j - T_i T_j \tag{7.1.13}$$

and

$$T_i \not\equiv T_j = T_i + T_j - 2T_i T_j \ .$$

Note, also, how extremely simple negation is, as compared to the Shannon inversion rule (theorem 4.1). Here, irrespective of the complexity of φ, simply

$$\bar{\varphi} = 1 - \varphi \ .$$

7.2 Transformation of Boolean Functions of Indicator Variables to Multilinear Form

The main practical reason for a considerable interest in Boolean functions written as polynomials with standard symbols $+$, $-$, and $\cdot$ for addition, subtraction, and multiplication, respectively, is the ease with which the expected value of a polynomial can be calculated, especially when all its variables are stochastically independent of each other. (More on this will follow in §§8 and 9.)

Using the fundamental formulas Eqs. (7.1.1) to (7.1.4), one can transform any Boolean function φ directly to a multilinear form. However, in many cases a DNF will be given. Then one can get along well with Eq. (7.1.2a), and, only on factoring out terms, will use be made of Eqs. (7.1.3) and (7.1.4).

Example 7.2.1 Two-out-of-three-function

From the well-known

$$\varphi = X_1 X_2 \vee X_1 X_3 \vee X_2 X_3 \ ,$$

by Eq. (7.1.2a)

$$\begin{aligned} \varphi &= 1-(1-X_1X_2)(1-X_1X_3)(1-X_2X_3) \\ &= 1-(1-X_1X_2-X_1X_3+X_1^2X_2X_3)(1-X_2X_3) \ . \end{aligned}$$

"Forgetting" the exponent in X_1^2 and also in squares that follow:

$$\begin{aligned} \varphi &= 1-1+X_2X_3+X_1X_2-X_1X_2X_3+X_1X_3 \\ &\quad -X_1X_2X_3-X_1X_2X_3+X_1X_2X_3 \\ &= X_1X_2+X_1X_3+X_2X_3-2X_1X_2X_3 \ . \qquad (7.2.1) \end{aligned}$$

It is obvious that, using this approach, one easily gets into severe problems with computational complexity. Typically, on transforming Eq. (4.4.21) with its DNF of 7 terms one has in the beginning $2^7 = 128$ terms. Here the algorithm of Satyaryanana and Prabhakar [19] may prove to be helpful. The following theorem leads the way to another solution of the above problem.

Theorem 7.2.1. In disjunctions of mutually disjoint functions the OR-operator ($\vee$) can be replaced by the plus sign ($+$).

Proof. If $\varphi_1\varphi_2 = 0$ from Eq. (7.1.2)

$$\varphi_1 \vee \varphi_2 = \varphi_1 + \varphi_2 \ .$$

If also $\varphi_1\varphi_3 = \varphi_2\varphi_3 = 0$, then

$$\begin{aligned}\varphi_1 \vee \varphi_2 \vee \varphi_3 &= (\varphi_1 + \varphi_2) + \varphi_3 - (\varphi_1 + \varphi_2)\varphi_3 \\ &= \varphi_1 + \varphi_2 + \varphi_3 - \varphi_1\varphi_3 - \varphi_2\varphi_3 \\ &= \varphi_1 + \varphi_2 + \varphi_3 \ .\end{aligned}$$

By induction (whose details are omitted here)

$$\bigvee_{\mathrm{i}} \varphi_{\mathrm{i}} = \sum_{\mathrm{i}} \varphi_{\mathrm{i}} \ , \quad \varphi_{\mathrm{j}}\varphi_{\mathrm{k}} = 0 \ , \quad j \neq k \ , \tag{7.2.2}$$

q.e.d.

Corollary 7.2.1. Any Boolean function φ is the sum of its minterms

$$\varphi = \sum_{\mathrm{i} \in \mathrm{I}_\varphi} \hat{M}_{\mathrm{i}} \ . \tag{7.2.2a}$$

Proof. Apply theorem 7.2.1 to the CDNF of φ!

Examples:
(a) From Eq. (4.4.18) or Eq. (4.4.25) for the two-out-of-three system

$$\varphi = X_1X_2 + X_1\bar{X}_2X_3 + \bar{X}_1X_2X_3 \ . \tag{7.2.3}$$

(b) From Eq. (4.4.23)

$$\varphi = X_6X_7 + X_1X_2X_3\bar{X}_6 + \ \dots \ + \bar{X}_1X_2X_3X_4\bar{X}_5X_6\bar{X}_7 \ . \tag{7.2.4}$$

(c) From Eq. (4.4.34)

$$\varphi = X_6X_7 + X_3X_5X_6\bar{X}_7 + \ \dots \ + X_1\bar{X}_2X_3X_4X_5\bar{X}_6\bar{X}_7 \ . \tag{7.2.5}$$

In general, we have for the important representation of Boolean functions (of indicator variables) by multilinear polynomials the following result.

Theorem 7.2.2. Every Boolean function φ of Boolean indicator variables $\mathbf{X} = (X_1, \dots, X_{\mathrm{n}})$ can be transformed to a multilinear form (of its literals):

$$\varphi(\mathbf{X}) = \varphi_{\mathrm{ML}}(\mathbf{X}) = c_0 + \sum_{\mathrm{i}=1}^{\mathrm{m}} \left(c_{\mathrm{i}} \prod_{\mathrm{j}=1}^{\mathrm{n_i}} \tilde{X}_{l_{\mathrm{i,j}}} \right) , \quad l_{\mathrm{i,j}} \neq l_{\mathrm{i,k}}{}^{\dagger} \tag{7.2.6}$$

† Note that it is not necessary that all $\prod_{\mathrm{j}} \tilde{X}_{l_{\mathrm{i,j}}}$ are different. If they are different, the c_{i} are integers.

with

$$n_i, l_{i,j} \in \{1, \ldots, n\}, \quad c_0 \in \{0, 1\}, c_i \in \mathbb{R}, \quad \tilde{X}_k \in \{X_k, \bar{X}_k\}, \quad m, n_i \in \mathbb{N}.$$

Proof. Corollary 7.2.1, since a sum of minterms is a multilinear polynomial.

Corollary 7.2.2. For every φ of the above type the following complement free multilinear polynomial form is also possible (with m, c_i, n_i generally differing from those of Eq. (7.2.6))

$$\varphi(\mathbf{X}) = c_0 + \sum_{i=1}^{m} \left(c_i \prod_{j=1}^{n_i} X_{l_{i,j}} \right); \quad l_{i,j} \neq l_{i,k}. \tag{7.2.7}$$

Proof. Using $\bar{X} = 1 - X$ to transform Eq. (7.2.6), after multiplying out one has a polynomial of the type Eq. (7.2.7).

Example 7.2.2 Two-out-of-three system

From Eqs (4.4.18) or (4.4.25) we found Eq. (7.2.3), i.e.

$$\varphi = X_1 X_2 + X_1 \bar{X}_2 X_3 + \bar{X}_1 X_2 X_3,$$

which is of the Eq. (7.2.6)-type with $c_0 = 0$, $c_1 = c_2 = c_3 = 1$.

Replacing $\bar{X}_i$ by $1 - X_i$ in Eq. (7.2.3), yields Eq. (7.2.1):

$$\begin{aligned} \varphi &= X_1 X_2 + X_1(1 - X_2)X_3 + (1 - X_1)X_2 X_3 \\ &= X_1 X_2 + X_1 X_3 - X_1 X_2 X_3 + X_2 X_3 - X_1 X_2 X_3 \\ &= X_1 X_2 + X_1 X_3 + X_2 X_3 - 2X_1 X_2 X_3 . \text{ — —} \end{aligned} \tag{7.2.8}$$

It is of – at least theoretical – interest, that Eq. (7.2.7) where all $\prod_j X_{l_{i,j}}$ are different is another canonical form, see the following theorem.

Theorem 7.2.3. The multilinear form with non-negated variables only (and integer coefficients)

$$\begin{aligned} X_\varphi = \varphi(X_1, \ldots, X_n) = a_0 &+ a_1 X_1 + a_2 X_2 + \ldots \\ &+ a_{1,2} X_1 X_2 + a_{1,3} X_1 X_3 + \ldots \\ &+ \ldots \\ &+ a_{1,2,\ldots,n} X_1 X_2 \ldots X_n \end{aligned} \tag{7.2.9}$$

is unique, i.e. it is another canonical form. (Found independently from [38].)

Proof. Let $a_{i,j,\ldots m}$ the first coefficient, in which Eq. (7.2.9) and a possibly different form of the same φ, namely

$$X'_\varphi = a'_0 + a'_1 X_1 + a'_2 X_2 + \ldots + a'_{1,2} X_1 X_2 + \ldots$$

differ. Then, setting $X_i = X_j = \ldots = X_m = 1$, and all the others to 0, we have

$$X_\varphi - X'_\varphi = a_{i,j,\ldots,m} - a'_{i,j,\ldots,m} .$$

Now, from $X'_\varphi = X_\varphi$ there follows the contradiction $a'_{i,j,\ldots,m} = a_{i,j,\ldots,m}$, q.e.d.

At this point it is appropriate to present a nice little algorithm [18] for finding Eq. (7.2.9) from any form of φ:

The Enzmann Algorithm

For

$$X_\varphi = \varphi(X_1, X_2, X_3, X_4, X_5, \ldots, X_{n-2}, X_{n-1}, X_n)$$

obviously

$$\begin{aligned}
a_0 &= \varphi(0, 0, 0, 0, 0, \ldots, 0, 0, 0) ,\\
a_0 + a_1 &= \varphi(1, 0, 0, 0, 0, \ldots, 0, 0, 0) ,\\
a_0 + a_2 &= \varphi(0, 1, 0, 0, 0, \ldots, 0, 0, 0) ,\\
a_0 + a_3 &= \varphi(0, 0, 1, 0, 0, \ldots, 0, 0, 0) ,\\
&\vdots\\
a_0 + a_1 + a_2 + a_{1,2} &= \varphi(1, 1, 0, 0, 0, \ldots, 0, 0, 0) ,\\
a_0 + a_1 + a_3 + a_{1,3} &= \varphi(1, 0, 1, 0, 0, \ldots, 0, 0, 0) ,\\
&\vdots\\
a_0 + a_1 + \ldots + a_{1,\ldots,n} &= \varphi(1, 1, 1, 1, 1, \ldots, 1, 1, 1) .
\end{aligned}$$

This is obviously a very simple system of linear equations to solve (no matrix inversion is necessary). The uniqueness of the solution together with its independence of the form, i.e. representation of φ, is another proof of theorem 7.2.3.

Example 7.2.3 Two-out-of-three system

As is well known, here

$$\varphi(X_1, X_2, X_3) = X_1 X_2 \vee X_1 X_3 \vee X_2 X_3 .$$

Hence, by the above equations

$$\begin{aligned}
a_0 &= \varphi(0, 0, 0) = 0 ,\\
a_0 + a_1 &= \varphi(1, 0, 0) = 0 \Rightarrow a_1 = 0 ,\\
a_0 + a_2 &= \varphi(0, 1, 0) = 0 \Rightarrow a_2 = 0 ,\\
a_0 + a_3 &= \varphi(0, 0, 1) = 0 \Rightarrow a_3 = 0 ,
\end{aligned}$$

$$a_0 + a_1 + a_2 + a_{1,2} = \varphi(1,1,0) = 1 \Rightarrow a_{1,2} = 1 \ ,$$

$$a_0 + a_1 + a_3 + a_{1,3} = \varphi(1,0,1) = 1 \Rightarrow a_{1,3} = 1 \ ,$$

$$a_0 + a_2 + a_3 + a_{2,3} = \varphi(0,1,1) = 1 \Rightarrow a_{2,3} = 1 \ ,$$

$$a_0 + a_1 + a_2 + a_3 + a_{1,2} + a_{1,3} + a_{2,3} + a_{1,2,3} = \varphi(1,1,1) = 1$$

$$\Rightarrow a_{1,2,3} = 1 - 3 = -2 \ .$$

Insertion in Eq. (7.2.9) yields the well-known result

$$\varphi = X_1 X_2 + X_1 X_3 + X_2 X_3 - 2 X_1 X_2 X_3 \ .$$

Mixing Boolean and Non-Boolean Notation

By theorem 7.2.1 the Shannon decomposition formula (Eq. (4.4.24)) can be rewritten as

$$\varphi(\mathbf{X}) = X_i \varphi(X_i = 1) + \bar{X}_i \varphi(X_i = 0) \ . \tag{7.2.10}$$

Working one's way down to a multilinear polynomial form of the Eq. (7.2.6)--type one will have both Boolean and usual arithmetical operators in intermediate steps. This should not produce any problems. On the contrary, the vanishing of the last OR ($\vee$) is the stop signal of the decomposition algorithm.

Example 7.2.4 (Example 4.4.4 redone)

From Eq. (4.4.27)

$$\varphi = X_6 \varphi_{1,1} + \bar{X}_6 X_1 \varphi_{1,2}$$

with (see Eq. (4.4.28) to Eq. (4.4.32))

$$\varphi_{1,1} = X_7 + \bar{X}_7 X_3 \varphi_{2,1} \ ,$$

$$\varphi_{2,1} = X_5 + \bar{X}_5 X_2 (X_1 + \bar{X}_1 X_4) \ ,$$

$$\varphi_{1,2} = X_2 (X_3 \vee X_4 X_7 \vee X_5 X_7) + \bar{X}_2 X_4 (X_7 \vee X_3 X_5)$$

$$= X_2 \varphi_{2,2} + \bar{X}_2 X_4 \varphi_{2,3} \ ,$$

$$\varphi_{2,2} = X_3 + \bar{X}_3 X_7 (X_4 + \bar{X}_4 X_5) \ ,$$

$$\varphi_{2,3} = X_7 + \bar{X}_7 X_3 X_5 \ .$$

Finally, instead of Eq. (4.4.33),

$$\varphi = X_6 \{X_7 + \bar{X}_7 X_3 [X_5 + \bar{X}_5 X_2 (X_1 + \bar{X}_1 X_4)]\}$$
$$+ \bar{X}_6 X_1 \{X_2 [X_3 + \bar{X}_3 X_7 (X_4 + \bar{X}_4 X_5)] + \bar{X}_2 X_4 (X_7 + \bar{X}_7 X_3 X_5)\} \ . \tag{7.2.11}$$

Speeding Up Transformations to a Multilinear Form

Intelligent combinations of the above methods for finding a φ_{ML} can substantially speed up this non-unique transformation process. For instance, in the process of the Shannon decomposition the transformation of $X_i \vee X_j \vee X_k \vee \ldots$ can be speeded up by using Eq. (7.1.2a), i.e. by using – in case inverse (negated) variables are allowed –

$$X_i \vee X_j \vee X_k \vee \ldots = 1 - \bar{X}_i \bar{X}_j \bar{X}_k \ldots$$

Starting from a DNF of φ the transformation to a φ_{ML} can be speeded up considerably by a proper bipartition of φ into φ_1 and φ_2:

$$\varphi(\mathbf{X}) = \varphi_1(\mathbf{X}) \vee \varphi_2(\mathbf{X}) = \varphi_1(\mathbf{X}) + \varphi_2(\mathbf{X}) - \varphi_1(\mathbf{X})\varphi_2(\mathbf{X}) \ . \tag{7.2.12}$$

Here "proper" means that it must be relatively easy to determine $\varphi_{1,ML}$ and $\varphi_{2,ML}$ and that the overall length of the last form of φ does not greatly exceed those found by the methods of §4.4.

Example 7.2.5 Simplified ARPA net (Example 4.4.2 redone)

Let

$$\varphi_1(\mathbf{X}) = X_6 X_7 \vee X_1 X_4 X_7 \vee X_3 X_5 X_6 \vee X_1 X_2 X_5 X_7 \vee X_2 X_3 X_4 X_6 \ , \tag{7.2.13}$$

$$\varphi_2(\mathbf{X}) = X_1 X_2 X_3 \vee X_1 X_3 X_4 X_5 \ . \tag{7.2.14}$$

Then with little effort the set addition algorithm of Abraham yields

$$\begin{aligned}\varphi_{1,ML} = {} & X_6 X_7 + X_1 X_4 \bar{X}_6 X_7 + X_3 X_5 X_6 \bar{X}_7 + X_1 X_2 \bar{X}_4 X_5 \bar{X}_6 X_7 \\ & + X_2 X_3 X_4 \bar{X}_5 X_6 \bar{X}_7 \ ,\end{aligned} \tag{7.2.15}$$

$$\varphi_{2,ML} = X_1 X_2 X_3 + X_1 \bar{X}_2 X_3 X_4 X_5 \ . \tag{7.2.16}$$

Because of $X^2 = X$ and $X\bar{X} = 0$, two of the possible ten terms of $\varphi_{1,ML}\varphi_{2,ML}$ vanish. The result is

$$\begin{aligned}\varphi_{1,ML}\varphi_{2,ML} = {} & X_1 X_2 X_3 X_6 X_7 + X_1 \bar{X}_2 X_3 X_4 X_5 X_6 X_7 \\ & + X_1 X_2 X_3 X_4 \bar{X}_6 X_7 + X_1 X_2 \bar{X}_3 X_4 X_5 \bar{X}_6 X_7 \\ & + X_1 X_2 X_3 X_5 X_6 \bar{X}_7 + X_1 \bar{X}_2 X_3 X_4 X_5 X_6 \bar{X}_7 \\ & + X_1 X_2 X_3 \bar{X}_4 X_5 \bar{X}_6 X_7 + X_1 X_2 X_3 X_4 \bar{X}_5 X_6 \bar{X}_7 \ .\end{aligned} \tag{7.2.17}$$

Hence $\varphi_{ML} = \varphi_{1,ML} + \varphi_{2,ML} - \varphi_{1,ML}\varphi_{2,ML}$ consists of $5 + 2 + 8 = 15$ terms, whereas Eq. (4.4.23) (with the $\vee$s replaced by $+$s) consists of only ten terms. — —

The above example will look much better when it will be part of an approximate probabilistic analysis in §8. There it will prove unnecessary to calculate all the terms of $\varphi_{1,M}\varphi_{2,M}$.

7.3 Coherence Revisited

The main results of §8 will hold true only for coherent systems. Therefore, here is the proper place to show how coherence can be expressed in the context of multilinear Boolean polynomials.

As an introductory example of a non-coherent function we have from Eq. (7.1.6)

$$X_1 \not\equiv X_2 = X_1(1-X_2)+(1-X_1)X_2 = X_1+X_2-2X_1X_2 \ . \tag{7.3.1}$$

From Eq. (7.2.7), on focusing interest on the dependence of φ on X_i, one can write

$$\varphi(\mathbf{X}) = X_i\varphi_i' + \varphi_i'' \ , \tag{7.3.2}$$

where φ_i', φ_i'' are independent of X_i. (φ_i', φ_i'' are different from φ', φ'' in Eq. (4.5.16).)

Obviously, φ_i' is the sum of all the terms containing X_i with X_i factored out, and φ_i'' is the sum of all the other terms of φ. More insight into the meaning of φ_i' and φ_i'' can be gained from Eq. (7.2.10), i.e. from

$$\varphi(\mathbf{X}) = X_i\varphi(X_i=1) + \bar{X}_i\varphi(X_i=0) \ . \tag{7.3.3}$$

With $\bar{X}_i = 1 - X_i$ Eq. (7.3.3) becomes

$$\varphi(\mathbf{X}) = X_i[\varphi(X_i=1) - \varphi(X_i=0)] + \varphi(X_i=0) \ . \tag{7.3.4}$$

Hence

$$\varphi_i'(X_1, \ldots, X_{i-1}, X_{i+1}, \ldots, X_n) = \varphi(X_i=1) - \varphi(X_i=0) \tag{7.3.5}$$

and

$$\varphi_i''(X_1, \ldots, X_{i-1}, X_{i+1}, \ldots, X_n) = \varphi(X_i=0) \ . \tag{7.3.6}$$

This completes the proof of the following theorem.

Theorem 7.3.1. Given the Boolean function φ of indicator variables $X_1, \ldots, X_n$ as

$$\varphi(\mathbf{X}) = X_i\varphi_i' + \varphi_i''; \qquad \varphi_i', \varphi_i'' \text{ independent of } X_i \ ,$$

then

(a) φ_i'' is Boolean,

(b) $\varphi_i' + \varphi_i''$ is Boolean,

(c) φ_i' is Boolean iff $\varphi(X)$ is monotonically non-decreasing in X_i. In general $\varphi_i' \in \{0, 1, -1\}$. — —

Also the Boolean difference of Eq. (6.2.2), transformed via Eq. (7.3.1), i.e.

$$\begin{aligned}\Delta_{X_i}\varphi &= \varphi(X_i = 0) \not\equiv \varphi(X_i = 1)\\ &= \varphi(X_i = 0) + \varphi(X_i = 1) - 2\varphi(X_i = 0)\varphi(X_i = 1)\end{aligned} \tag{7.3.7}$$

can be used to illuminate monotonicity. Inserting from Eqs. (7.3.5), (7.3.6) into Eq. (7.3.7) yields

$$\Delta_{X_i}\varphi = \varphi_i'' \not\equiv (\varphi_i' + \varphi_i'') = \varphi_i'' + \varphi_i' + \varphi_i'' - 2\varphi_i''(\varphi_i' + \varphi_i'') \ .$$

Since φ_i'' is Boolean, by the idempotence rule (7.1.4),

$$\Delta_{X_i}\varphi = \varphi_i'(1 - 2\varphi_i'') = \varphi_i' - 2\varphi_i'\varphi_i'' \ . \tag{7.3.8}$$

Lemma 7.3.1. If φ_i' is Boolean, then

$$\varphi_i'\varphi_i'' = 0 \ .$$

Proof. Since $\varphi_i' + \varphi_i''$ and φ_i'' are Boolean, $\varphi_i' = 1 \Rightarrow \varphi_i'' = 0$, q.e.d.

Theorem 7.3.2. If a Boolean function $\varphi(X_1, \ldots, X_n)$ is monotonically non-decreasing in X_i, then the Boolean difference of φ with respect to X_i is the formal partial derivative of φ given in multilinear polynomial form.

Proof. By Eq. (7.3.2), formally, i.e. forgetting the discreteness of X_i,

$$\frac{\partial\varphi}{\partial X_i} = \varphi_i' \ .$$

By Eq. (7.3.8), theorem 7.3.1c and lemma 7.3.1, for the φ of theorem 7.3.2

$$\Delta_{X_i}\varphi = \varphi_i' \ . \tag{7.3.9}$$

Hence, in the case of Boolean φ_i',

$$\Delta_{X_i}\varphi = \frac{\partial\varphi}{\partial X_i} \ , \tag{7.3.10}$$

q.e.d.

Finally, it is shown how unateness can be expressed in standard algebraic notation: By definition 4.5.5 φ is unate, if DNFs φ_i' and φ_i'' (without X_i) can be found such that

$$\varphi = \tilde{X}_i\varphi_i' \vee \varphi_i'' \ , \quad \tilde{X}_i \in \{X_i, \bar{X}_i\} \ . \tag{7.3.11}$$

By Eqs. (7.1.1)–(7.1.3)

$$\varphi = \tilde{X}_i \varphi_i' + \varphi_i'' - \tilde{X}_i \varphi_i' \varphi_i'' = \tilde{X}_i \varphi_i' \overline{\varphi_i''} + \varphi_i''$$
$$= \tilde{X}_i \varphi_i''' + \varphi_i'' \ , \qquad \varphi_i''' := \varphi_i' \overline{\varphi_i''} \ , \tag{7.3.12}$$

where both, φ_i'' and φ_i''' are Boolean.

Exercises

7.1
Calculate the shortest multilinear polynomial form of

$$\varphi = (X_1 \not\equiv X_2) \not\equiv X_3 \ .$$

(a) by using Eq. (7.1.6) twice,
(b) by applying the ENZMANN algorithm.
(c) Use the result of (a) for a new proof of the associative law (6.1.10), i.e. of

$$(X_1 \not\equiv X_2) \not\equiv X_3 = X_1 \not\equiv (X_2 \not\equiv X_3) \ . \tag{E7.1}$$

7.2
Use Eq. (7.3.10) to calculate

$$\Delta_{X_1}(X_1 X_2 \vee X_1 X_3 \vee X_2 X_3) \ .$$

7.3
Why can the multilinear form of a coherent Boolean function not contain a constant term $c_0 \neq 0$?

7.4
Prove that for coherent Boolean functions in the complement free canonical polynomial form, Eqs. (7.2.7), and (7.2.9), respectively,

$$c_1 + c_2 + \ldots + c_m = 1 \ ,$$

and

$$a_1 + a_2 + \ldots + a_{1,2} + a_{1,3} + \ldots + a_{1,2\ldots,n} = 1 \ ,$$

respectively.

8 Stochastic Theory of Boolean Functions

§§8 and 9 are devoted to the stochastics (probabilistics) of Boolean analysis. Hence, it is advisable to first refresh ones knowledge of basic probability theory via the appendix §11. As to practical applications, §8 is devoted to the determination of probabilities of various states of given systems, where states will be described by Boolean vectors. In principle, code words of binary codes are included. The new, namely the probabilistics, aspect consists of the – properly modelled – uncertainty of the actual values of one or more Boolean (binary) variables. (The most primitive experiment in this field is the well-known tossing of a coin.) In §8.1 the concepts connected with the probability of a binary state are discussed. Then, in §8.2 the probability of the value 1 of a Boolean function is determined. §8.3 is concerned with approximate calculations of this probability, and in §8.4 moments of the random value of a Boolean function of random variables are derived.

8.1 Probability of a Binary State

If a state S_i is defined to be a collection (set) of elementary states s_j, i.e.

$$S_i = \{s_j \mid j \in I_i\} \ , \qquad I_i \subset \{1, \ldots, 2^n\} \ ,$$

then, because of the disjointness of the s_j, the probability of S_i is simply

$$P\{S_i\} = \sum_{j \in I_i} P\{s_j\} \ . \tag{8.1.1}$$

In Boolean analysis s_j is equivalent to a Boolean n-tuple

$$\mathbf{X}_j = (X_{j,1}, \ldots, X_{j,n}) \ ; \quad X_{j,k} \in \{0, 1\} \ .$$

Hence, with $\mathbf{X}$ for the general state vector, and random events $a_k := (X_k = X_{j,k})$:

$$P\{s_j\} = P\{\mathbf{X} = \mathbf{X}_j\} := P\left\{\bigcap_{k=1}^{n} (X_k = X_{j,k})\right\} \ . \tag{8.1.2}$$

If the components of $\mathbf{X}$ are all (stochastically) independent of each other, by the product rule, Eq. (11.2.13),

$$P\{s_j\} = \prod_{k=1}^{n} P\{X_k = X_{j,k}\} \ . \tag{8.1.3}$$

Here nothing of a possible (Boolean) function of **X** need be known. This changes when using the other concept of state, namely the binary state of the value X_φ of a Boolean function φ. By Eq. (11.3.4)

$$P\{X_\varphi=1\}=E\{X_\varphi\} \; . \tag{8.1.4}$$

If $X_\varphi(t)=0$ from 0 to L_φ, and 1 thereafter, then by Eq. (11.3.7)

$$P\{X_\varphi(t)=1\}=P\{L_\varphi \leqq t\}=F_{L_\varphi}(t) \; . \tag{8.1.5}$$

After these introductory remarks the time dependent probability of a single binary state is investigated, i.e. $P\{X(t)=1\}$ for different t.

For ease of notation let X_i be the indicator of a reparable unit where $X=1$ denotes the "good" state; see Fig. 2.1. Hence, in the reliability context

$$P\{X_i=1\}=A_i(t)$$

being the availability of component i.

To get meaningful results under fairly general assumptions, it is assumed that the time intervals, where $X=1$ and $X=0$, respectively, define an *alternating renewal process*; see §11.4. In that case time intervals of the (random) length L (for life) follow (and are followed by) time intervals of the (random) length D (for down time); see Fig. 8.1.1. For $A_i(t)$ the following recursion relation (a linear integral equation) holds:

$$A_i(t)=\int_0^t f_{L+D}(\tau)A_i(t-\tau)d\tau+1-F_L(t) \; . \tag{8.1.6}$$

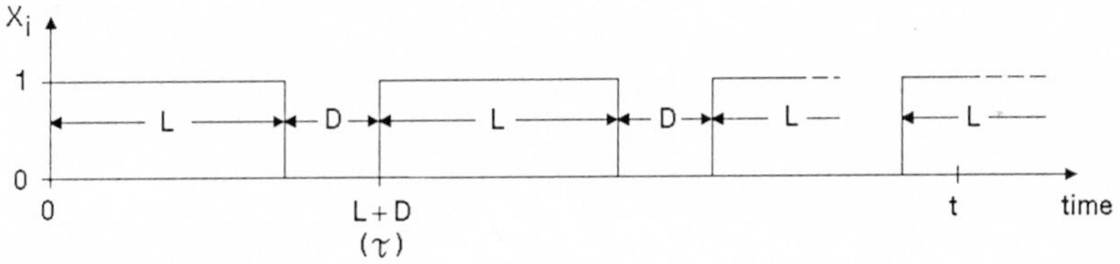

Fig. 8.1.1. Sample function $X_i(t)$

The proof is similar to that of Eq. (11.3.11) and will be treated in the solution of exercise 8.6. A totally different approach is treated in §11.4. To show the equality of the results, Eq. (8.1.6) is "solved" via the Laplace transform (see §11.5).

Using the convolution theorem, $\mathscr{L}\{1\}=1/s$, and the integration rule, Eq. (8.1.6), becomes

$$A_i^*(s)=f_L^*(s)\,f_D^*(s)A_i^*(s)+\frac{1}{s}-\frac{1}{s}\,f_L^*(s) \; .$$

The extraction of $A_i^*(s)$ yields

$$A_i^*(s)=\frac{1-f_L^*(s)}{s[1-f_L^*(s)\,f_D^*(s)]}\,, \tag{8.1.7}$$

which equals Eq. (11.4.12).

Example 8.1.1 Fully exponential case

Let

$$f_L(t)=\lambda\exp(-\lambda t)\,,\qquad f_D(t)=\mu\exp(-\mu t)\,. \tag{8.1.8}$$

Then by example 11.5.1

$$f_L^*(s)=\frac{\lambda}{\lambda+s}\,,\qquad f_D^*(s)=\frac{\mu}{\mu+s}\,. \tag{8.1.8a}$$

Insertion in Eq. (8.1.7) yields

$$A_i^*(s)=\frac{s(\lambda+s)(\mu+s)}{(\lambda+s)s[(\lambda+\mu)s+s^2]}=\frac{\mu+s}{s(\lambda+\mu+s)}\,. \tag{8.1.9}$$

To ease the backward transform this is split in two terms:

$$A_i^*(s)=\frac{1}{\lambda+\mu+s}+\frac{\mu}{s(\lambda+\mu+s)}\,. \tag{8.1.9a}$$

Finally, using example 11.5.1 and the integration rule of the Laplace transform

$$A_i(t)=\exp[-(\lambda+\mu)t]+\frac{\mu}{\lambda+\mu}\{1-\exp[-(\lambda+\mu)t]\}$$

$$=\frac{\mu}{\lambda+\mu}+\frac{\lambda}{\lambda+\mu}\exp[-(\lambda+\mu)t]\,. \tag{8.1.10}$$

Note that $A_i(0)=1$ and $A_i(\infty)=\mu/(\lambda+\mu)$.

It is of theoretical and practical interest to try to expand the above results for a single $X_i(t)$ to $X_\varphi(t)$. How far can one proceed? The following example shows the graphical construction of $X_\varphi(t)$ in a very simple case.

Example 8.1.2 Two-out-of-three time function

Figure 8.1.2 shows the superposition of the sample functions of three indicator processes (useful in reliability theory) superposed according to

$$X_\varphi(t)=X_1(t)X_2(t)\vee X_1(t)X_3(t)\vee X_2(t)X_3(t)\,.$$

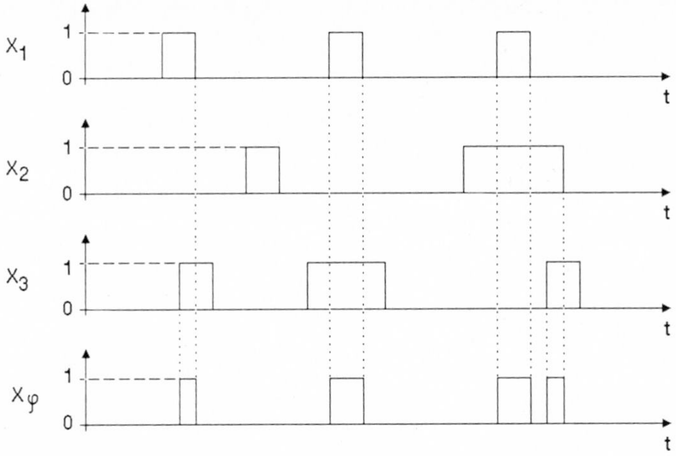

Fig. 8.1.2. Two-out-of-three system; behaviour in time

Since the superposed process $\{X_\varphi(t)\}$ is, in general, not the indicator process of an alternating renewal process, the results derived for $\{X_i(t)\}$ above are not directly applicable. However, in the case of *s*-independent $X_i(t)$ for every moment *t* one can use from Eq. (8.1.7)

$$p_i(t) := P\{X_i(t)=1\} = \mathscr{L}^{-1}\left\{\frac{1-f_L^*(s)}{s[1-f_L^*(s)\,f_D^*(s)]}\right\}$$

in

$$p_\varphi(t) := P\{X_\varphi(t)=1\} \ .$$

Details of how to find $p_\varphi(t)$ from $p_1(t), \ldots, p_n(t)$ are discussed in §8.2. In the stationary case the above notation will be simplified as follows:

$$p_i := P\{X_i=1\} \ , \quad p_\varphi = P\{X_\varphi=1\} \ . \tag{8.1.11}$$

8.2 Probability of the Value 1 of a Boolean Function

The probabilities with which a given Boolean function assumes the values 0 and 1 are certainly quantities of general interest in stochastic Boolean functions theory. Here two main approaches to calculate these probabilities are pursued: (1) calculation of $E\{X_\varphi\}$ instead of $P\{X_\varphi=1\}$ (see §8.2.1); (2) calculation of probabilities of parts of a partition of the event $X_\varphi=1$ in the framework of homogeneous Markov processes (see §8.2.2). Both approaches have specific advantages and disadvantages. Approach (1) is excellent for mutually *s*-independent X_i in

$$X_\varphi := \varphi(X_1, \ldots, X_n) \ ,$$

but becomes awkward once probabilities or expectations respectively of the type

$$P\{(X_i=1)\cap(X_j=1)\cap\ldots\}=E\{X_iX_j\ldots\}$$

must be known, since, in practice, such information is rarely easy to get. Approach (2) assumes that all system states last for an exponentially distributed time. This is a strong restriction. However, in this case very detailed questions on the behaviour of a system modelled by φ can be answered. Typically, for a three-components system, where the probabilities of all the elementary states can be calculated,

$$P\{(X_1=1)\cap(X_2=1)\}$$

can be found as

$$P\{(X_1=1)\cap(X_2=1)\}=P\{(X_1=1)\cap(X_2=1)\cap(X_3=0)\}$$
$$+P\{(X_1=1)\cap(X_2=1)\cap(X_3=1)\}.$$

8.2.1 Expected Value of a Boolean Function

It was mostly for the stochastic theory of Boolean indicator functions and of Boolean indicator processes (§9) that the typical Boolean operators were replaced by the usual addition, subtraction and multiplication in §7. This was motivated by the fact that – as will be shown subsequently in detail – usually the probability of the state $X_\varphi=1$ is not so easy to calculate for non trivial $\varphi(\mathbf{X})$, such that one is tempted to try to use the expected value $E\{X_\varphi\}$ instead of $P\{X_\varphi=1\}$. In fact – as was already pointed out in [15] – for $X_\varphi=\varphi(\mathbf{X})$ by definition of $E\{\cdot\}$ (§11)

$$E(X_\varphi\}=1\cdot P\{X_\varphi=1\}+0\cdot P\{X_\varphi=0\}=P\{X_\varphi=1\}\ . \tag{8.2.1}$$

However, it is not clear how to treat Boolean operators inside φ. Contrarily, if φ is a polynomial (see Eq. (7.2.6)), viz.

$$X_\varphi=\varphi(\mathbf{X})=c_0+\sum_{i=1}^{m}\left(c_i\prod_{j=1}^{n_i}\tilde{X}_{l_{i,j}}\right);\quad \tilde{X}_k\in\{X_k,\bar{X}_k\}\ , \tag{8.2.2}$$

and if the X_j are all stochastically independent of each other, then by Eq. (11.3.5)

$$E\{X_\varphi\}=c_0+\sum_{i=1}^{m}\left(c_i\prod_{j=1}^{n_i}E\{\tilde{X}_{l_{i,j}}\}\right) \tag{8.2.3}$$

or, with

$$\tilde{p}:=P\{\tilde{X}=1\}=E\{\tilde{X}\}\ , \tag{8.2.4}$$

$$p_\varphi=c_0+\sum_{i=1}^{m}\left(c_i\prod_{j=1}^{n_i}\tilde{\mathbf{p}}_{l_{i,j}}\right);\quad \tilde{\mathbf{p}}_k=\begin{cases}p_k,\ \tilde{X}_k=X_k\\ \bar{p}_k:=1-p_k,\ \tilde{X}_k=\bar{X}_k\ .\end{cases} \tag{8.2.5}$$

As a first trivial example Eq. (7.1.2) yields for $\varphi = X_1 \vee X_2$

$$p_\varphi = p_1 + p_2 - p_1 p_2 \ . \tag{8.2.6}$$

Example 8.2.1 Two-out-of-three majority function

Equation (7.2.3) yields

$$p_\varphi = p_1 p_2 + p_1 \bar{p}_2 p_3 + \bar{p}_1 p_2 p_3 \ . \tag{8.2.7}$$

Equation (7.2.8) yields the same p_φ as

$$p_\varphi = p_1 p_2 + p_1 p_3 + p_2 p_3 - 2 p_1 p_2 p_3 \ . \text{——} \tag{8.2.7a}$$

Comment. The multilinear form is not the only one for which

$$p(\varphi = 1) = \varphi(\mathbf{p}) \ , \tag{8.2.8}$$

meaning that in $\varphi(\mathbf{X})$ the vector $\mathbf{X}$ can be replaced by the vector

$$\mathbf{p} := (p_1, \ldots, p_n) \ ,$$

or, more generally, $\tilde{\mathbf{X}}$ by $\tilde{\mathbf{p}}$.

The following theorem is obvious.

Theorem 8.2.1. In any form of $\varphi(\mathbf{X})$, where on the way to a multilinear polynomial the idempotence rule cannot be applied any more, $\mathbf{X}$ can be replaced by $\mathbf{p}$ to yield

$$p_\varphi = P\{\varphi(\mathbf{X}) = 1\} \text{ (for stochastically independent components of } \mathbf{X}) \ . \text{——}$$

A typical example is the nested form of the final result of the Shannon decomposition, see Eq. (7.2.11).

Another example is the nested form derived from a tree-type syntax diagram. In reliability theory this corresponds to a reliability block diagram of the series/parallel type with all edges named differently. For better motivation, the following little theorem is preceded by an introductory example.

Example 8.2.2 Redundant computer system

Let Fig. 8.2.1 be the reliability block diagram (see §3.8) of a computer system consisting of two simplex units (of high reliability), three duplicated units (of standard reliability) and one triplicated unit (of low reliability, e.g. because of wear and tear). This diagram then corresponds to the fault tree (a true tree here) of Fig. 8.2.2 and the syntax diagram of Fig. 8.2.3, the latter being pictures of special DNFs.

Obviously

$$X_\varphi = X_1 \vee X_2 \vee X_3 X_4 \vee X_4 X_5 \vee X_5 X_6 \vee X_7 X_8 \vee X_9 X_{10} X_{11} \ .$$

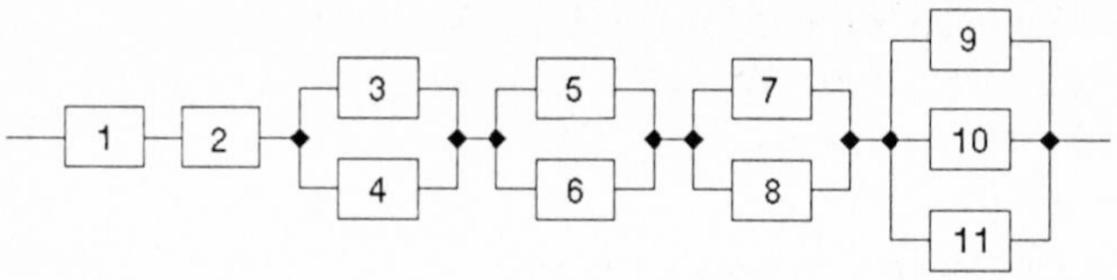

Fig. 8.2.1. Reliability block diagram of a typical redundant system

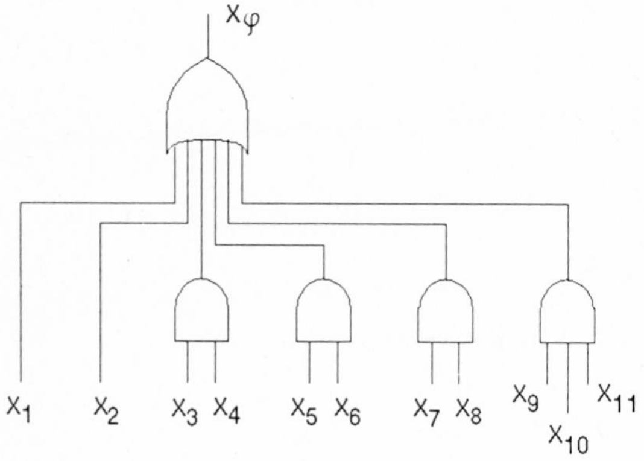

Fig. 8.2.2. Fault tree corresponding to Fig. 8.2.1

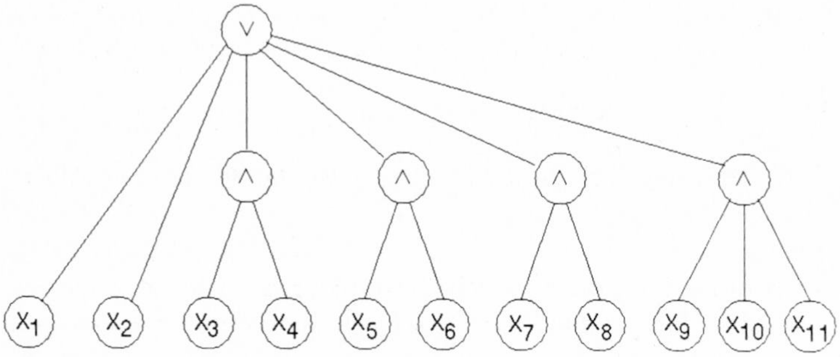

Fig. 8.2.3. Syntax diagram corresponding to the DNF of Fig. 8.2.2

Applying Eqs. (7.1.1a), (7.1.2a)

$$X_\varphi = 1-(1-X_1)(1-X_2)(1-X_3X_4)(1-X_4X_5)(1-X_5X_6)$$
$$\times(1-X_7X_8)(1-X_9X_{10}X_{11}) \ .$$

Obviously, by theorem 8.2.1, here

$$p_\varphi = 1-(1-p_1)(1-p_2)(1-p_2p_4)(1-p_4p_5)(1-p_5p_6)$$
$$\times(1-p_7p_8)(1-p_9p_{10}p_{11}) \ .$$

Theorem 8.2.2. If a syntax diagram of a Boolean function φ with the operators AND, OR, and NOT ($\wedge$, $\vee$, $^{-}$) is a tree, then after applying Eqs. (7.1.1a), (7.1.2a), (7.1.3) (to get rid of $\wedge$, $\vee$, $^{-}$), every X_i can be replaced by $p_i = P\{X_i = 1\}$.

Proof. In the primary nested form of this φ (with Boolean operators $\wedge$, $\vee$, $^{-}$) each X_i will appear at most once. This will not change on applying Eqs. (7.1.1a), (7.1.2a), (7.1.3). Hence, on factoring out the new (standard algebra) form of φ to a polynomial form, no powers of any variable nor pairs X_i, $\bar{X}_i$ will show up. Hence, by theorem 8.2.1, already in this nested form any X_i can be replaced by p_i, and any $\bar{X}_i$ by $\bar{p}_i$, q.e.d.

Comment. In theorem 8.2.2 the word "syntax" can, of course, be replaced by the word "logic", and the sequence "syntax diagram of a Boolean" by "fault tree".

Special Applications of p_φ

p_φ can also be used to express the probability with which a given (permanent) fault can influence the desired operation of a (combinational) switching circuit. In that case let

$$\varphi := \tilde{\psi}_\alpha = \tilde{\varphi} \not\equiv \tilde{\varphi}_\alpha$$

be the indicator of wrong outputs of a circuit that should implement $\tilde{\varphi}$, but suffers from the (permanent i.e. stuck) fault α, as discussed in §6.3, and p_φ the corresponding probability. Of course, a value of p_φ which is definitely positive means that the permanent fault shows up in an intermediate fashion with probability p_φ. Only, if p_φ is practically zero, α will be tolerable in practice.

It is interesting to note that not only any DDNF gives easy access to the number $N_{\hat{M}}(\varphi)$ of minterms of a given Boolean function φ but also p_φ. Putting all p_i to 1/2 makes the values of the probabilities of the elementary states (given by argument vectors $\underline{X}_i$ of φ) equal:

$$P\{\underline{X} = \underline{X}_i\} = 1/2^n \ ,$$

so that, with $\underline{2^{-1}} := \left(\frac{1}{2}, \frac{1}{2}, \ldots, \frac{1}{2}\right)$,

$$N_{\hat{M}}(\varphi) = 2^n \varphi(\underline{2^{-1}}) \ . \tag{8.2.9}$$

Example 8.2.3 Two-out-of-three majority function

From Eq. (8.1.7)

$$N_{\hat{M}}(\varphi) = 2^3 \cdot \varphi\left(\frac{1}{2}, \frac{1}{2}, \frac{1}{2}\right) = 8 ; \quad \left(\frac{1}{4} + \frac{1}{4} + \frac{1}{4} - 2\frac{1}{8}\right) = 4 \ .$$

This checks with Eq. (4.3.7). — —

8.2.2 Probabilities of Arguments (States) in the Homogeneous Markov Model

In the Markov model† we have states as introduced with the state (transition) diagrams of §3.6. For n variables $X_1, \ldots, X_n$ describing (indicating) the binary states of the n components of a system, characterized by the Boolean function $\varphi(X_1, \ldots, X_n)$, there are, independent of φ, 2^n elementary system states. A given φ tells for which states $\varphi = 1$ and $\varphi = 0$ (e.g. "good" and "bad" elementary states) respectively. Then, according to the detailed problems to be solved, as few as possible states S_i (disjoint sets of elementary states) are defined. Finally, these S_i should constitute a partition of the 2^n elementary states.

Definition 8.2.1. Let $S(t)$ be the "general" *system state* at time t. (The term state variable seems to be misleading.) — —

Now, in the Markov method one is primarily‡ interested in state probabilities of the type

$$P_i(t) := P\{S(t) = S_i\} \ . \tag{8.2.10}$$

Other probabilities, especially $P\{X_\varphi = 1\}$, are sums of appropriate P_is.

Obviously $S(t + \Delta t)$ can have evolved from any state at t: According to the rule (11.2.16) of *total probability* for the m states of the partition of definition 8.2.1

$$P\{S(t+\Delta t) = S_j\} = \sum_{i=1}^{m} P\{[S(t+\Delta t) = S_j] \cap [S(t) = S_i]\}$$

$$= \sum_{i=1}^{m} P\{S(t+\Delta t) = S_j \mid S(t) = S_i\} \, P\{S(t) = S_i\} \tag{8.2.11}$$

(see Fig. 8.2.4). Note that this equation does not make use of the so-called Markov property, viz. the independence of complex conditional probabilities depending on several past states from all but the most recent of the latter. The "deep" question concerning the existence of the conditional probabilities of Eq. (8.2.11) for non-Markovian processes is not discussed here.

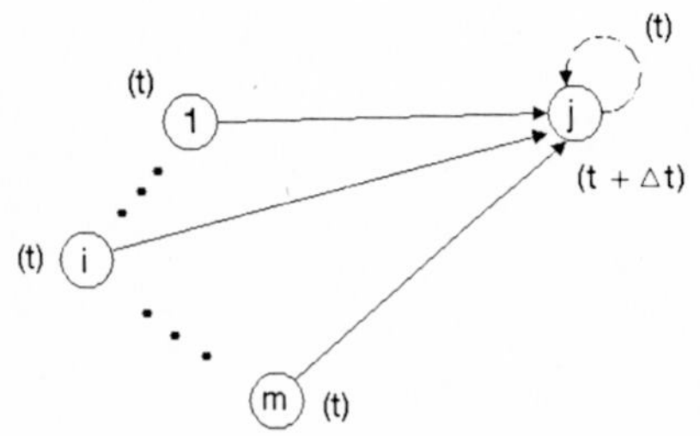

Fig. 8.2.4. Trivial state transition diagram. The dotted line is usually omitted

† In this text only continuous time Markov chains (with a finite number of states) are considered.
‡ In elementary analyses, i.e. in a first step.

It does not make much sense to indicate the possible stay (between t and $t+\Delta t$) in S_j by an extra edge since this possibility is trivial. (With clocked switching automata this is different, of course.)

With Eq. (8.2.10) and the so-called *conditional transition-probability*

$$P_{j|i}(\Delta t) := P\{S(t+\Delta t) = S_j | S(t) = S_i\} \tag{8.2.12}$$

(being independent of t for the time-homogeneous case), Eq. (8.2.11) is condensed to

$$P_j(t+\Delta t) = \sum_{i=1}^{m} P_{j|i}(\Delta t) P_i(t) \; . \tag{8.2.13}$$

Only the time-homogeneous case will be treated here.

Definition 8.2.2. $\lambda_{i,j}$ is the *transition rate* (a conditional probability density) of state transitions from S_i to S_j. — —

Now, the simplest plausible assumption as to $P_{j|i}(\Delta t)$ is the first-order Taylor approximation

$$P_{j|i}(\Delta t) = \begin{cases} \lambda_{i,j}\Delta t, & j \neq i \\ 1+\lambda_{i,i}\Delta t, & j = i \; , \end{cases} \tag{8.2.14}$$

where the $\lambda_{i,j}$ are constant.

By the normation rule, Eqs. (11.2.17), (8.2.14)

$$\sum_{j=1}^{m} P_{j|i}(\Delta t) = 1 + \Delta t \sum_{j=1}^{m} \lambda_{i,j} = 1 \; .$$

Hence

$$\sum_{j=1}^{m} \lambda_{i,j} = 0 \; ; \quad \lambda_{i,i}{}^{\dagger} = - \sum_{\substack{j=1 \\ j \neq i}}^{m} \lambda_{i,j} \; . \tag{8.2.15}$$

Since, plausibly, $\lambda_{i,j} \geqq 0$ for $j \neq i$, by Eq. (8.3.15) $\lambda_{i,i} < 0$. So, for sufficiently small Δt, both the right hand sides of Eq. (8.2.14) are between 0 and 1; as probabilities should be; and conditional probabilities too.

Inserting Eq. (8.2.14) in Eq. (8.2.13) yields, after shifting $P_j(t)$ to the l.h.s.,

$$P_j(t+\Delta t) - P_j(t) = \sum_{i=1}^{m} \lambda_{i,j} P_i(t) \Delta t \; .$$

Division by Δt and $\Delta t \to 0$ yield, with a dot for the time derivative, the system of linear differential equations

$$\dot{P}_j(t) = \sum_{i=1}^{m} \lambda_{i,j} P_i(t) \; ; \quad j = 1, \ldots, m \; , \tag{8.2.16}$$

† $\lambda_{i,i}$ should not be considered as a transition rate, but only as an abbreviation for the r.h.s. of this equation.

which, for constant $\lambda_{i,j}$, can be solved by standard methods [20]. Note that the equations are linearly dependent and that the normation equation

$$P_1 + P_2 + \ldots + P_m = 1 \tag{8.2.17}$$

has to be used instead of one of Equations (8.2.16)!

The following examples will allow for some very interesting interpretations and comparisons.

Example 8.2.4 A two-state system

The simplest example of Eqs. (8.2.16) and (8.2.17) is

$$\dot{P}_1(t) = -\lambda_{1,2} P_1(t) + \lambda_{2,1} P_2(t) \;, \tag{8.2.18}$$

$$\dot{P}_2(t) = \lambda_{1,2} P_1(t) - \lambda_{2,1} P_2(t) \;, \tag{8.2.19}$$

$$P_1(t) + P_2(t) = 1 \;. \tag{8.2.20}$$

For the steady state solution this triple of equations can be simplified to

$$0 = -\lambda_{1,2} P_1 + \lambda_{2,1} P_2 \;, \qquad P_1 + P_2 = 1$$

[where Eq. (8.2.19) was not used] with the obvious solution

$$P_1 = \lambda_{2,1}/(\lambda_{1,2} + \lambda_{2,1}) \;, \qquad P_2 = \lambda_{1,2}/(\lambda_{1,2} + \lambda_{2,1}) \;. \tag{8.2.21}$$

For the general solution a typical initial value is $P_1(0) = 1$, whence $P_2(0) = 0$. Using the Laplace transform (see §11.5) Eqs. (8.2.18) and (8.2.20) become

$$sP_1^*(s) - 1 = -\lambda_{1,2} P_1^*(s) + \lambda_{2,1} P_2^*(s) \;, \tag{8.2.22}$$

and

$$P_1^*(s) + P_2^*(s) = \frac{1}{s} \;, \tag{8.2.23}$$

respectively. Replacing $P_2^*(s)$ in Eq. (8.2.22) by

$$P_2^*(s) = \frac{1}{s} - P_1^*(s) \tag{8.2.24}$$

from Eq. (8.2.23) yields

$$P_1^*(s) = \left(\frac{\lambda_{2,1}}{s} + 1\right) \Big/ (\lambda_{1,2} + \lambda_{2,1} + s)$$

$$= \frac{\lambda_{2,1} + s}{s(\lambda_{1,2} + \lambda_{2,1} + s)} = \frac{\lambda_{2,1}}{s(\lambda_{1,2} + \lambda_{2,1} + s)} + \frac{1}{\lambda_{1,2} + \lambda_{2,1} + s} \;. \tag{8.2.25}$$

Using Eqs. (8.1.9a) and (8.1.10) formally the inverse Laplace transform of Eq. (8.2.25) is very simple. The result is

$$P_1(t)=\frac{\lambda_{2,1}}{\lambda_{1,2}+\lambda_{2,1}}+\frac{\lambda_{1,2}}{\lambda_{1,2}+\lambda_{2,1}}\exp[-(\lambda_{1,2}+\lambda_{2,1})t]\ . \tag{8.2.26}$$

Checks. (1) Obviously $P_1(0)=1$; (2) $P_1(\infty)$ is the r.h.s. of Eq. (8.2.21), i.e. the correct stationary value.

In reliability theory $P_1(t)$ equals $R(t)$, i.e. the reliability of the unit under consideration. In case of no repair $\lambda_{2,1}=0$ and $P_1(t)$ is the survivor function; i.e. with L for the life of the unit under consideration

$$P_1(t)=1-F_L(t)\ , \tag{8.2.27}$$

so that

$$F_L(t)=1-\exp(-\lambda_{1,2}t)\ , \tag{8.2.28}$$

where $\lambda_{1,2}$ is the so-called failure rate. Using Eq. (11.3.8)

$$E(L)=1/\lambda_{1,2}\ . \tag{8.2.29}$$

Discussion. Looking once again at Eq. (8.1.10) reveals that for a single unit the renewal theory approach for exponentially distributed lives and down times coincides with the (homogeneous) Markov approach. To make this coincidence complete one should convince oneself that $\lambda_{1,2}$ and $\lambda_{2,1}$ equal λ and μ of §8.1. — —

Example 8.2.5 A special three-state system

The system to be investigated can be modelled by the state transition graph of Fig. 8.2.5. The given interpretation in terms of reliability aspects is not the only one possible. A likewise very plausible interpretation is in terms of a simple service system with a queue with, at the most, three customers and with interarrival and service rates that depend on system state [54].

The connection with Boolean functions is given simply by the fact that, with

$$X_i=\begin{cases}1, & \text{component } i \text{ is good}\\ 0, & \text{else}\end{cases}$$

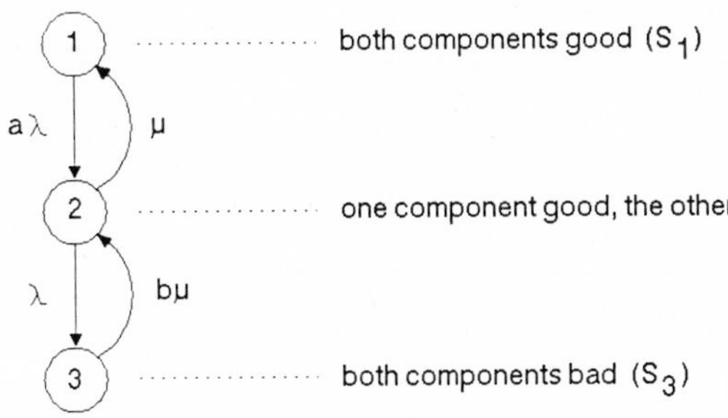

Fig. 8.2.5. Three-state system with intepretation as a (possibly failing) 1-out-of-2 system. (See also Fig. 3.6.1.)

and

$$X_\varphi = \begin{cases} 1, & \text{system is good} \\ 0, & \text{else} \end{cases}$$

we have

$$X_\varphi = X_1 \vee X_2 = \begin{cases} 1, & \text{for states 1 and 2} \\ 0, & \text{for state 3} \end{cases} \tag{8.2.30}$$

The equations for an analysis of this Markov model system are, by Eqs. (8.2.16), (8.2.17),

$$\dot{P}_1(t) = -a\lambda P_1(t) + \mu P_2(t) \ , \tag{8.2.31}$$

$$\dot{P}_2(t) = a\lambda P_1(t) - (\lambda + \mu) P_2(t) + b\mu P_3(t) \ , \tag{8.2.32}$$

$$P_1(t) + P_2(t) + P_3(t) = 1 \ . \tag{8.2.33}$$

(The equation for $\dot{P}_3(t)$ is omitted, since it is not used here.)

Steady State Solution

With $P_i := P_i(\infty)$; $\dot{P}_i(\infty) = 0$ Eq. (8.2.31) becomes

$$P_2 = \frac{a\lambda}{\mu} P_1 \ .$$

Summing Eqs. (8.2.31), (8.2.32) results in

$$P_3 = \frac{\lambda}{b\mu} P_2 = \frac{a\lambda^2}{b\mu^2} P_1 \ .$$

Now, by Eq. (8.2.33)

$$P_1 = \left(1 + \frac{a\lambda}{\mu} + \frac{a\lambda^2}{b\mu^2}\right)^{-1} = \frac{b\mu^2}{a\lambda^2 + ab\lambda\mu + b\mu^2} \ . \tag{8.2.34}$$

Hence

$$P_2 = \frac{ab\lambda\mu}{a\lambda^2 + ab\lambda\mu + b\mu^2} \tag{8.2.35}$$

and

$$P_3 = \frac{a\lambda^2}{a\lambda^2 + ab\lambda\mu + b\mu^2} \ . \tag{8.2.36}$$

P_3 is the unavailability of the system. If

$$\lambda = \frac{1}{\text{MTBF}} \ , \qquad \mu = \frac{1}{\text{MTTR}} \tag{8.2.37}$$

(with MTBF and MTTR being mean time before failures and mean time to repair for any of the two equal components) the unavailability of a single reparable component is (in the stationary state)

$$U = \frac{\mathrm{MTTR}}{\mathrm{MTBF}+\mathrm{MTTR}} = \frac{\lambda}{\lambda+\mu} . \tag{8.2.38}$$

With stochastically independent components, by the product rule, Eq. (11.2.13), of probability calculus

$$P_3 = U^2 = \frac{\lambda^2}{\lambda^2+2\lambda\mu+\mu^2} . \tag{8.2.39}$$

Comparing this result with Eq. (8.2.36) reveals: For $a = b = 2$ and only in that case, both components are stochastically independent (of each other).

Returning to the introductory remarks of this §8.2, it is obvious how to calculate $P\{X_\varphi = 1\}$. It is simply the sum of all those (mutually exclusive) states S_i whose state vectors $\mathbf{X}_i$ yield $\varphi(\mathbf{X}_i) = 1$. In the present example system availability is simply

$$A_\varphi = P_1 + P_2 .$$

Had state 2 been split in two elementary states as in Fig. 3.6.2, then the same A_φ would be

$$A_\varphi = P_{00} + P_{01} + P_{10} .$$

General Solution

The general solution of Eqs. (8.2.31)–(8.2.33) is possible with the aid of the Laplace transform (see §11.5). However, the details are so tedious that they are regarded as a non-appropriate exercise in algebraic manipulations for this text. After all, in exercise 8.5 the interested reader is guided to find $P_1^*(s)$. However, the inverse Laplace transform of $P_1^*(s)$ is not easy to determine.

In §9.3 a special solution of Eqs. (8.2.31)–(8.2.33) is presented in the time domain. There, too, tedious algebra will "block" the way to any results.

8.3 Approximate Probability of the Value 1

Typically, in fault-tree analysis of reliability technology, practitioners are strongly interested in approximate analyses, since they often have to process Boolean functions of many variables with many terms.

There are two major ways to reduced computational complexity while maintaining an acceptable degree of accuracy. One is not to aim at a multilinear polynomial form of φ (in its variables or literals), which is pursued in §8.2.1. The other is to shorten the DNF of φ and calculate the DDNF or a different type of polynomial only for the rest φ_r of φ, which is discussed in §8.2.2.

8.3.1 Principle of Inclusion–Exclusion

Let a_i be the following random event (see the Appendix §11) with $\hat{T}_i$ a Boolean term:

$$a_i=(\hat{T}_i=1) \ .$$

Now, if $\hat{T}_i$ is a term $\hat{T}'_j$ of a DDNF, then by the additivity axiom, Eq. (11.2.4), of probability calculus, from

$$X_\varphi=\varphi(X_1,\ldots,X_n)=\bigvee_{j=1}^{m'}\hat{T}_j \ , \tag{8.3.1}$$

there follows [by the extension Eq. (11.2.10) of Eq. (11.2.4)]

$$p_\varphi:=P\{X_\varphi=1\}=\sum_{j=1}^{m'}P\{\hat{T}'_j=1\} \ ,$$

e.g. with X_φ for the two-out-of-three majority function, Eq. (4.4.18a)

$$p_\varphi=P\{X_1X_2\bar{X}_3=1\}+P\{X_1\bar{X}_2X_3=1\}+P\{X_2X_3=1\} \ . \tag{8.3.2}$$

However, in practice, instead of the DDNF Eq. (8.3.1), rather a "simple" DNF

$$X_\varphi=\bigvee_{i=1}^{m}\hat{T}_i \tag{8.3.3}$$

is given. Then, instead of Eq. (11.2.10), we must use the cumbersome formula, Eq. (11.2.9).

Example 8.3.1 Two-out-of-three majority function

The well-known function

$$X_\varphi=X_1X_2 \vee X_1X_3 \vee X_2X_3$$

yields (for *s*-dependent X_1, X_2, X_3)

$$\begin{aligned}p_\varphi&=P\{X_1X_2=1\}+P\{X_1X_3=1\}+P\{X_2X_3=1\}\\&\quad-P\{(X_1X_2=1)\cap(X_1X_3=1)\}-P\{(X_1X_2=1)\cap(X_2X_3=1)\}\\&\quad-P\{(X_1X_3=1)\cap(X_2X_3=1)\}\\&\quad+P\{(X_1X_2=1)\cap(X_1X_3=1)\cap(X_2X_3=1)\}\\&=P\{X_1X_2=1\}+P\{X_1X_3=1\}+P\{X_2X_3=1\}\\&\quad-P\{X_1X_2X_3=1\}-P\{X_1X_2X_3=1\}\\&\quad-P\{X_1X_2X_3=1\}+P\{X_1X_2X_3=1\}\\&=P\{X_1X_2=1\}+P\{X_1X_3=1\}+P\{X_2X_3=1\}\\&\quad-2P\{X_1X_2X_3=1\} \ .\end{aligned} \tag{8.3.4}$$

Comment. Even this small example shows one of the typical drawbacks of the inclusion–exclusion (IE) principle: the cancelling of terms. Here [19] seems to lead a way out; and certainly also the ENZMANN algorithm of §7.2 does so.

Obviously, the computational procedure and hence the computational complexity of the IE principle is very similar to Eq. (7.1.2a). However, the recursive version of the IE principle given by

$$P\left\{\bigcup_{i=1}^{k+1} a_i\right\} = P\left\{\bigcup_{i=1}^{k} a_i\right\} + P\{a_{k+1}\} - P\left\{a_{k+1} \cap \bigcup_{i=1}^{k} a_i\right\} , \tag{8.3.5}$$

which is trivially true by Eq. (11.2.7), may be easier to manage; especially as to programming on a computer.

The most important point of the IE principle is probably that it lends so easily to upper and lower bounds on $P\left\{\bigcup_i a_i\right\}$ via the Bonferroni inequalities.

Theorem 8.3.1 (Bonferroni inequalities). If the sums in the general summation formula (11.2.9) are successively called $\Sigma_1, \Sigma_2, \ldots, \Sigma_m$, then

$$p_m := P\left\{\bigcup_{i=1}^{m} a_i\right\} \begin{cases} \leqq \Sigma_1 - \Sigma_2 + - \ \ldots \ + \Sigma_{2k-1};\ 2k-1 \leqq m \\ \geqq \Sigma_1 - \Sigma_2 + - \ \ldots \ - \Sigma_{2k};\ 2k \leqq m \end{cases} k = 1, 2, \ldots \tag{8.3.6}$$

Proof. [26] (for those who can read German, also [17]).

Corollary 8.3.1. If p_m of Eq. (8.3.6) is approximated by $\Sigma_1 - \Sigma_2 + - \ \ldots \ \Sigma_j$, then the approximation error is at most Σ_{j+1}.

Proof. If $j = 2k$, or $j = 2k - 1$ respectively, then p_m must lie in an interval of length Σ_{2k+1} and Σ_{2k}, respectively (being the upper r.h.s. of Eq. (8.3.6) minus the lower one), q.e.d.

8.3.2 Approximation with a Given Error Bound

Let a DNF of φ be given, which is denoted by φ for simplicity. Further, let

$$\varphi = \varphi_r \vee \varphi_d , \tag{8.3.7}$$

where φ_r and φ_d are DNF's such that any term $\hat{T}_i$ of φ reappears either in φ_r or in φ_d. (An optimal choice of φ_r and φ_d is discussed below.) For any choice of φ_r and φ_d by Eqs. (7.1.2) and (7.1.5)

$$\varphi = \varphi_r + \varphi_d - \varphi_r \varphi_d , \tag{8.3.8}$$

and

$$\varphi = \varphi_r + \bar{\varphi}_r \varphi_d , \tag{8.3.9}$$

respectively. Since φ_r, $\bar{\varphi}_r$, and φ_d are Boolean:

$$\varphi_r \varphi_d \geqq 0 , \quad \bar{\varphi}_r \varphi_d \geqq 0 .$$

Hence

$$\varphi_r \leqq \varphi \leqq \varphi_r + \varphi_d \ , \tag{8.3.10}$$

and, by the definition of the expectation operator,

$$E\{\varphi_r\} \leqq E\{\varphi\} \leqq E\{\varphi_r + \varphi_d\} = E\{\varphi_r\} + E\{\varphi_d\} \ . \tag{8.3.11}$$

Obviously, $E\{\varphi_d\}$ is the maximum (positive) error for an approximation of $E\{\varphi\} = p_\varphi$ by $E\{\varphi_r\}$.

Clearly, if φ_d is not very short, the calculation $E\{\varphi_d\}$ becomes a difficult task. Hence, in the spirit of the first Bonferroni inequality, i.e.

$$P\left\{\bigcup_i a_i\right\} \leqq \sum_i P\{a_i\} \ , \tag{8.3.12}$$

the function φ_d

$$\varphi_d := \bigvee_{i \in I_d} \hat{T}_i \tag{8.3.13}$$

(to be deleted from φ to yield the rest φ_r) is replaced by (the non-Boolean)

$$\varphi_d' := \sum_{i \in I_d} \hat{T}_i \ , \tag{8.3.14}$$

where by Eq. (8.3.12)

$$E\{\varphi_d\} \leqq E\{\varphi_d'\} \ , \tag{8.3.15}$$

such that Eq. (8.3.11) is transformed to

$$E\{\varphi_r\} \leqq E\{\varphi\} \leqq E\{\varphi_r\} + E\{\varphi_d'\} \ . \tag{8.3.16}$$

If one desires to achieve

$$E\{\varphi_d'\} \leqq e_{max} \ ,$$

one calculates all the $E\{\hat{T}_i\}$ of φ, puts them in ascending order, and adds as many of them as possible, as long as their sum is at the most e_{max}. The $\hat{T}_i$ found this way are those of an optimal φ_d. Finally $E\{\varphi_r\}$ has to be calculated, e.g. by any of the methods discussed in this book. More details can be found in [42].

8.4 Moments of Boolean Functions

Expectations of powers higher than 1 of Boolean indicators have obviously not yet found many applications. By Eq. (8.2.1) the importance of $E\{X_i\}$ and $E\{X_\varphi\}$ is apparent. Hence, in this short section only higher moments will be considered.

By the idempotence law $X^m = X$,

$$E\{X^m\} = E\{X\} = P\{X=1\} \ . \tag{8.4.1}$$

Specifically,

$$E\{X_\varphi^m\} = P\{X_\varphi = 1\} = p_\varphi \ .$$

As to central moments, only the variance is considered here. By Eq. (11.3.14)

$$\begin{aligned}\sigma_X^2 &= E\{X^2\} - (E\{X\})^2 = E\{X\}(1 - E\{X\}) \\ &= P\{X=1\}(1 - P\{X=1\}) = P\{X=1\}P\{X=0\} \ .\end{aligned} \tag{8.4.2}$$

Another quantity of some practical interest is the random (non-Boolean) number

$$N_1 = \sum_{i=1}^{n} X_i \ ; \quad X_i = 0, 1 \ ; \quad n = \text{const}, \tag{8.4.3}$$

see exercise 11.3. N_1 can be the number of failed components of an n components system or the number of erroneous bits of a binary (n, k) code. By definition of the expectation functional,

$$E\{N_1\} = \sum_{i=1}^{n} E\{X_i\} \ , \tag{8.4.4}$$

and by Eq. (11.3.14)

$$\begin{aligned}\sigma_{N_1}^2 &= E\{N_1^2\} - (E\{N_1\})^2 \\ &= \sum_{i=1}^{n} E\{X_i\} + 2\sum_{j=1}^{n-1} \sum_{j=i+1}^{n} E\{X_i X_j\} - \left(\sum_{i=1}^{n} E\{X_i\}\right)^2 \ .\end{aligned} \tag{8.4.5}$$

For s-independent X_is, by Eq. (11.3.5)

$$E\{X_i X_j\} = E\{X_i\} E\{X_j\}$$

can be inserted in Eq. (8.4.5). For equal

$$E\{X_i\} = P\{X_i = 1\} = p \ ; \quad i = 1, \ldots, n \tag{8.4.6}$$

one finds

$$E\{N_1\} = np \ . \tag{8.4.7}$$

Furthermore, for equal and s-independent X_is

$$\sigma_{N_1}^2 = np + n(n-1)p^2 - n^2 p^2 = np(1-p) \ . \tag{8.4.8}$$

Exercises

8.1
Why is it that, if multilinear forms of Boolean functions $\varphi(X_1, \ldots, X_n)$ were considered as real-valued on the unit n-cube, the value of the function would be a value between 0 and 1?
Hint. "Point of" means "point inside or on the surface of".

8.2
Prove the Bonferroni inequalities, Eq. (8.3.6), for $m=3$.

8.3
Figure E8.1 shows the state transition graph of the $M/M/1$-queue. Calculate

(a) the steady state probabilities of the system states and
(b) the mean queue length.

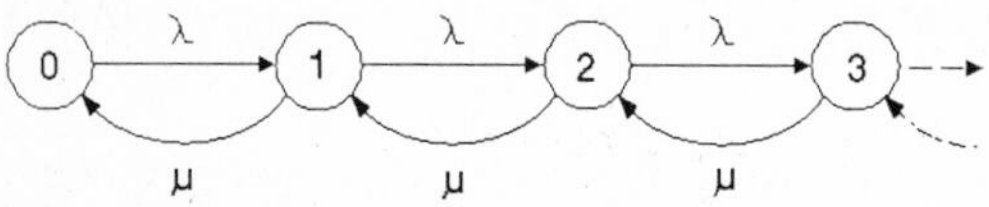

Fig. E8.1. $M/M/1$-queueing system. λ is the arrival rate, μ is the service rate ($M/M/1$ stands for Markov arrival rate/Markov service rate/1 server.)

Hint. The index of the system state is the number of customers in the system.

8.4
Given P_1 and P_2 of the 1-out-of-2 system of Fig. 8.2.5 for the case of stochastically independent X_i. How big are $p_i := P\{X_i = 1\}$, $i = 1, 2, 3$?

8.5
Find $P_1^*(s)$ for $P_1(t)$ of the system of Fig. 8.2.5 in the special case of $P_1(0) = 1$.

8.6
Derive the linear integral Eq. (8.1.6).

9 Stochastic Theory of Boolean Indicator Processes

This chapter is devoted to the stochastic theory of Boolean functions in time, i.e. to binary processes depending on other such processes via a Boolean function. The main concern will be with the mean duration of binary states defined by one of the two values of a given Boolean function. In the Markov model this analysis will be refined in that the value of the given Boolean function will not determine the state in question, but specific subspaces of its arguments. Feedback as is typical for digital automata will not be considered.

The type of processes to be discussed was shown pictorially in Fig. 8.1.2. However, in §8 only point events in time were considered, whereas the concern of this chapter is mainly with random lengths of time intervals with binary processes. For the single variable, i.e. for $\{X_i(t)\}$, some results of renewal theory are given in §11.4; e.g. the c.d.f. of the time between m renewals, or the mean number of renewals in an interval of length t, the so-called renewal function. For the logical (Boolean) superposition of several $\{X_i(t)\}$ to yield $\{X_\varphi(t)\}$, only very few of these results are easily derivable. After all, mean durations of certain states can be found, even though not even the corresponding variances are known, not to mention corresponding distribution functions.

In order to gradually increase the level of sophistication, a basic result for §9.2 is only derived in §9.3. However, at the proper place a convincing heuristic argument is given.

9.1 Mean Duration of States in the Markov Model

Since this text is not at all a monograph on Markov processes, only two small points concerning the mean duration of states will be discussed here; more can be found in [20]. First, as to the duration D_i of a single state i as defined in the Markov model differential Eqs. (8.2.16), we show the following.

Theorem 9.1.1. With D_i the duration of state i (in the Markov model), using $\lambda_{i,i}$ of Eq. (8.2.15),

$$E\{D_i\} = -1/\lambda_{i,i} \; . \tag{9.1.1}$$

Proof. In a Markov system the future behaviour is independent of how the present state was reached. (This is proved elsewhere.) Therefore, to get $E\{D_i\}$, one can start at time 0 with probability one in state i, exclude ("prohibit") any returns to state i and determine $E\{D_i\}$ in this special situation. Then, by Eqs. (8.2.16) and (8.2.15)

$$\dot{P}_i(t)=\lambda_{i,i}P_i(t)$$

with the solution (for $P_i(0)=1$)

$$P_i(t)=\exp(\lambda_{i,i}t) \; .$$

Obviously, with $F_{D_i}(t)$ being the *probability distribution function* of D_i,

$$P_i(t)=P\{D_i>t\} =: 1-F_{D_i}(t) \; . \tag{9.1.2}$$

Hence, by Eq. (11.3.8), Eq. (9.1.1) is true, q.e.d.

Special Time-Dependent Solutions of Markov Equations

Now, the above method of interpreting a Markov state probability as a survivor function will be applied to a couple of states.

Example 9.1.1 (Example 8.2.5 contd.)

Differentiating Eq. (8.2.31) and inserting from Eqs. (8.2.31) and (8.2.32) one gets the 2nd order equation

$$\ddot{P}_1(t) + [(a+1)\lambda+(b+1)\mu]\dot{P}_1(t)+(a\lambda^2+ab\lambda\mu+b\mu^2)P_1(t)=b\mu^2 \tag{9.1.3}$$

with the general solution

$$P_1(t)=P_1+\beta_1\exp(-\gamma_1 t)+\beta_2\exp(-\gamma_2 t) \; . \tag{9.1.4}$$

We restrict the further investigation for brevity to the case $a=b=2$. Then for the case (initial condition) $P_1(0)=1$

$$P_1(t)=\frac{1}{(\lambda+\mu)^2}\{\mu^2+\lambda^2\exp[-2(\lambda+\mu)t]+2\lambda\mu\exp[-(\lambda+\mu)t]\} \tag{9.1.5}$$

and

$$P_2(t)=\frac{1}{(\lambda+\mu)^2}\{2\lambda\mu-2\lambda^2\exp[-2(\lambda+\mu)t]+2\lambda(\lambda-\mu)\exp[-(\lambda+\mu)t]\} \; . \tag{9.1.6}$$

To obtain the mean time of the 1-state of the OR-function (i.e. the mean time) the system S_φ of Fig. 8.2.5 is in either of the states 1 and 2, we model the system behaviour so that the time the system is in states 1 or 2 becomes the life of a non-repairable system S'_φ. Then, with $F_L(t)$ for the distribution function of the life of S'_φ, the mean duration of S_φ in states 1 or 2 is, by Eq. (11.3.8),

$$E\{D_{\varphi,1}\}=\int_0^\infty [1-F_L(t)]dt=\int_0^\infty \bar{F}_L(t)dt \; . \tag{9.1.7}$$

Changing S_φ to S'_φ is very easy. Set $b=0$ and $P_2(0)=1$.[†] This way the starting state of S'_φ is exactly the state from which one begins the counting of $D_{\varphi,1}$, i.e. the situation $X_\varphi=1$ of Eq. (8.2.30).

To insert in

$$\bar{F}_L(t) := P_1(t) + P_2(t) \tag{9.1.8}$$

one finds for $b=0$ and $P_2(0)=1$ from Eq. (9.1.5, 6), after some elementary steps,

$$P_1(t) = \frac{\mu}{\gamma_2 - \gamma_1}[\exp(-\gamma_1 t) - \exp(-\gamma_2 t)] , \tag{9.1.9}$$

and

$$P_2(t) = \frac{1}{\gamma_2 - \gamma_1}[(2\lambda - \gamma_1)\exp(-\gamma_1 t) - (2\lambda - \gamma_2)\exp(-\gamma_2 t)] , \tag{9.1.10}$$

where

$$\gamma_{1,2} = \frac{1}{2}(3\lambda + \mu) \pm \frac{1}{2}\sqrt{\lambda^2 + 6\lambda\mu + \mu^2} . \tag{9.1.11}$$

Finally, using

$$\int_0^\infty \exp(-\alpha t)dt = 1/\alpha , \quad \alpha > 0 ,$$

the mean time during which the OR-function is 1 is simply (see exercise 9.5)

$$E\{D_{\varphi,1}\} = \frac{1}{\lambda} + \frac{\mu}{2\lambda^2} . \tag{9.1.12}$$

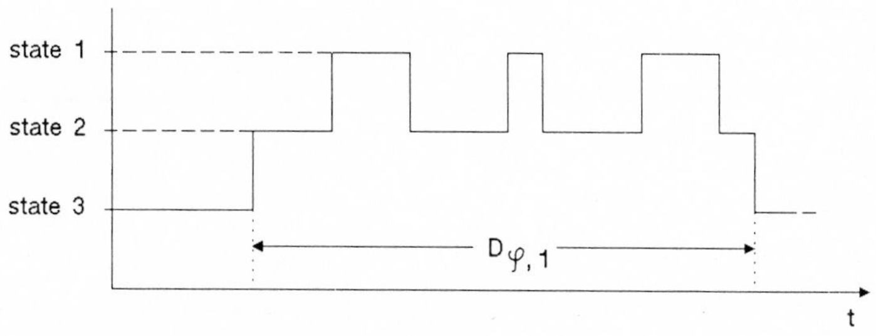

Fig. 9.1.1. Sample value of duration of time, where $X_\varphi=1$ for the one-out-of-two system

[†] This way S'_φ always starts in S_2. Then it either "goes" to S_1 or to S_3. From S_1 it eventually returns to S_2, etc.

$1/\lambda$ is the mean time a single component is in the 1-state. Hence for $\mu \gg \lambda$ the OR-function, i.e. the one-out-of-two system can stay much longer in the state $X_\varphi = 1$ (consisting of the Markov-states 1 and 2) than can the single components.

Figure 9.1.1 shows a sample value of $D_{\varphi,1}$; there are three changes from state 2 to state 1 and back again.

Now we try to derive results for mean durations of binary states in non-Markovian cases.

9.2 Mean Duration of Boolean Functions' Values

Let $D_{i,c}$ be the duration (time) the indicator process $\{X_i(t)\}$ spends in the c-state, $c \in \{0, 1\}$, and let

$$\nu_i = \frac{1}{E\{D_{i,o}\} + E\{D_{i,1}\}} \tag{9.2.1}$$

be the *mean frequency* (rate) of starting (changing to) a state of a definite type (either 0-type or 1-type, not both).

Definition 9.2.1. The Boolean *sensitivity function* φ_{si} of φ with respect to X_i is a function that shows explicitly how φ depends on X_i. It corresponds to the partial derivative in real functions analysis. More precisely, with

$$X_\varphi = \varphi(\mathbf{X}) = X_i \varphi_i' + \varphi_i''$$

(as in Eq. (7.3.2)) φ_i' is the sensitivity function φ_{si} of φ with respect to X_i in cases, where φ_i' is Boolean:

$$\varphi_{si} = \varphi_i' \; . — — \tag{9.2.2}$$

Obviously, by theorem 7.3.1, for $\varphi_i' = 1$ one has $\varphi_i'' = 0$, and, consequently, $\varphi = X_i$. For $\varphi_i' = 0$, φ does not depend on X_i. For $\varphi_i' = -1$ we have $\varphi_i'' = 1$, so that $\varphi = \bar{X}_i$.

Definition 9.2.2. Let

$$p_{\varphi,i} := P\{\varphi_{si} = 1\} \tag{9.2.3}$$

be the *sensitivity probability* of φ with respect to X_i.

Lemma 9.2.1. For Boolean φ_i'

$$p_{\varphi,i} = P\{\varphi_i' = 1\} = E\{\varphi_i'\} \; . \tag{9.2.3a}$$

Proof. Equation (11.3.4).

Theorem 9.2.1. For a monotonously non-decreasing (isotonous) Boolean function $\varphi(X_1, \ldots, X_n)$ the mean frequency of the recurrence of a definite state (either 0 or

1, but not both) of the associated binary process $\{X_\varphi(t)\}$ is

$$\nu_\varphi := \frac{1}{E\{D_{\varphi,0}\} + E\{D_{\varphi,1}\}} = \sum_{i=1}^{n} p_{\varphi,i}\nu_i \ . \tag{9.2.4}$$

Proof. (see proof of theorem 9.3.2 for the time-variable case).

Heuristics of Proof. Analogous to the rule of total probability one can imagine the existence of a rule of total frequency, where the conditional probability, that the system behaves as the single component C_i i.e. that $X_\varphi = X_i$, is $p_{\varphi,i}$.

Before giving an example, let us elaborate a little bit on the importance of Eq. (9.2.4). As is shown elsewhere [21], in case of stationarity

$$p_i := P\{X_i = 1\} = \frac{E\{D_{i,1}\}}{E\{D_{i,0}\} + E\{D_{i,1}\}} \ ; \qquad i = 1, \ldots, n,^\dagger \tag{9.2.5}$$

so that

$$E\{D_{i,1}\} = p_i/\nu_i \tag{9.2.6}$$

and

$$E\{D_{i,0}\} = \bar{p}_i/\nu_i \ , \tag{9.2.7}$$

where

$$\bar{p}_i := 1 - p_i \ . \tag{9.2.8}$$

This means that from ν_φ and p_φ one can easily calculate the (often highly desirable) mean state durations $E\{D_{\varphi,0}\}$ and $E\{D_{\varphi,1}\}$ of 0-state and 1-state of the binary random process $\{X_\varphi(t)\}$ being the logical superposition (by φ) of the binary processes $\{X_i(t)\}$; $i = 1, \ldots, n$.

With the abbreviations (the rates)

$$\lambda_i := \frac{1}{E\{D_{i,1}\}} \ ; \qquad \mu_i := \frac{1}{E\{D_{i,0}\}} \tag{9.2.9}$$

Eqs. (9.2.6) and (9.2.7) can be combined to

$$\nu_i = p_i\lambda_i = \bar{p}_i\mu_i \ . \tag{9.2.10}$$

Example 9.2.1 AND-function

Let

$$\varphi = X_1 X_2 \ . \tag{9.2.11}$$

† For an alternating renewal process compare this with Eq. (11.4.18).

Here by Eq. (9.2.2)

$$\varphi_1' = X_2 \;, \qquad \varphi_2' = X_1 \;, \tag{9.2.12}$$

so that

$$p_{\varphi,1} = p_2 \;, \qquad p_{\varphi,2} = p_1 \;. \tag{9.2.13}$$

Inserting in Eq. (9.2.4) yields

$$\nu_\varphi = p_2 \nu_1 + p_1 \nu_2 \;, \tag{9.2.14}$$

or, using Eq. (9.2.11),

$$\nu_\varphi = p_1 p_2 \lambda_1 + p_1 p_2 \lambda_2 \;.$$

Since (for *s*-independent) X_1, X_2:

$$p_\varphi = p_1 p_2 \;, \tag{9.2.15}$$

we have

$$\lambda_\varphi = \lambda_1 + \lambda_2 \;. \tag{9.2.16}$$

This is a very important result in reliability theory, where the φ of Eq. (9.2.9) could model an irredundant two-component system. Equation (9.2.14) states that the component failure rates must be added to the yield system failure rate.

Example 9.2.2 OR function (Example 9.1.1 contd.)

Let

$$X_\varphi = X_1 \vee X_2 \;.$$

From Eq. (7.1.2)

$$\begin{aligned} X_\varphi &= X_1 + X_2 - X_1 X_2 \\ &= X_1(1 - X_2) + X_2 \Rightarrow \varphi_1' = 1 - X_2 \\ &= X_2(1 - X_1) + X_1 \Rightarrow \varphi_2' = 1 - X_1 \;. \end{aligned}$$

Here

$$p_{\varphi,1} = 1 - p_2 = \bar{p}_2 \;, \qquad p_{\varphi,2} = 1 - p_1 = \bar{p}_1 \;.$$

Hence, by Eq. (9.2.4)

$$\nu_\varphi = \bar{p}_2 \nu_1 + \bar{p}_1 \nu_2 \;.$$

Replacing the v_i according to Eq. (9.2.6)

$$v_\varphi = \frac{\bar{p}_2 p_1}{E\{D_{1,1}\}} + \frac{\bar{p}_1 p_2}{E\{D_{2,1}\}}$$

$$= \frac{p_1}{E\{D_{1,1}\}} + \frac{p_2}{E\{D_{2,1}\}} - p_1 p_2 \left(\frac{1}{E\{D_{1,1}\}} + \frac{1}{E\{D_{2,1}\}} \right) . \tag{9.2.17}$$

By Eq. (9.2.9), Eq. (9.2.15) is transformed to

$$v_\varphi = p_1 \lambda_1 + p_2 \lambda_2 - p_1 p_2 (\lambda_1 + \lambda_2) . \tag{9.2.18}$$

Alternatively, from the first r.h.s. of Eq. (9.2.17), using Eq. (9.2.9):

$$v_\varphi = \frac{\bar{p}_2 \bar{p}_1}{E\{D_{1,0}\}} + \frac{\bar{p}_1 \bar{p}_2}{E\{D_{2,0}\}} = \bar{p}_1 \bar{p}_2 (\mu_1 + \mu_2) . \tag{9.2.19}$$

Finally, with (see Eq. (8.2.6))

$$p_\varphi = p_1 + p_2 - p_1 p_2 = 1 - \bar{p}_1 \bar{p}_2 ,$$

the mean durations of the function's values 1 and 0 are, by Eqs. (9.2.6) and (9.2.7)

$$E\{D_{\varphi,1}\} = \frac{p_1 + p_2 - p_1 p_2}{p_1 \lambda_1 + p_2 \lambda_2 - p_1 p_2 (\lambda_1 + \lambda_2)} \tag{9.2.20}$$

and

$$E\{D_{\varphi,0}\} = \frac{1}{\mu_1 + \mu_2} . \tag{9.2.21}$$

Alternatively, with rates (mean frequencies) only, remembering that, by Eq. (9.2.16), Eq. (9.2.5) is transformed to

$$p_i = \frac{\mu_i}{\lambda_i + \mu_i} , \qquad \bar{p}_i = \frac{\lambda_i}{\lambda_i + \mu_i} , \tag{9.2.22}$$

one gets

$$\lambda_\varphi = \frac{\lambda_1 \lambda_2 (\mu_1 + \mu_2)}{\lambda_1 \mu_2 + \lambda_2 \mu_1 + \mu_1 \mu_2} , \tag{9.2.23}$$

and (trivially by Eq. (9.2.9))

$$\mu_\varphi = \mu_1 + \mu_2 . \tag{9.2.24}$$

It is interesting to note that, for equal components, Eq. (9.2.23) checks with Eq. (9.1.12).

Example 9.2.3 Two-out-of-three majority function

From Eq. (7.2.8)

$$X_\varphi = X_1X_2 + X_1X_3 + X_2X_3 - 2X_1X_2X_3 \ . \tag{9.2.25}$$

Here, the φ_i' of (7.3.2) and the corresponding $p_{\varphi,i}$ are

$$\begin{aligned} \varphi_1' &= X_2 + X_3 - 2X_2X_3 \ , \quad & p_{\varphi,1} &= p_2 + p_3 - 2p_2p_3 \ , \\ \varphi_2' &= X_1 + X_3 - 2X_1X_3 \ , \quad & p_{\varphi,2} &= p_1 + p_3 - 2p_1p_3 \ , \\ \varphi_3' &= X_1 + X_2 - 2X_1X_2 \ , \quad & p_{\varphi,3} &= p_1 + p_2 - 2p_1p_2 \ . \end{aligned} \tag{9.2.26}$$

By Eqs. (9.2.4) and (9.2.17)

$$\begin{aligned} v_\varphi &= (p_2 + p_3 - 2p_2p_3)p_1\lambda_1 + (p_1 + p_3 - 2p_1p_3)p_2\lambda_2 \\ &\quad + (p_1 + p_2 - 2p_1p_2)p_3\lambda_3 \\ &= p_1p_2(\lambda_1 + \lambda_2) + p_1p_3(\lambda_1 + \lambda_3) + p_2p_3(\lambda_2 + \lambda_3) \\ &\quad - 2p_1p_2p_3(\lambda_1 + \lambda_2 + \lambda_3) \ . \end{aligned} \tag{9.2.27}$$

From v_φ and the p_φ of Eq. (8.2.7a) λ_φ and μ_φ can be easily found.

The comparisons of Eq. (8.2.6) with Eq. (9.2.18) and of Eq. (8.2.7a) with Eq. (9.2.27) point to the following theorem:

Theorem 9.2.2. If the Boolean function $\varphi(X_1, \ldots, X_n)$ is given in the multilinear polynomial from Eq. (7.2.6), then for *s*-independent $X_1, \ldots, X_n$

$$v_\varphi = \sum_{i=1}^{m} \left[c_i \left(\prod_{j=1}^{n_i} \tilde{p}_{l_{i,j}} \right) \sum_{j=1}^{n_i} \tilde{\lambda}_{l_{i,j}} \right] , \tag{9.2.28}$$

where

$$\tilde{p}_k = \begin{cases} p_k & \text{for } \tilde{X}_k = X_k \\ \bar{p}_k = 1 - p_k & \text{for } \tilde{X}_k = \bar{X}_k \end{cases} \tag{9.2.29}$$

and

$$\tilde{\lambda}_k = \begin{cases} \lambda_k & \text{for } \tilde{X}_k = X_k \\ -\mu_k & \text{for } \tilde{X}_k = \bar{X}_k \ . \end{cases} \tag{9.2.30}$$

Prior to the proof of this theorem some helpful lemmas are derived.

Lemma 9.2.2. If in Eq. (7.3.2), i.e. in

$$X_\varphi = X_i\varphi_i' + \varphi_i'' \tag{9.2.31}$$

X_i is *s*-independent from the rest of the variables, then

$$p_\varphi = p_ip_{\varphi,i} + p\{\varphi_i'' = 1\} \ , \tag{9.2.32}$$

and hence

$$p_{\varphi,i} = \frac{\partial}{\partial p_i} p_\varphi(p_1, \ldots, p_n) \ . \tag{9.2.33}$$

The proof is obvious.

A trivial consequence of this lemma is the following:

Lemma 9.2.3. If p_φ is given as a multilinear polynomial of the p_is, then $p_{\varphi,i}$ is the sum of all terms of p_φ, which contain p_i and where p_i has been replaced by 1. — —

Similar results follow from any representation

$$X_\varphi = X_i \psi_i' + \bar{X}_i \psi_i'' + \psi_i''' \ , \tag{9.2.34}$$

which is, of course, not unique. Even ψ_i', ψ_i'', ψ_i''' need not be Boolean.

Example 9.2.4 Two-out-of-three function

From the CDNF by corollary 7.2.1 on the one hand

$$X_\varphi = X_1 X_2 X_3 + \bar{X}_1 X_2 X_3 + X_1 \bar{X}_2 X_3 + X_1 X_2 \bar{X}_3 \ . \tag{9.2.35}$$

Here

$$\psi_1' = X_2 X_3 + \bar{X}_2 X_3 + X_2 \bar{X}_3 = X_3 + X_2 \bar{X}_3 \ ,$$

$$\psi_1'' = X_2 X_3 \ , \quad \psi_1''' = 0 \ . \tag{9.2.36}$$

On the other hand, after merging the first two terms of Eq. (9.2.35)

$$X_\varphi = X_2 X_3 + X_1 \bar{X}_2 X_3 + X_1 X_2 \bar{X}_3 \ , \tag{9.2.37}$$

so that now

$$\psi_1' = \bar{X}_2 X_3 + X_2 \bar{X}_3 \ , \quad \psi_1'' = 0 \ , \quad \psi_1''' = X_2 X_3 \ . \text{— —} \tag{9.2.38}$$

From Eq. (9.2.34), obviously, for *s*-independent X_is and Boolean ψ_i', ψ_i'', ψ_i'''

$$p_\varphi = p_i P\{\psi_i' = 1\} + \bar{p}_i P\{\psi_i'' = 1\} + P\{\psi_i''' = 1\} \ , \tag{9.2.39}$$

and by Eq. (9.2.33) for any choice of ψ_i', ψ_i''

$$p_{\varphi,i} = P\{\psi_i' = 1\} - P\{\psi_i'' = 1\} \ . \tag{9.2.40}$$

Proof of theorem 9.2.2: Inserting Eq. (9.2.10) in Eq. (9.2.4) yields the two alternatives

$$\nu_\varphi = \sum_{i=1}^{n} p_{\varphi,i} \begin{cases} p_i \lambda_i \\ \bar{p}_i \mu_i \ . \end{cases} \tag{9.2.41}$$

Choose $p_i\lambda_i$ for those terms of $p_{\varphi,i}$ which belong to $P\{\psi_i'=1\}$ and $\bar{p}_i\mu_i$ for those belonging to $-P\{\psi_i''=1\}$. The result of this is that the i-th addend of the sum in Eq. (9.2.41) consists of all the terms of p_φ, which contain p_i or $\bar{p}_i$ and the former are multiplied by λ_i whilst the latter are multiplied by $-\mu_i$. Finally from the n addends of v_φ of Eq. (9.2.41) the common terms are extracted yielding for each term $p_i p_j \ldots \bar{p}_k \bar{p}_e \ldots$ of p_φ the new term

$$p_i p_j \ldots \bar{p}_k \bar{p}_l \ldots (\lambda_i + \lambda_j + \ldots - \mu_k - \mu_l - \ldots) , \tag{9.2.42}$$

q.e.d. (For a different proof see [22].)

Equation (9.2.27) is an example for theorem 9.2.2. Yet it does not show its full power since the transformation of a polynomial with p_is and $\bar{p}_i$s to one with p_is only can be rather cumbersome. Therefore, the following example is added.

Example 9.2.5 (Example 9.2.3 redone)

From Eq. (8.2.7) in the case of the two-out-of-three function

$$p_\varphi = p_1 p_2 + p_1 \bar{p}_2 p_3 + \bar{p}_1 p_2 p_3 \; . \tag{9.2.43}$$

By theorem 9.2.2

$$v_\varphi = p_1 p_2(\lambda_1+\lambda_2) + p_1\bar{p}_2 p_3(\lambda_1-\mu_2+\lambda_3) + \bar{p}_1 p_2 p_3(-\mu_1+\lambda_2+\lambda_3) \; . \tag{9.2.44}$$

Check. Using Eq. (9.2.10), this result should be equal to Eq. (9.2.27). The first terms of both results are equal. Furthermore

$$\begin{aligned} p_1\bar{p}_2 p_3(\lambda_1-\mu_2+\lambda_3) &= p_1(1-p_2)p_3(\lambda_1+\lambda_3) - p_1 p_3 \bar{p}_2 \mu_2 \\ &= p_1 p_3(\lambda_1+\lambda_3) - p_1 p_2 p_3(\lambda_1+\lambda_3) - p_1 p_3 p_2 \lambda_2 \\ &= p_1 p_3(\lambda_1+\lambda_3) - p_1 p_2 p_3(\lambda_1+\lambda_2+\lambda_3) \; . \end{aligned}$$

Similarly (after permuting indices)

$$\bar{p}_1 p_2 p_3(-\mu_1+\lambda_2+\lambda_3) = p_2 p_3(\lambda_2+\lambda_3) - p_1 p_2 p_3(\lambda_1+\lambda_2+\lambda_3) \; ,$$

q.e.d.

9.3 Mean Frequency of Changes of Functions' Values

In certain applications, typically in reliability technology, rather than the mean duration of a state (fixed value of X_φ) the mean frequency of changes of state, e.g. changes of X_φ from 0 to 1, are of practical interest. In the stationary case discussed in §9.2 the mean frequency v_φ and the mean duration of states $X_\varphi=0$ and 1 are related by Eq. (9.2.4). In the non-stationary case a similar relation is not quite obvious, even though we cannot think of a frequency of point events [34] without thinking of intervals between neighbouring points. The main result of this section will be a formula for determining v_φ also in case of s-dependent X_is. We start with a plausible lemma.

Lemma 9.3.1. The mean point density of unidirectional changes of state of an orderly [32] alternating point process i (PPi) with the indicator process given by

$$X_i(t) = \left\{ \begin{array}{ll} 0, & \text{for state "0"} \\ 1, & \text{for state "1"} \end{array} \right\} \text{ of point process } i$$

is the temporal joint probability density

$$v_i(t) = \lim_{\Delta t \to 0} \frac{1}{\Delta t} P\{[X_i(t)=0] \cap [X_i(t+\Delta t)=1]\} \ . \tag{9.3.1}$$

(See also [31].)

Proof. With

$$N_i(t_1, t_2) := \{\text{Number of changes of } X_i(\tau) \text{ from 0 to 1 in the time interval from } t_2 \text{ to } t_2\} \tag{9.3.2}$$

we have for any orderly *PPi* by Korolyuk's theorem [32]

$$P\{[X_i(t)=0] \cap [X_i(t+\Delta t)=1]\} = P\{N_i(t, t+\Delta t) \geqq 1\} + 0(\Delta t). \tag{9.3.3}$$

Furthermore, due to orderliness:

$$\begin{aligned} P\{N_i(t, t+\Delta t) \geqq 1\} &= \sum_{j=1}^{\infty} j\, P\{N_i(t, t+\Delta t) = j\} + 0(\Delta t) \\ &=: E\{N_i(t, t+\Delta t)\} + 0(\Delta t) \ . \end{aligned} \tag{9.3.4}$$

Clearly, as a mean point density

$$\lim_{\Delta t \to 0} \frac{1}{\Delta t} E\{N_i(t, t+\Delta t)\} = v_i(t) \ . \tag{9.3.5}$$

Hence, dividing Eqs. (9.3.4) and (9.3.3) by Δt and letting $\Delta t \to 0$ one has Eq. (9.3.1), q.e.d.

Now it is easy to prove the following theorem.

Theorem 9.3.1. With the notation of lemma 9.3.1

$$v_i(t) = \lim_{\Delta t \to 0} \frac{1}{\Delta t} E[\bar{X}_i(t) X_i(t+\Delta t)] \ . \tag{9.3.6}$$

Proof: Trivially, with

$$\bar{X} := 1 - X \quad \text{for} \quad X \in \{0, 1\}: \tag{9.3.7}$$

$$(X_i = 0) \Leftrightarrow (\bar{X}_i = 1) \ . \tag{9.3.8}$$

Hence the following equation between random events holds:

$$\{[X_i(t) = 0] \cap [X_i(t + \Delta t) = 1]\} = \{\bar{X}_i(t) X_i(t + \Delta t) = 1\} \ . \tag{9.3.9}$$

Further, by Eq. (11.3.4)

$$P\{\bar{X}_i(t) X_i(t + \Delta t) = 1\} = E[\bar{X}_i(t) X_i(t + \Delta t)] \ . \tag{9.3.10}$$

In conjunction with Eq. (9.3.1) this yields Eq. (9.3.6), q.e.d.

Comment. Sometimes, instead of $\bar{X}(t)X(t+\Delta t)$, initially $X(t)X(t+\Delta t)$ is given. By (see Eq. (9.3.7))

$$X(t)X(t + \Delta t) = [1 - \bar{X}(t)]X(t + \Delta t) = X(t + \Delta t) - \bar{X}(t)X(t + \Delta t) \tag{9.3.11}$$

the desired product $\bar{X}(t)X(t + \Delta t)$ is easily gained. — —

The key result of §9 follows in a trivial way: Since theorem 9.3.1 is true also for the index φ instead of i, we have

$$\nu_\varphi(t) = \lim_{\Delta t \to 0} \frac{1}{\Delta t} E[\bar{X}_\varphi(t) X_\varphi(t + \Delta t)] \ . \tag{9.3.12}$$

Let us try to use this result for practical applications!

In the subsequent analysis it will be slightly more practical to write (for any index)

$$\nu(t)\Delta t + 0(\Delta t) = E[\bar{X}(t)X(t + \Delta t)] \ . \tag{9.3.13}$$

If both $X_\varphi = \varphi(\mathbf{X})$ and $\bar{X}_\varphi = \bar{\varphi}(\mathbf{X})$ are multilinear polynomials of the literals $X_1, \bar{X}_1, X_2, \bar{X}_2, \ldots, X_n, \bar{X}_n$, the notation of Eq. (9.3.13) will show very efficiently which products of terms of φ or $\bar{\varphi}$ will vanish with $\Delta t \to 0$ and which will not. This is explained better with an example.

For better reading, we use henceforth the notation

$$A(t) = P\{\bar{X}(t) = 1\} = E[\bar{X}(t)] \quad \text{(availability)} \ , \tag{9.3.14}$$

$$U(t) = P\{X(t) = 1\} = E[X(t)]$$
$$= 1 - A(t) \quad \text{(unavailability)} \ . \tag{9.3.15}$$

Note that in fault-tree analysis the (mean) frequency of changes of X_φ from 0 to 1 is the failure frequency of the system under investigation.

Example 9.3.1 One-out-of-two: G system

In the most simple case of redundancy, the fault-tree function is

$$X_\varphi = X_1 X_2 \;, \tag{9.3.16}$$

and (as is easily checked)

$$\bar{X}_\varphi = \bar{X}_1 + \bar{X}_2 - \bar{X}_1 \bar{X}_2 \;. \tag{9.3.17}$$

Hence

$$\begin{aligned}\bar{X}_\varphi(t) X_\varphi(t+\Delta t) = {} & [\bar{X}_1(t) X_1(t+\Delta t)] X_2(t+\Delta t) \\ & + [\bar{X}_2(t) X_2(t+\Delta t)] X_1(t+\Delta t) \\ & - [\bar{X}_1(t) X_1(t+\Delta t)][\bar{X}_2(t) X_2(t+\Delta t)] \;,\end{aligned} \tag{9.3.18}$$

and by Eqs. (9.3.13)–(9.3.15) for *s*-independnet $X_i(t)$

$$\begin{aligned}\nu_\varphi(t)\Delta t + 0(\Delta t) = {} & \nu_1(t)\Delta t\, U_2(t+\Delta t) + \nu_2(t)\Delta t\, U_1(t+\Delta t) \\ & - \nu_1(t)\Delta t\, \nu_2(t)\Delta t \;,\end{aligned}$$

such that, after dividing by Δt and then letting $\Delta t \to 0$,

$$\nu_\varphi(t) = \nu_1(t) U_2(t) + \nu_2(t) U_1(t) \;. \tag{9.3.19}$$

This result is plausible, since, for $U_1(t)$, $U_2(t) \ll 1$ the system failure frequency is much smaller than the failure frequency of any of its components. Furthermore, this result checks perfectly with Eq. (9.2.14) derived for the stationary case, once $U_i(t)$ is replaced by p_i.

Example 9.3.2 Two-out-of-three system

By Eq. (7.2.8)

$$X_\varphi = X_1 X_2 + X_1 X_3 + X_2 X_3 - 2 X_1 X_2 X_3 \tag{9.3.20}$$

and, obviously,

$$\bar{X}_\varphi = \bar{X}_1 \bar{X}_2 + \bar{X}_1 \bar{X}_3 + \bar{X}_2 \bar{X}_3 - 2 \bar{X}_1 \bar{X}_2 \bar{X}_3 \;. \tag{9.3.21}$$

Here the product of the first terms yields by Eq. (9.3.13)

$$\begin{aligned}& E[\bar{X}_1(t) X_1(t+\Delta t) \bar{X}_2(t) X_2(t+\Delta t)] \\ & \quad = [\nu_1(t)\Delta t + 0(\Delta t)][\nu_2(t)\Delta t + 0(\Delta t)] \;,\end{aligned} \tag{9.3.22}$$

which, after the division by Δt as prescribed in Eq. (9.3.12), vanishes with Δt. Consequently, using Eq. (9.3.13) and gathering 6 individual $0(\Delta t)$ to the one appearing at the l.h.s. of the following equation, for stochastically independent $X_i(t)$

$$\begin{aligned} v(t)\Delta t + 0(\Delta t) = {} & v_1(t)\Delta t E[\bar{X}_2(t)X_3(t+\Delta t)] + v_2(t)\Delta t E[\bar{X}_1(t)X_3(t+\Delta t)] \\ & + v_1(t)\Delta t E[\bar{X}_3(t)X_2(t+\Delta t)] + v_3(t)\Delta t E[\bar{X}_1(t)X_2(t+\Delta t)] \\ & + v_2(t)\Delta t E[\bar{X}_3(t)X_1(t+\Delta t)] + v_3(t)\Delta t E[\bar{X}_2(t)X_1(t+\Delta t)] \ . \end{aligned} \tag{9.3.23}$$

Division by Δt gives for $\Delta t \to 0$ and stochastically independent X_i or $\bar{X}_i$, respectively (using the notation of Eqs. (9.3.14) and (9.3.15))

$$\begin{aligned} v_\varphi(t) &= v_1(t)A_2(t)U_3(t) + v_2(t)A_1(t)U_3(t) \\ &\quad + v_1(t)A_3(t)U_2(t) + v_3(t)A_1(t)U_2(t) \\ &\quad + v_2(t)A_3(t)U_1(t) + v_3(t)A_2(t)U_1(t) \\ &= v_1(t)[A_2(t)U_3(t) + A_3(t)U_2(t)] \\ &\quad + v_2(t)[A_1(t)U_3(t) + A_3(t)U_1(t)] \\ &\quad + v_3(t)[A_1(t)U_2(t) + A_2(t)U_1(t)] \ . \end{aligned} \tag{9.3.24}$$

This result checks with Eq. (9.2.27), if $v_i(t)$ is replaced by $p_i\lambda_i$, $A_i(t)$ by $1-p_i$, and $U_i(t)$ by p_i. — —

Now let us discuss briefly how a polynomial form of $\bar{\varphi}$ can be found. If φ is a (multilinear) polynomial of the type (7.2.6) then, trivially, an acceptable polynomial is

$$\bar{\varphi}(\mathbf{X}) = 1 - \varphi(\mathbf{X}) \ . \tag{9.3.25}$$

For instance in the case of the two-out-of-three system, instead of Eq. (7.2.8) one can use

$$\bar{X}_\varphi = 1 - X_1X_2 - X_1X_3 - X_2X_3 + 2X_1X_2X_3 \ . \tag{9.3.26}$$

In $E[\bar{X}_\varphi(t)X_\varphi(t+\Delta t)]$ this would initially lead to terms like

$$E[X_i(t)X_i(t+\Delta t)X_j(t)X_k(t+\Delta t)] \ .$$

Here it would be necessary to replace $X_i(t)X_i(t+\Delta t)$ according to Eq. (9.3.11).

To show that this little trick works, let us do a little example.

Example 9.3.3 One-out-of-two system (Example 9.3.1 redone)

In the case of a one-out-of-two system from Eq. (9.3.16), in contrast to Eq. (9.3.17)

$$\bar{X}_\varphi = 1 - X_1X_2 \ . \tag{9.3.27}$$

Now, by Eq. (9.3.11),

$$\begin{aligned}\bar{X}_\varphi(t)X_\varphi(t+\Delta t) &= X_1(t+\Delta t)X_2(t+\Delta t)\\ &-[X_1(t+\Delta t)-\bar{X}_1(t)X_1(t+\Delta t)][X_2(t+\Delta t)-\bar{X}_2(t)X_2(t+\Delta t)]\\ &=\bar{X}_1(t)X_1(t+\Delta t)X_2(t+\Delta t)+\bar{X}_2(t)X_2(t+\Delta t)X_1(t+\Delta t)\\ &-\bar{X}_1(t)X_1(t+\Delta t)\bar{X}_2(t)X_2(t+\Delta t)\ ,\end{aligned} \tag{9.3.28}$$

which equals Eq. (9.3.18), such that again the $v_\varphi(t)$ of Eq. (9.3.19) will result.

A More Elegant Way to Determine the Time-Specific Failure Frequency for *S*-Independent *X*s

As is obvious from the last few examples, the above "brute force" approach generally leads to a formidable amount of algebraic manipulations, even for small-size problems. The following theorem, an extension of theorem 9.2.1, shows a way out for *s*-independent X_is.

Theorem 9.3.2. The frequency of changes of $\varphi(\mathbf{X})$ from 0 to 1 is for the n stochastically independent components X_i of $\mathbf{X}$

$$v_\varphi(t)=\sum_{i=1}^{n} v_i(t)p_{\varphi,i}(t)\ , \tag{9.3.29}$$

where, as an obvious generalization of lemma 9.2.7,

$$p_{\varphi,i}(t):=P\{\varphi_i'(t)=1\} \tag{9.3.30}$$

with

$$\varphi_i'(t)=\frac{\partial}{\partial X_i}\varphi(\mathbf{X})\ ;\qquad \mathbf{X}=\mathbf{X}(t)\ , \tag{9.3.31}$$

being the formal† partial derivative with respect to X_i of the Boolean function $\varphi(\mathbf{X})$, i.e. φ_i' is defined by

$$\varphi = X_i\varphi_i'+\varphi_i''\ , \tag{9.3.32}$$

where φ_i' and φ_i'' don't depend on X_i.

Proof. By the rule of total probability, with $N_i(t_1, t_2)$ from Eq. (9.3.2), and $N_\varphi(t_1, t_2)$ the same counting variable for the system characterized by φ:

$$\begin{aligned}P\{N_\varphi(t, t+\Delta t)=1\}=\sum_{i=1}^{n} &P\{N_\varphi(t, t+\Delta t)=1\,|\,N_i(t, t+\Delta t)=1\}\\ &\cdot P\{N_i(t, t+\Delta t)=1\}\ .\end{aligned} \tag{9.3.33}$$

† Clearly, Boolean functions are not differentiable in the usual sense.

From Eqs. (9.3.3) and (9.3.4) we have (for indices φ and i)

$$P\{N(t, t+\Delta t) = 1\} = v(t)\Delta t + 0(\Delta t) . \tag{9.3.34}$$

Further, the conditional probability in Eq. (9.3.33) is obviously the probability that, given a change of X_i, φ will behave in the same way (between t and $t+\Delta t$). As can be concluded from Eq. (9.3.32) this is equivalent to having $\varphi_i' = 1$ (between t and $t+\Delta t$); in short

$$\lim_{\Delta t \to 0} P\{N_\varphi(t, t+\Delta t) = 1 | N_i(t, t+\Delta t) = 1\} = P\{\varphi_i'(t) = 1\} =: p_{\varphi,i}(t) . \tag{9.3.35}$$

Hence, inserting Eqs. (9.3.34), (9.3.35) in Eq. (9.3.33), dividing by Δt and letting $\Delta t \to 0$ yields Eq. (9.3.29), q.e.d.

Examples are readily found amongst the applications of Eq. (9.2.4), where the dependence on t is easily introduced.

9.4 The Distribution of Residual Life Times

In a number of applications of Boolean functions modelling the time between reaching a given set of states and the state $X_\varphi = 1$ is of some interest. For instance, when repairs must be postponed it can be of vital interest to know the probability with which a fault tolerant system will fail at the latest τ time units after a tolerable faulty state; see the following introductory example.

Example 9.4.1 One-out-of-two: G system

A one-out-of-two: G system with two components, which are hot standby for each other, will fail together with the second failing component. Let both of them have lives L_1 and L_2, respectively. Let residual life start on the first failure of a component. Obviously, if $L_i < L_j$ then $L_j - L_i$ is the system's residual life L_r. Then, for s-independent L_1 and L_2

$$F_{L_r}(\tau) = 1 - \int_0^\infty \{f_{L_1}(t)[1 - F_{L_2}(t+\tau)] + f_{L_2}(t)[1 - F_{L_1}(t+\tau)]\}\, dt . \tag{9.4.1}$$

Proof. First, it is best to exchange c.d.fs for survivor functions $\bar{F} := 1 - F$. This changes Eq. (9.4.1) to

$$\bar{F}_{L_r}(\tau) = \int_0^\infty f_{L_1}(t)\bar{F}_{L_2}(t+\tau)\, dt + \int_0^\infty f_{L_2}(t)\bar{F}_{L_1}(t+\tau)\, dt .$$

Now, similar to the Example of Eq. (11.3.11), for s-independent L_i and L_j

$$\begin{aligned} P\{(L_r > \tau) \cap (L_i \leqq L_j)\} &= \sum_{k=1}^{\infty} f_{L_i}(k\Delta t)\Delta t \bar{F}_{L_j}(k\Delta t + \tau) + 0(\Delta t) \\ &= \int_0^\infty f_{L_i}(t)\bar{F}_{L_j}(t+\tau)\, d\tau , \end{aligned}$$

where $(i, j) \in \{(1, 2), (2, 1)\}$. Finally,

$$\begin{aligned} P\{L_r > \tau\} &= P\{(L_r > \tau) \cap [(L_1 \leqq L_2) \cup (L_2 \leqq L_1)]\} \\ &= P\{(L_r > \tau) \cap (L_1 \leqq L_2)\} + P\{(L_r > \tau) \cap (L_2 \leqq L_1)\} \ , \end{aligned}$$

q.e.d.

In case of cold standby of component 2 for component 1, trivially $L_r = L_2$.

In the general theory, developed in [57], a basic concept is that of the residual life set $S_{r,j}$, which is a set j of components, the last of which fails at $L_{r,j}$. The system fails at L, where the Boolean function φ describes the redundancy structure (e.g. the fault tree) of the given system. Then, as to residual life set j

$$L_r = L_\varphi - L_{r,j} \ ; \quad j = 1, \ldots, m \ . \tag{9.4.2}$$

In example 9.4.1 there are (for hot standby) two residual life sets, namely component 1 and component 2.

Since residual life can start at any moment t, it makes sense to define a probability density $p_L(t, \tau)$ such that

$$\bar{F}_{L_r}(\tau) = \int_0^\infty p_L(t, \tau)\, dt \ . \tag{9.4.3}$$

It can be shown [57] that

$$p(t, \tau) = \frac{\partial}{\partial t} P\{a(t, t', \tau)\}_{/t'=t} \ , \tag{9.4.4}$$

with the random event

$$a(t, t', \tau) = \bigcup_{j=1}^{m} [(L_{r,j} \leqq t) \cap (L_\varphi > t' + \tau)] \ . \tag{9.4.5}$$

Obviously, as equations of random events with X_i and X_φ as fault-free input and output, respectively:

$$(L_{r,j} \leqq t) = \left(\bigwedge_{i \in I_{S_{r,j}}} X_i(t) = 1 \right) \ , \tag{9.4.6}$$

where $I_{S_{r,j}}$ is the index set of $S_{r,j}$, and

$$(L_\varphi > t' + \tau) = (X_\varphi(t' + \tau) = 0) = (\bar{X}_\varphi(t' + \tau) = 1) \ . \tag{9.4.7}$$

Hence

$$a(t, t', \tau) = \left\{ \bigvee_{j=1}^{m} [\bar{X}_\varphi(t' + \tau) \bigwedge_{i \in I_{S_{r,j}}} X_i(t)] = 1 \right\} \ . \tag{9.4.8}$$

For $t' \geqq t$ and no repairs $X_i(t) = 1$ induces $X_i(t'+\tau) = 1$. Therefore, Eq. (9.4.8) can be simplified by replacing $\bar{X}_\varphi(t'+\tau)$ by

$$\bar{X}_\varphi(t'+\tau)/_{X_k=1} \; ; \quad k \in I_{S_{r,j}} \; . \tag{9.4.9}$$

Another example will show the usefulness of this approach.

Example 9.4.2 Two-out-of-three-system

As is well known from §7, for $X_i = 1$, iff component i is bad, here

$$\bar{X}_\varphi = \bar{X}_1 \bar{X}_2 + \bar{X}_1 \bar{X}_3 + \bar{X}_2 \bar{X}_3 - 2\bar{X}_1 \bar{X}_2 \bar{X}_3 \; . \tag{9.4.10}$$

The only plausible $m = 3$ residual life sets are

$$S_{r,j} = \{C_j\} \; ; \quad j = 1, 2, 3 \; . \tag{9.4.11}$$

Hence, by Eqs. (9.4.8) and (9.4.9) and Eq. (7.1.6), aiming at $a(t, t', \tau)$:

$$\begin{aligned} \bigvee_{j=1}^{3} \bar{X}_\varphi(t'+\tau)/_{X_j=1} X_j(t) &= X_1(t)\bar{X}_2(t'+\tau)\bar{X}_3(t'+\tau) \vee X_2(t)\bar{X}_1(t'+\tau)\bar{X}_3(t'+\tau) \\ &\quad \vee X_3(t)\bar{X}_1(t'+\tau)\bar{X}_2(t'+\tau) \\ &= 1 - [1 - X_1(t)\bar{X}_2(t'+\tau)\bar{X}_3(t'+\tau)] \\ &\quad \times [1 - X_2(t)\bar{X}_1(t'+\tau)\bar{X}_3(t'+\tau)] \\ &\quad \times [1 - X_3(t)\bar{X}_1(t'+\tau)\bar{X}_2(t'+\tau)] \; . \end{aligned} \tag{9.4.12}$$

On simplification of the r.h.s. to a polynomial, use can be made of

$$X_i(t)\bar{X}_i(t'+\tau) = 0 \; , \quad t' \geqq t \; , \quad \tau > 0 \; . \tag{9.4.13}$$

This is true because in the case of no repairs X_i never returns to 0, once it has changed from 0 (good state) to 1 (bad state). Therefore, by Eqs. (9.4.4), (9.4.8), (8.2.1), and (8.1.5)

$$\begin{aligned} p(t, \tau) &= E\{X_1(t)\bar{X}_2(t'+\tau)\bar{X}_3(t'+\tau)\} + E\{X_2(t)\bar{X}_1(t'+\tau)\bar{X}_3(t'+\tau)\} \\ &\quad + E\{X_3(t)\bar{X}_1(t'+\tau)\bar{X}_2(t'+\tau)\} \\ &= F_{L_1}(t)\bar{F}_{L_2}(t'+\tau)\bar{F}_{L_3}(t'+\tau) + F_{L_2}(t)\bar{F}_{L_1}(t'+\tau)\bar{F}_{L_3}(t'+\tau) \\ &\quad + F_{L_3}(t)\bar{F}_{L_1}(t'+\tau)\bar{F}_{L_2}(t'+\tau) \; . \end{aligned} \tag{9.4.14}$$

Furthermore,

$$\begin{aligned} \frac{\partial}{\partial t} p(t, \tau) &= f_{L_1}(t)\bar{F}_{L_2}(t'+\tau)\bar{F}_{L_3}(t'+\tau) + f_{L_2}(t)\bar{F}_{L_1}(t'+\tau)\bar{F}_{L_3}(t'+\tau) \\ &\quad + f_{L_3}(t)\bar{F}_{L_1}(t'+\tau)\bar{F}_{L_2}(t'+\tau) \; . \end{aligned} \tag{9.4.15}$$

Finally, by Eq. (9.4.3)

$$\bar{F}_{L_r}(\tau) = \int_0^\infty [f_{L_1}(t)\bar{F}_{L_2}(t+\tau)\bar{F}_{L_3}(t+\tau) + f_{L_2}(t)\bar{F}_{L_1}(t+\tau)\bar{F}_{L_3}(t+\tau) + f_{L_3}(t)\bar{F}_{L_1}(t+\tau)\bar{F}_{L_2}(t+\tau)]\, dt \ , \qquad (9.4.16)$$

which is a very plausible result, since $L_r > \tau$, if on a failure of component i both other components do not fail for at least τ further units of time. Details are given in exercise 9.4.

Exercises

9.1
How long is the mean time of a state of the $M/M/1$ queueing system in which $i > 0$ customers are waiting for service?

9.2
Review example 4.4.3 (Shannon decomposition of the two-out-of-three majority function). Show that, similar to theorem 8.2.1, it is not always necessary first to find a multilinear polynomial form of φ (or p_φ) in order to calculate v_φ. In the case of the Shannon decomposition one can also multiply the innermost terms of the raw (non-polynomial) result of the decomposition algorithm with the sums of λ_is and $-\mu_j$s of theorem 9.2.2. Give the corresponding details for the two-out-of-three majority function.

9.3
How could Eq. (9.2.28) look like in case of equal components, i.e. for

$$p_i = p \ , \quad \lambda_i = \lambda \ , \quad \mu_i = \mu \ ; \quad i = 1, \ldots, n \ ?$$

9.4
Provide for an elementary derivation of Eq. (9.4.16).

9.5
Prove (9.1.12).

10 Some Algorithms and Computer Programs for Boolean Analysis

Boolean functions' analysis is also possible with computers. Sometimes, the sheer size of the problem, i.e. the number of the Boolean variables involved, demands computer-aided analysis and/or synthesis.

In this chapter some algorithms pertinent to this field are explained and implemented in PASCAL and partly (where special insight or much higher computational speed may be gained) in an assembly language. For example in [39], the rightmost 1 of the binary number N is deleted (and then also easily counted) by

$$N \textit{ and } (N-1) \quad ,^{\dagger}$$

where *and* is the bitwise AND operator and $N-1$ is easily found by *dec* N,‡ with *dec* for decrement.

Clearly, the huge field of computer applications to Boolean analysis is not covered by the contents of this chapter, yet the selection will, hopefully, help in many situations of practical interest.

Note that, as in the vast majority of computer outputs, only upper case letters are used in the formulation of the following algorithms. For a detailed understanding of the computer codes given the reader is supposed to have a working knowledge of PASCAL [23], [24]. However, in most cases flowcharts will make an understanding of the algorithms independent from the mastering of a special programming language.

The difficult field of the complexity of algorithms' is not discussed here. The reader will, hopefully, find useful information, if needed, in [39], [40].

Finally, it should be stressed that this chapter was written with the non-expert programmer in mind. The author, being such a person himself, would not therefore be surprised if, even in PASCAL, experts find better programs. His main intention is to help the non-expert solve small and medium-sized problems alone and discuss large problems with programming experts.

Warning. The correctness of the PASCAL procedures and programs of this chapter cannot be guaranteed, even though most of them were run on several computers with different compilers and different examples.

---

† Clearly N and $N-1$ would be inside two registers, say $R1$ and $R2$, and a proper assembly language formulation would be $R1$ *and* $R2$.

‡ Rather $R2 := \textit{dec } R1$

10.1 Computing Values of a Boolean Function

The computation of values of a given function φ depends strongly on its representation. Fortunately, the compilers of high level computer languages help. They get along with all sorts of Boolean algebra expressions. Except for the length of the operators, algebraic forms of Boolean functions are most easily transformed to computer code. For example, in PASCAL, with

$$XI \triangleq X_i$$

the well-known two-out-of-three majority function

$$\varphi = X_1 X_2 \vee X_1 X_3 \vee X_2 X_3 \tag{10.1.1}$$

simply becomes (after defining VAR PHI, X1, X2, X3: BOOLEAN)

```
PHI:=X1 AND X2 OR X1 AND X3 OR X2 AND X3 .
```

Of course, in high-level languages Boolean (indicator) variables are of the data type "Boolean". If necessary, a transformation to the data type "integer" must be executed, e.g. by

```
IF XI THEN YI:=1 ELSE YI:=0
```

(In PASCAL it is not necessary to write explicitly IF XI=TRUE.)

Trivially, having a multilinear polynomial form of a Boolean function, one can stay in the integer numbers' domain. Then

$$\varphi = X_1 X_2 + X_1 X_3 + X_2 X_3 - 2X_1 X_2 X_3$$

is in PASCAL (with VAR PHI, X1, X2, X3: INTEGER)

```
PHI:=X1*X2+X1*X3+X2*X3-2*X1*X2*X3 .
```

Note that this may need considerably more processing time than the Boolean notation version! Note also that here a transformation of φ to the data type "Boolean" may be necessary.

Other problems are connected with the computation of a complete function table or truth table.[†] The first problem is the optimal choice of the sequence of the Boolean argument n-tuples **X**. A simple solution is the design of a binary counter observing the fact that incrementing a binary number is equivalent to complementing all digits from the right (least significant bit) up to and including the first 0.

[†] Details in §10.2.1.

Example 10.1.1 Binary counting

$$\begin{array}{r} 1100111 \\ + \quad\quad 1 \\ \hline 1101000 \end{array} \qquad \begin{array}{r} 100 \\ + \quad 1 \\ \hline 101 \end{array}$$

The logical details are given by the flow chart of Fig. 10.1.1.

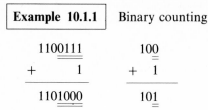

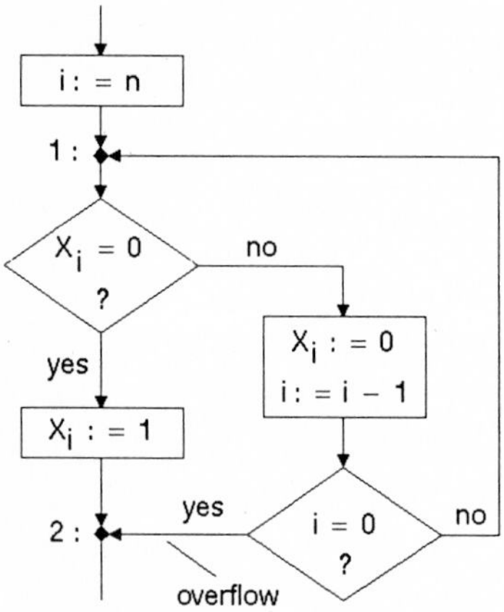

Fig. 10.1.1. Procedure for simulating a binary counter. (The successor of 1 . . . 1 is 0 . . . 0.) As to the flowcharts, consult §3.9

The short program of Table 10.1.1 is a PASCAL implementation of such a counter starting with an n-tuple and producing the next one.

Now, computer scientists have good reasons for regarding the excessive use of GOTO jumps as harmful. Therefore, a GOTO-free version of the above counting procedure is added. The relevant control flow graph of Fig. 10.1.2 looks simpler than that of Fig. 10.1.1. However, in the case of the highest possible n-tuple (1 . . . . 1) the algorithm will only leave the loop if some $X_i=0$ is found, preferably, $X_0=0$. Table 10.1.2 gives a PASCAL program.

Table 10.1.1. Low level PASCAL program for a binary counter

```
I:=N;
1:IF X[I]=0 THEN X[I]:=1
            ELSE
            BEGIN
              X[I]:=0;
              I:=I-1;
              IF I=0 THEN GOTO 2
                     ELSE GOTO 1
            END;
2:...
```

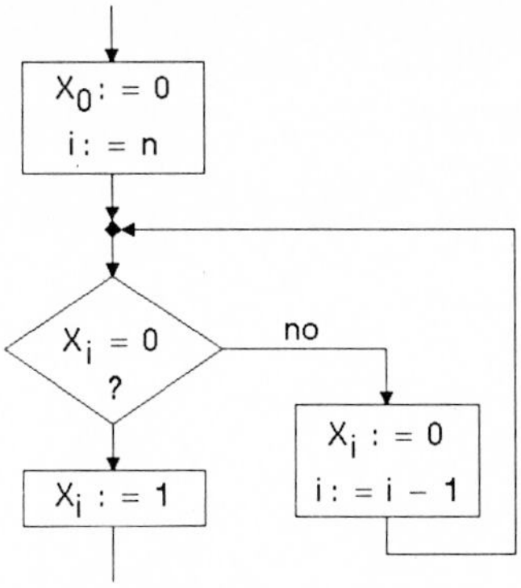

Fig. 10.1.2. Simplified version of binary counter

Table 10.1.2. PASCAL implementation for Fig. 10.1.2

```
X[0]:=0 {FOR A STOP WHEN INCREMENTING 1 . . . 1}
I:=N;
WHILE X[I]=1 DO
     BEGIN
        X[I]:=0;
        I:=I−1
     END;
X[I]:=1;
```

Table 10.1.3. PASCAL procedure "COUNT"

```
PROCEDURE COUNT (VAR X:LIST1);†
   VAR I:INTEGER;
   BEGIN {N-TUPLE TO BE INCREMENTED IS IN X[1 . . N] WITH DEFINITE VALUES}
         {ADD X[0]:=0 IF N-TUPLE COULD BE 1 . . . 1}
      I:=N;
      WHILE X[I]=1  DO
          BEGIN
             X[I]:=0;
             I:=I−1
          END;
      X[I]:=1
   END;
```

† In the main program there should be definitions of the following kind:
CONST N= . . . ;
TYPE LIST 1=ARRAY [O . . N] OF INTEGER;

Table 10.1.3 contains the complete procedure COUNT which will be used here later on. The simplest output routine for the counter is

```
FOR I:=1 TO N DO
WRITE (X[I]) ;
```

For $N=5$ this will be 11010, if $X_1=1$, $X_2=1$, $X_3=0$, $X_4=1$, $X_5=0$.

For a Boolean array to be changed by pseudo counting (to get a new n-tuple) Table 10.1.4 contains the adapted procedure.

Alternatively, one can use successive integer divisions (without a residue) to get the binary digits of a number which was incremented by the usual (integer) addition of 1. Table 10.1.5 gives the details of a simple routine for conversion from decimal to binary.

Note that, in the case of Table 10.1.5, first the most significant bit is produced, and this will be bit N−I from the left (see the second REPEAT statement).

Also, primitive polynomials can be used to produce binary n-tuples in the way pseudo random numbers are produced [25].

An extremely simple higher languages method to get Boolean n-tuples is a nesting of FOR loops as in Table 10.1.6. Note that this may be a memory space as well as a time-consuming way to count.

Table 10.1.4. PASCAL PROCEDURE "PSEUDOCOUNT"†

```
PROCEDURE PSEUDOCOUNT (VAR X:LIST1B);
   VAR I:INTEGER;
   BEGIN {N-TUPLE TO BE INCREMENTED IS IN X[1..N] WITH BOOLEAN VALUES}
         {ADD X[0]:=FALSE IF N-TUPLE COULD BE TRUE ... TRUE}
      I:=N;
      WHILE X[I] DO
          BEGIN
             X[I]:=FALSE;
             I:=I-1
          END;
      X[I]:=TRUE
   END;
```

† In the main program there should be definitions of the following kind:
CONST N= ...;
TYPE LIST 1 B=ARRAY [O .. N] OF BOOLEAN;

Table 10.1.5. Transformation of an integer from decimal to binary. Binary equivalent of NUMB is put in array X[1 .. N] initially filled with zeros

```
VAR NUMB, S, I:INTEGER;
BEGIN
   S:=1;
   I:=0;
   REPEAT S:=S*2;
          I:=I+1
   UNTIL S>NUMB;
   REPEAT S:=S DIV 2;
          I:=I-1;
          X[N-I]:=NUMB DIV S;
          IF X[N-I]=1 THEN NUMB:=NUMB MOD S
   UNTIL  S=1
END
```

Table 10.1.6. Nested "binary" FOR loops to simulate a binary counter

```
VAR X1, X2, . . . : INTEGER;
BEGIN
  FOR X1:=0 TO 1 DO
      FOR X2:=0 TO 1 DO
            . . . . . . . . . . . . . . . . . . . . .
            FOR XN:=0 TO 1 DO
               {USE OF X1, X2, . . . , XN}
END
```

Of course, for the evaluating of $\varphi(X_1, \ldots, X_n)$ it would be better to work with BOOLEAN instead of INTEGER X1, . . . , XN. Table 10.1.7 shows this alternative for PASCAL.

Clearly, it is not very practical to add or delete lines in the program of Table 10.1.7 depending on the actual value of N. In PASCAL one can use instead the recursive procedure of Table 10.1.8, the name of which is RECURSION.

Table 10.1.7. Nested "Boolean" FOR loops. (In PASCAL TRUE "follows" FALSE)

```
VAR X1, X2, . . . : BOOLEAN;
BEGIN
  FOR X1:=FALSE TO TRUE DO
     FOR X2:=FALSE TO TRUE DO
         . . . . . . . . . . . . . . . . . . . . . .
         FOR XN:=FALSE TO TRUE DO
              {USE OF X1, X2, . . . , XN}
END
```

Table 10.1.8. Recursion for nesting Boolean FOR loops; as a nesting of two procedures

```
PROCEDURE RECURSION;
  CONST N= ; {INSERT AN INTEGER VALUE OF N}
  VAR X:ARRAY[1 . . N] OF BOOLEAN;
  PROCEDURE BOOLFOR(DEPTH:INTEGER);
    VAR COUNT:BOOLEAN;
    BEGIN
    FOR COUNT:=FALSE TO TRUE DO
      BEGIN X[DEPTH]:=COUNT;
         IF DEPTH<N THEN BOOLFOR (DEPTH+1)
                    ELSE . . . {USE OF X[1], . . . , X[N]}
      END
    END;
  BEGIN
  BOOLFOR(1);
  END;
```

n-Tuples of Weight *k*

Sometimes it can be desirable to process the 2^n binary n-tuples rather in the order of a rising Hamming weight w_H, i.e. the number of 1s. If a DNF of φ with non-negated variables has no terms of order† lower than m, it is obvious that the binary n-tuples being the equivalent of minterms of φ will have

$$w_H \geqq m \ .$$

More precisely, in that case the

$$\binom{n}{0}+\binom{n}{1}+\ldots+\binom{n}{m-1}$$

n-tuples with $w_H < m$ need not be produced. Table 10.1.9 indicates in which order one can produce binary n-tuples $\mathbf{X}$ with $w_H = k$. Further details will be given in the solutions of exercise 10.2.

Table 10.1.9. Binary n-tuples with $w_H = k$

No.	1	2	3	...	$k-1$	k	$k+1$	...	$n-k$	$n-k+1$	...	$n-1$	n
1	1	1	1	...	1	1	0	...	0	0	...	0	0
2	1	1	1	...	1	0	1	...	0	0	...	0	0
⋮													
$n-k$	1	1	1	...	1	0	0	...	0	0	...	1	0
$n-k+1$	1	1	1	...	1	0	0	...	0	0	...	0	1
⋮	.....												
	0	1	1	...	1	1	1	...	0	0	...	0	0
	.....												
⋮	0	1	1	...	1	1	0	...	0	0	...	0	1
	.....												
	0	0	0	...	0	0	0	...	1	1	...	1	0
	0	0	0	...	0	0	0	...	1	1	...	0	1
	0	0	0	...	0	0	0	...	1	1	...	1	1
	.....												
	0	0	0	...	0	0	0	...	1	0	...	1	1
$\binom{n}{k}$	0	0	0	...	0	0	0	...	0	1	...	1	1

Ternary Evaluation

Occasionally some of the X_is of $\mathbf{X} = (X_1, \ldots, X_n)$ are given definite values (0 or 1) and one would like to know if $\varphi = 0$ or $\varphi = 1$ or if now φ is a "true" function of some or all of the remaining X_is. (In §10.4 this is an important issue.) Let us denote the case "φ neither 0 nor 1" by the symbol d for don't care. Then it makes sense to rename all the non-fixed variables (the above "remaining" X_is) by d. This way $\varphi(\mathbf{X})$ is changed to a single-variable ternary function with the "values" 0, 1, and d, both for the variable and the function.

† Order equals length, i.e. number of literals.

Example 10.1.2 Two-out-of-three majority function

If in Eq. (10.1.1), i.e. in

$$X_\varphi = \varphi(X_1, X_2, X_3) = X_1 X_2 \vee X_1 X_3 \vee X_2 X_3$$

$X_1 = 1$, then

$$X_\varphi = X_2 \vee X_3 \ .$$

Hence, if X_2 or X_3 are ds, defining in this case $X_\varphi = d$, makes sense. If also $X_2 = 1$, then

$$X_\varphi = 1 \vee X_3 = 1 \ . — —$$

Now it is necessary to define (simple) rules for a Boolean algebra with the operands 0, 1, and d. Clearly,

$$0 \vee d = d \ , \quad 1 \vee d = 1 \ , \quad 0 \wedge d = 0 \ , \quad 1 \wedge d = d \ . \tag{10.1.2}$$

However, $X_i \bar{X}_j$ would initially yield $d\bar{d}$, and $d\bar{d} = 0$ is not a desirable rule. Hence, to simplify the following, we assume once again that φ be given as a DNF. Then the rule

$$\tilde{X}_i \leftarrow d \ , \quad \tilde{X}_i \in \{X_i, \bar{X}_i\} \tag{10.1.3}$$

for the transformation from m-variable binary to 1-variable ternary makes sense, and (10.1.2) can be augmented by the idempotence laws

$$d \vee d = d \ , \quad d \wedge d = d \ . \tag{10.1.4}$$

A little problem still remains to be solved for evaluating by a computer, say

$$\varphi(0, d, d, 1, 1, d, 0, 0) = d \ .$$

How can one handle d? To avoid cumbersome symbol manipulations, d is given a simple numerical value, say 3. Hence, with a double apostrophe for the numerical pseudovalue of the term $\hat{T}_i$, for a given **X**, whose components are either 0 or 1 or 3:

$$\hat{T}_i'' = \prod_{j=1}^{n_i} \tilde{X}_{l_{ij}} \ ; \quad \bar{X}_k = 1 - X_k \ .$$

For instance, $\hat{T}_i = X_1 \bar{X}_2 X_3$ yields for $X_1 = 1$, $X_2 = d$, $X_3 = d$

$$\hat{T}_i'' = 1 . (1-3) . 3 = -6 \mathrel{\hat{=}} \hat{T}_i = d \ .$$

A DNF can be evaluated according to the flowchart of Fig. 10.1.3.

Table 10.1.10 shows the procedure "FUNEVA" for a Boolean function's evaluation according to Fig. 10.1.3. The m terms $\hat{T}_1'', \ldots, \hat{T}_m''$ of the pseudo DNF are assumed to be given in array $T[1 \,..\, M]$. To avoid problems with PASCAL symbolism $\hat{T}_i''$ is replaced by T_i or, rather, by $T[I]$.

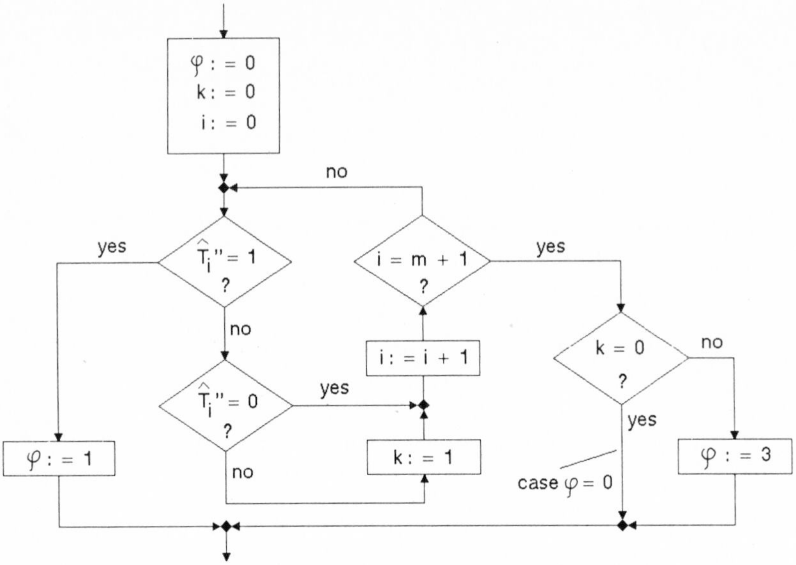

Fig. 10.1.3. Evaluation of a Boolean DNF with all its m terms $\hat{T}_1, \ldots, \hat{T}_m$ either 0, or 1, or d. ($\varphi=3$ corresponds to $\varphi=d$, with k as the indicator variable)

Table 10.1.10. Implementation of the algorithm of Fig. 10.1.3 as the procedure FUNEVA

```
CONST M= . . . ;
TYPE LIST2=ARRAY[1 . . M]OF INTEGER;
. . . . . . . . . . . . . . . . . . . . . . . . . . . . . . . .
PROCEDURE FUNEVA (T:LIST2; VAR PHI:INTEGER);
   VAR I, K:INTEGER;
   BEGIN
   I:=1;
   K:=0;
   PHI:=0;
   REPEAT IF T[I]=1 THEN PHI:=1
                         ELSE IF T[I]<>0 THEN K:=1; {T[I]=3}
            I:=I+1
   UNTIL (PHI=1) OR (I=M+1);
   IF (PHI=0) AND (K=1) THEN PHI:=3
   END;
```

It is certainly interesting to note how simple the logic of the assembly level flowchart can be implemented in PASCAL. Specifically, the repeat loop is left as soon as the first term, which is 1, has been found or when no such term could be found.

In closing, note that $d \triangleq 2$ would be a bad choice, since $d=2$ could lead to the following wrong result:

$$X_1 \leftarrow d \ , \ X_2 \leftarrow d \Rightarrow \bar{X}_1 \bar{X}_2 = \bar{d}\bar{d} \triangleq (1-2)(1-2)=1 \ ,$$

whereas

$$\bar{X}_1 \bar{X}_2 = d$$

would be correct.

10.2 Canonical Representations of a Boolean Function

Since Boolean functions typically allow for numerous representations some computer aid for the production of canonical forms is highly desirable.

The most elementary canonical form of a Boolean function, the CDNF is easily found using results of §10.1; see §10.2.1.

§10.2.2 contains an implementation of the ENZMANN algorithm for finding the canonical multilinear polynomial form of φ.

Comment. Even though a DNF with all the prime implicants of a Boolean function is also a canonical representation, I am sure that the better known algorithms for providing all these terms, the so-called *complete basis* of φ, have been implemented many times. Corresponding reports can certainly be found in the libraries of many laboratories of computer engineering.

10.2.1 The Canonical Disjunctive Normal Form

To find a minterm $\hat{M}_i$ of φ it is only necessary to select an n-tuple $\mathbf{X}_i$ with $\varphi(\mathbf{X}_i) = 1$ and to replace the components $X_{i,j}$ of $\mathbf{X}_i$ according to

$$X_{i,j} \Rightarrow \begin{cases} X_j \ , & \text{for} \quad X_{i,j} = 1 \\ \bar{X}_j \ , & \text{for} \quad X_{i,j} = 0 \ , \end{cases} \quad j = 1, \ldots, n \ .$$

Finally, all the $\hat{M}_i$ must be ORed to yield the CDNF of φ.

The systematic determination of a new minterm via the procedure "MINDET" using the procedure "PSEUDOCOUNT" of Table 10.1.4 is illustrated in Fig. 10.2.1. Table 10.2.1 shows a PASCAL implementation of MINDET. In non-PASCAL

Table 10.2.1. PASCAL version of procedure "MINDET"

```
PROCEDURE MINDET (VAR X: LIST1B; VAR PHI: BOOLEAN);
PROCEDURE PSEUDOCOUNT (VAR X: LIST1B);
......... (Table 10.1.4)
BEGIN {MINDET}
   IF NOT X[0] THEN
      REPEAT
         PSEUDOCOUNT (X);
         PHI:= ...                    {INSERT GIVEN FUNCTION PHI}
      UNTIL (PHI=TRUE) OR (X[0]=TRUE)
END
```

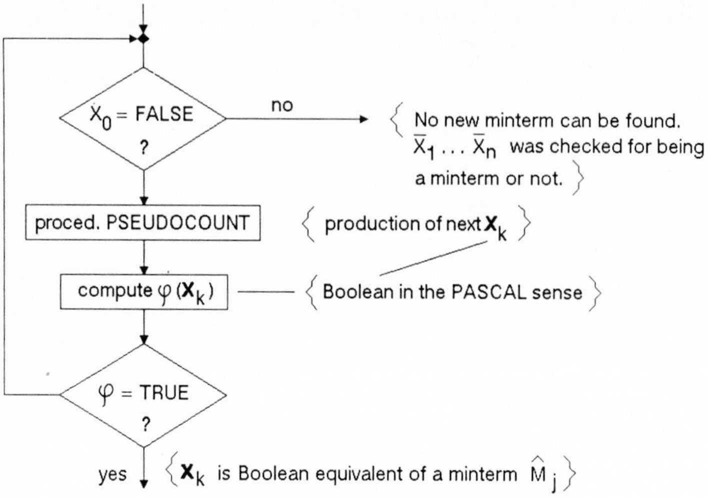

Fig. 10.2.1. Procedure MINDET for the determination of a new minterm. The n-tuple $\underline{\text{true}}$ = (true, . . . , true) is the last one of PSEUDOCOUNT

Boolean notation MINDET starts with checking the n-tuple (0, . . . , 0, 1) and ends with checking (0, . . . , 0, 0). Hence, in Table 10.2.1 the initial values of LIST1B should be $n+1$ times FALSE.

The printing of a complete table of minterms, with X_i and $\bar{X}_i$ replaced by 1 and 0, respectively, can be done by means of the program of Table 10.2.2.

Table 10.2.2. Program to produce a table form of a CDNF

```
PROGRAM CDNF_TAB (OUTPUT);
CONST N= ...; {INSERT NUMBER OF VARIABLES OF PHI}
TYPE LIST1B:ARRAY [0 .. N] OF BOOLEAN;
VAR J: INTEGER; X:LIST1B; PHI:BOOLEAN;
PROCEDURE MINDET (VAR X:LIST1B; VAR PHI:BOOLEAN);
...... {Table 10.2.1}
BEGIN {TABLE OF THE CDNF OF PHI}
FOR J:= 0 TO N DO X[J]:= FALSE;
WHILE X[0]=FALSE DO
     BEGIN
     MINDET (X; PHI) ;
     IF PHI:= TRUE THEN
        BEGIN
              FOR I:= 1 TO N DO
                 IF X[I] THEN WRITE ('1')
                          ELSE WRITE ('0');
              WRITELN
        END
     END
END.
```

Clearly a possibly more comprehensive table will result with the following substitutions:

WRITE('1') → WRITE('X', I) ,
WRITE('0') → WRITE('−X', I)

with the minus sign to denote negation.

Example 10.2.1 Two-out-of-three majority function

With $N = 3$ and

PHI := X[1] AND X[2] OR X[1] AND X[3] OR X[2] AND X[3]

the computer net output is

011	or	$-X_1$	X_2	X_3
101	or	X_1	$-X_2$	X_3
110	or	X_1	X_2	$-X_3$
111	or	X_1	X_2	X_3

This corresponds obviously to the CDNF

$$\varphi = \bar{X}_1 X_2 X_3 \vee X_1 \bar{X}_2 X_3 \vee X_1 X_2 \bar{X}_3 \vee X_1 X_2 X_3 \ . \text{——}$$

10.2.2 The Canonical Multilinear Polynomial Form

Now the ENZMANN algorithm of section 7.2 is implemented. As shown in section 7.2 the canonical multilinear polynomial form of $\varphi(X_1, \ldots, X_n)$ is

$$\begin{aligned} X_\varphi = a_0 &+ \sum_{i_1=1}^{n} a_{i_1} X_{i_1} + \sum_{i_1=1}^{n-1} \sum_{i_2=i_1+1}^{n} a_{i_1,i_2} X_{i_1} X_{i_2} \\ &+ \sum_{i_1=1}^{n-2} \sum_{i_2=i_1+1}^{n-1} \sum_{i_3=i_2+1}^{n} a_{i_1,i_2,i_3} X_{i_1} X_{i_2} X_{i_3} + \ldots , \end{aligned} \tag{10.2.1}$$

where $a_0 \in \{0, 1\}$, a_{i_1}, a_{i_1,i_2}, $a_{i_1,i_2,i_3}, \ldots$ positive or negative integers. In order to determine $a_{i_1,\ldots,i_j}$ one has to set

$$X_k = \begin{cases} 1, & \text{for } k \in I_j = \{i_1, i_2, \ldots, i_j\};\ i_1 < i_2 < \ldots < i_j \ , \\ 0, & \text{else} \end{cases}$$

and has to realize that then φ assumes the (Boolean) value

$$\begin{aligned} X_\varphi = a_0 &+ \sum_{i_1 \in I_j} a_{i_1} + \sum_{\substack{i_1 < i_2 \\ i_1, i_2 \in I_j}} a_{i_1,i_2} + \sum_{\substack{i_1 < i_2 < i_3 \\ i_1, i_2, i_3 \in I_j}} a_{i_1,i_2,i_3} + \ldots \\ &+ \sum_{\substack{i_1 < i_2 < \ldots < i_{j-1} \\ i_1, i_2, \ldots, i_{j-1} \in I_j}} a_{i_1,i_2,\ldots,i_{j-1}} + a_{i_1,i_2,\ldots,i_j} \ . \end{aligned} \tag{10.2.2}$$

Obviously in Eq. (10.2.1)

$$a_0 = \varphi(\mathbf{0}) := \varphi(0, \ldots, 0) \tag{10.2.3}$$

and

$$X_\varphi = \underset{(1)}{\varphi(0}, \ldots, 0, \underset{(\mathrm{i}_1)}{1}, 0, \ldots, 0, \underset{(\mathrm{i}_2)}{1}, 0, \ldots, 0, \ldots, \underset{(\mathrm{i}_j)}{1}, 0, \ldots, \underset{(\mathrm{n})}{0}) =: \varphi(\mathbf{X}_\mathrm{j}) \,.$$

Hence Eq. (10.2.2) can be rewritten as

$$a_{\mathrm{i}_1,\mathrm{i}_2,\ldots,\mathrm{i}_\mathrm{j}} = \varphi(\mathbf{X}_\mathrm{j}) - \varphi(\mathbf{0}) - \sum_{\mathrm{i}_1 \in \mathrm{I}_\mathrm{j}} a_{\mathrm{i}_1} - \sum_{\substack{\mathrm{i}_1 < \mathrm{i}_2 \\ \mathrm{i}_1, \mathrm{i}_2 \in \mathrm{I}_\mathrm{j}}} a_{\mathrm{i}_1,\mathrm{i}_2}$$

$$- \ldots - \sum_{\substack{\mathrm{i}_1 < \mathrm{i}_2 < \ldots < \mathrm{i}_{\mathrm{j}-1} \\ \mathrm{i}_1, \ldots, \mathrm{i}_{\mathrm{j}-1} \in \mathrm{I}_\mathrm{j}}} a_{\mathrm{i}_1,\ldots,\mathrm{i}_\mathrm{j}} \,.$$

This is the basis of a recursive solution of the problem of determining the coefficients of the canonical multilinear polynomial form (CMPF) of φ. Initially

$$a_\mathrm{i} = \underset{(1)}{\varphi(0}, \ldots, 0, \underset{(\mathrm{i})}{1}, 0, \ldots, \underset{(\mathrm{n})}{0}) - \varphi(\mathbf{0}) \,, \tag{10.2.4}$$

$$a_{\mathrm{i},\mathrm{j}} = \underset{(1)}{\varphi(0}, \ldots, 0, \underset{(\mathrm{i})}{1}, 0, \ldots, 0, \underset{(\mathrm{j})}{1}, 0, \ldots, \underset{(\mathrm{n})}{0}) - \varphi(\mathbf{0}) - a_\mathrm{i} - a_\mathrm{j} \,, \tag{10.2.5}$$

$$a_{\mathrm{i},\mathrm{j},\mathrm{k}} = \underset{(1)}{\varphi(0}, \ldots, 0, \underset{(\mathrm{i})}{1}, 0, \ldots, 0, \underset{(\mathrm{j})}{1}, 0, \ldots, 0, \underset{(\mathrm{k})}{1}, 0, \ldots, \underset{(\mathrm{n})}{0}) - \varphi(\mathbf{0}) - a_\mathrm{i} - a_\mathrm{j} - a_\mathrm{k}$$

$$- a_{\mathrm{i},\mathrm{j}} - a_{\mathrm{i},\mathrm{k}} - a_{\mathrm{j},\mathrm{k}} \tag{10.2.6}$$

etc.

In reliability theory, where X_φ is often the output variable (top event) of a fault tree, as a rule, terms of more than 3rd order are negligible, since the corresponding probabilities are negligibly small. Hence one can use a relatively primitive algorithm to implement Eqs. (10.2.4)–(10.2.6). In the following example φ is determined up to and including all 2nd order terms.

Example 10.2.2 (Example 7.2.3 redone)

Table 10.2.3 contains the PASCAL implementation of a very elementary way of solving Eqs. (10.2.3)–(10.2.6). Clearly $a_{\mathrm{i},\ldots}$ is replaced by A[I, . . .]. Since PHI (φ) is Boolean originally (when evaluated) one needs PHII for the integer value of φ.

Output:

A 1, 2 = 1

A 1, 3 = 1

A 2, 3 = 1

Note that the printing of coefficients $a_{i,\ldots}$ that are zero is suppressed. — —

Table 10.2.3. PASCAL program to calculate a 2nd order "approximation" of the CMPF of φ

```
PROGRAM CMPF_APP2 (OUTPUT);
CONST N = 3;
TYPE LISTB = ARRAY[1 .. N] OF BOOLEAN;
VAR I, J, PHII:INTEGER;
    C1:ARRAY[0 .. N]OF INTEGER;           {CI IS AN I-TH ORDER POLYNOMIAL}
    C2:ARRAY[1 .. N, 2 .. N]OF INTEGER;  {COEFFICIENT}
     X:LISTB;
FUNCTION PHI (X:LISTB)  :BOOLEAN;
BEGIN
PHI:= X[1] AND X[2] OR X[1] AND X[3] OR X[2] AND X[3]
END;
BEGIN
  FOR I:= 1 TO N DO X[I]:= FALSE;
  IF PHI (X) THEN BEGIN C1[0]:= 1; WRITELN ('A0 = 1') END
               ELSE C1[0]:= 0;
  FOR I:= 1 TO N DO
    BEGIN X[I]:= TRUE;
      IF PHI(X)THEN PHII := 1  ELSE PHII:= 0;
      C1[I]:= PHII-C1[0];
      IF C1[I] < > 0 THEN WRITELN('A', I, '=', C1[I]);
      X[I]:= FALSE
    END;
  FOR I:= 1 TO N-1 DO
    BEGIN
      X[I]:= TRUE;
      FOR J:= I+1 TO N DO
        BEGIN X[J]:= TRUE;
        IF PHI (X) THEN PHII:= 1 ELSE PHII:= 0;
        C2[I, J]:= C1[0]-C1[I]-C1[J];
        IF C2[I, J] < > 0 THEN WRITELN ('A', I, ',', J, '=', C2[I, J]) ;
        X[J]:= FALSE
        END;
      X[I]:= FALSE
    END;
END.
```

Finally, one should never forget that "approximating" Boolean functions, e.g. by cutting off higher order terms of a CMPF, is by itself nonsense. For instance, in the last example cutting off terms of 3rd (and higher) order yields

$$\varphi^*(X_1, X_2, X_3) = X_1 X_2 + X_1 X_3 + X_2 X_3$$

with

$$\varphi^*(1, 1, 1) = 3 \ ,$$

which is certainly in no way an "approximately Boolean" value. It is only in connection with probabilities as in §8.3 that cutting off higher order terms of a DNF can make sense.

10.3 Probability of a Given Value of a Boolean Function

By Eq. (7.2.2a) the Boolean function (describing e.g. a technical system with n components) is

$$X_\varphi = \varphi(X_1, \ldots, X_n) = \sum_{i \in I_\varphi} \hat{M}_i \;, \tag{10.3.1}$$

where $\hat{M}_i$ is minterm i (by any numbering) and I_φ is the set of the indices of the minterms of φ. Further, with simplified indexing, let

$$\hat{M}_i = \bigwedge_{k=1}^{n} \tilde{X}_k \;; \quad \tilde{X}_k \in \{X_k, \bar{X}_k\} \tag{10.3.2}$$

let Q_i be the probability of an elementary system 1-state:

$$Q_i := P\{\hat{M}_i = 1\} \;, \tag{10.3.3}$$

and let, as in Eq. (8.1.11), p_i be the probability of the 1-state of component i:

$$p_i := P\{X_i = 1\} \;, \quad i \in \{1, \ldots, n\} \;. \tag{10.3.4}$$

Then, with simplified indexing, for s-independent $X_1, \ldots, X_n$

$$Q_i = \prod_{k=1}^{n} \tilde{p}_k \;, \quad \tilde{p}_k = \begin{cases} p_k \;, & \text{for } \tilde{X}_k = X_k \\ \bar{p}_k = 1 - p_k \;, & \text{for } \tilde{X}_k = \bar{X}_k \;. \end{cases} \tag{10.3.5}$$

Now, the algorithm "SUMPES" computes p_φ, the probability of $X_\varphi = 1$ as the sum of the probabilities of the elementary states with $X_\varphi = 1$:

$$p_\varphi = \sum_{i \in I_\varphi} Q_i \tag{10.3.6}$$

Table 10.3.1. PASCAL version of procedure "PROBMIN"

```
CONST N= . . . ;
TYPE LIST1B = ARRAY[0 . . N] OF BOOLEAN;
      LISTR = ARRAY[1 . . N] OF REAL;
PROCEDURE PROBMIN (X:LIST1B; P:LISTR; VAR Q: REAL);
VAR I:INTEGER;
BEGIN
  Q:= 1;
  FOR I:=1 TO N DO
        IF X[I] THEN Q:= Q*P[I]
                ELSE Q:= Q*(1-P[I])
END;
```

for given $p_1, \ldots, p_n$. (In reliability theory p_i is the unavailability of component i and p_φ is system unavailability.) Figure 10.3.1 shows the flowchart of the whole algorithm.

In Fig. 10.3.2 one can find the details of the procedure PROBMIN (probability of a minterm). (The procedure MINDET is discussed amply in §10.2.1.) Table 10.3.1 contains a PASCAL program for Fig. 10.3.2.

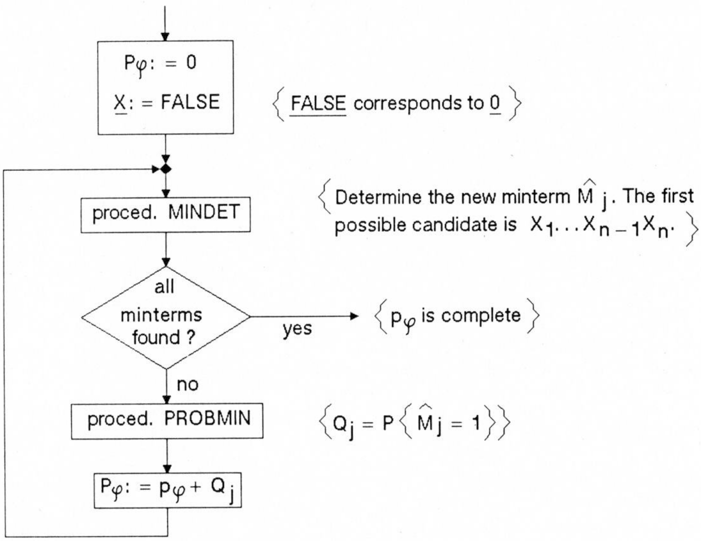

Fig. 10.3.1. Flowchart of a possible procedure SUMPES

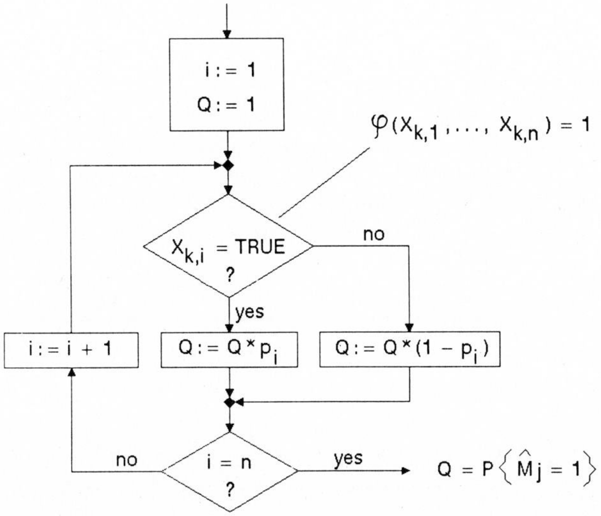

Fig. 10.3.2. Flowchart of the procedure PROBMIN

Discussion. The above approach to the determination of p_φ is not pursued in more detail since in §10.4 far more efficient alternatives will be presented. Rather, in this section the first two Bonferroni inequalities, Eq. (8.3.6), yielding an upper and a lower bound, respectively, will be implemented.

An Upper Limit

Given

$$\varphi(\mathbf{X}) = \bigvee_{i=1}^{m} \hat{T}_i\,, \quad \hat{T}_i = \bigwedge_{j=1}^{n_i} \tilde{x}_{l_{i,j}}\,, \quad l_{i,j} \in \{1, \ldots, n\},\ \tilde{X}_k \in \{X_k, \bar{X}_k\}$$

an $\underline{\text{upper}}$ $\underline{\text{l}}\text{imit}$ for $P\{X_\varphi = 1\}$ is (for s-independent X_j),

$$p_{ul} := \sum_{i=1}^{m} \prod_{j=1}^{n_i} \tilde{p}_{l_{i,j}}\,. \tag{10.3.7}$$

The PASCAL notation for this is that of Table 10.3.2.

Table 10.3.2 Computing a polynomial value for Eq. (10.3.7)

```
PUL:= 0;
FOR I:= 1 TO M DO
    BEGIN PP:= 1;
          FOR J:= 1 TO N[I] DO PP:= PP*PL[I, J];
          PUL:= PUL+PP
    END
```

A Lower Limit

The determination of the $\underline{\text{l}}\text{ower}$ $\underline{\text{l}}\text{imit}$

$$p_{ll} := p_{ul} - \sum_{i=1}^{m-1} \sum_{j=i+1}^{m} P\{\hat{T}_i \hat{T}_j = 1\} \tag{10.3.8}$$

is not quite as trivial as that of p_{ul}. In Table 10.3.3 the value of the double sum of Eq. (10.3.8) is printed as "MAXIMUM ERROR". Note that if $\hat{T}_i$ and $\hat{T}_j$ have common literals $\tilde{X}_k$, prior to calculating $P\{\hat{T}_i \hat{T}_j = 1\}$ by replacing in $\hat{T}_i \hat{T}_j$ every X by p, the idempotence rule

$$\tilde{X}_k \tilde{X}_k = \tilde{X}_k$$

must be applied.

Comments. (1) The program of Table 10.3.3 implicitly assumes that all $\tilde{p}_k = p_k$. If this is not the case, generalizations will be obvious from §10.5. (2) The problem of applying the idempotence rule is solved by asking if an index k appears in the index set S[I] of $\hat{T}_i$ or in the index set $S[J]$ of $\hat{T}_j$.

Table 10.3.3. Difference between an upper and a lower bound on p_φ

```
PROGRAM MAXERROR (INPUT, OUTPUT);
CONST N = ...;  M = ...; {THERE ARE N VARIABLES AND M TERMS}
TYPE INDEXSET = SET OF 1 ... N; {SET OF INDICES OF A TERM TI}
VAR I, J, K:INTEGER; Q, DIF:REAL;
        P: ARRAY [1 .. N] OF REAL; {P[I] = PROB. OF X[I] = 1}
        S: ARRAY [1 .. M] OF INDEXSET;
BEGIN
....... {INPUT P AND S}
   DIF:= O;
   FOR I:= 1 TO M-1 DO
      FOR J:= I+1 TO M DO
            BEGIN
               Q := 1;
               FOR K:= 1 TO N DO
                           IF K IN (S[I] + S[J]) THEN
                           Q := Q*P[K];
               DIF := DIF + Q
            END;
   WRITE ('MAXIMUM ERROR = ', DIF:10)
END.
```

10.4 Algorithms for Making the Terms of a Given DNF Disjoint

§§8 and 9 give several good reasons for a strong interest in multilinear polynomial forms of Boolean functions. As was shown in §7, also DDNFs can be regarded (after replacing $\vee$ by $+$) as such polynomials.

In the last few years many algorithms have been proposed – mostly in the IEEE Trans. Reliability – for producing DDNFs. The most promising three of these are implemented in §§4.4.2, 3, 4. Only in the case of the set-addition algorithm of §4.4.2 is the starting point a (any) DNF of the given $\varphi(\mathbf{X})$; in the other two cases any preprocessing to get a DNF is superfluous.

The determination of the average case computational complexity of the algorithms of that section is all but trivial, because, depending on the field of application, the type of "typical" Boolean function may vary considerably. Hence, questions of computational complexity are not discussed at any length here; see [39], [40].

As to the practical application of a DDNF, remember that – with s-independent X_is –

$$p_\varphi = P\{\varphi = 1\}$$

is found by replacing (in a DDNF of φ) any X_i by p_i and $\bar{X}_i$ by $1-p_i$. Further, to find the frequency of changes of the value of φ from j to $1-j, j\in\{0, 1\}$ one can replace any term $X_i \bar{X}_j \ldots$ of a DDNF by

$$p_i(1-p_j) \ldots (1/E\{D_{0,1,i}\} - 1/E\{D_{1,0,j}\} + - \ldots) .$$

In this section only a fairly detailed implementation of the binary decision tree algorithm will be presented since the other two of the above mentioned algorithms have been reported on elsewhere:

- the set addition algorithm in the original paper [6] of ABRAHAM,
- the Shannon decomposition algorithm in [41] by CORYNEN, and in [33].

The reader is warned not to underestimate the implementation problems implied since not only the main ideas must be implemented but also certain obvious optimizations, lest one runs, even for relatively low n (number of variables) and/or m (number of terms of polynomials), into severe computational complexity problems.

Implementing the Binary Decision Tree Algorithm of §4.4.4

First of all, in order to be able to use the "ternary evaluation procedure" at the end of §10.1, the Boolean function $\varphi(\mathbf{X})$, for which a short DDNF is to be found, is supposed to be given as a DNF (in an external integer file) with the notation

O for OR ($\vee$) and $+i$ for X_i as well as $-i$ for $\bar{X}_i$

Table 10.4.1. PASCAL program DECISION_TREE for the decision tree algorithm to transform $\varphi(X_1, \ldots, X_n)$ from a DNF to a DDNF

```
PROGRAM DECISION_TREE (INFILE, OUTPUT);
CONST N= ...; {N2=2N; X(N+K) IS X(K) NEGATED}
TYPE SETI=SET OF 1 .. N2; {INDEX SET}
     LIST=↑NODE;
     NODE=RECORD
               TERM: SETI; {INDICES OF DNF TERM}
               CARD: INTEGER; {LENGTH OF A TERM}
               AFTER: LIST
          END;
     LITERALS=ARRAY[1 .. N] OF INTEGER; {INDICES OF ALL LITERALS}
VAR START: LIST;
     I, DIMENSION: INTEGER;
     ORDER: LITERALS; {VARIABLES ARE GIVEN VALUES}
                      {0 OR 1 IN THE ORDER OF THIS}
                      {ARRAY 'ORDER'}
     ARGUMENT: LITERALS;  {ARGUMENT N-VECTOR WITH}
                          {COMPONENTS 0, OR 1, OR 3†}
PROCEDURE INREAD; {READING EXTERNAL DNF-FILE AND}
                  {PRODUCING A SIMPLY CONCATENATED LIST OF}
                  {INDEX SETS OF RISING LENGTH}
     VAR INFILE: FILE OF INTEGER;
          ALLOCATE: LIST;     {NEW ELEMENT OF LIST}
          BEFORE: LIST;       {PREDECESSOR OF NEW ELEMENT}
          NUMBER: INTEGER;    {NUMBER IN THE FILE}
          LENGTH: INTEGER;    {LENGTH OF TERM}
          LITSET: SETI;       {SET OF LITERAL INDICES IN A TERM}
```

For example, $X_1X_2 \vee X_1\bar{X}_2X_3 \vee \bar{X}_1X_2X_3$ appears in the input file as

$+1+2\text{o}+1-2+1\text{o}-1+2+3\text{o}$

with a final zero for closing-up.

Furthermore, it is assumed that all indices of $\mathbf{X}=(X_1, \ldots, X_n)$ appear in the DNF, otherwise some trivial renumbering should be practised.

The detailed program of Table 10.4.1 is based on the following main ideas, which determine the main steps of the underlying algorithm:

(1) The process of assigning 0s or 1s to the different X_i should tend to yield $X_\varphi = 1$, i.e. $\hat{T}_j = 1$ as soon as possible, since in this way short terms of the DDNF are found. Hence, the procedure INREAD (not READ, which is a PASCAL standard procedure) transforms the input file to a linked list of index sets (of DNF terms) of rising cardinality.

(2) It facilitates programming if the decision variables; i.e. those variables, whose values are 0 for the left-hand son and 1 for the right-hand son, respectively, only depend on the depth of the decision tree. An optimal sequencing of these variables (in the spirit the preceding step 1) is fixed in the procedure SUCCESSION.

Table 10.4.1. (continued)

```
PROCEDURE SEARCH (VAR BEFORE: LIST; LENGTH: INTEGER);
                  {ORDERING THE INPUT FILE}
   LABEL 1;
   BEGIN
      BEFORE:= NIL;
      IF START <> NIL THEN
        IF START↑. CARD <= LENGTH THEN
          BEGIN
               BEFORE:= START;
               WHILE BEFORE↑. AFTER <> NIL DO
                  IF BEFORE↑. AFTER↑. CARD <= LENGTH
                    THEN BEFORE:= BEFORE↑.AFTER
                    ELSE GOTO 1
          END
1: END;
BEGIN {MAIN BODY OF PROCEDURE INREAD}
   RESET (INFILE, 'INTO.DAT'); {NON-STANDARD PASCAL PROCEDURE}
                               {WHICH CONNECTS}
                               {'INTO.DAT' WITH 'INFILE'.}
                               {INFILE IS READ LATER ON.}
   START:= NIL;
   LENGTH:= 0;
   LITSET:= [ ];
   DIMENSION:= 0;
   BEFORE:= NIL;
   WHILE NOT EOF (INFILE) DO
      BEGIN
         READ (INFILE, NUMBER); {PASCAL STANDARD PROC.}
```

Table 10.4.1. (continued)

```
            IF NUMBER > 0
            THEN BEGIN {ADD NEW LITERAL FROM FILE}
                        LITSET := LITSET + [NUMBER];
                        LENGTH := LENGTH + 1;
                        IF (NUMBER < = N) AND (NUMBER > DIMENSION)
                           THEN DIMENSION := NUMBER
                           ELSE IF (NUMBER-N > DIMENSION)
                                     THEN DIMENSION := NUMBER-N
                 END
              ELSE BEGIN {INSERT NEW TERM IN LIST}
                        NEW (ALLOCATE); {PASCAL STANDARD PROCEDURE}
                        ALLOCATE↑. TERM := LITSET;
                        ALLOCATE↑. CARD := LENGTH;
                        SEARCH (BEFORE, LENGTH);
                        IF BEFORE = NIL
                             THEN BEGIN {INSERT AT THE HEAD}
                                        ALLOCATE↑. AFTER := START;
                                        START := ALLOCATE
                                  END;
                             ELSE BEGIN {INSERT BEHIND BEFORE}
                                        ALLOCATE↑. AFTER := BEFORE↑.AFTER
                                        BEFORE↑. AFTER := ALLOCATE
                                  END;
                             LITSET := [  ];
                             LENGTH := 0
                        END
   END {FILE IS TRANSFORMED TO LINKED LIST}
END; {PROCEDURE INREAD}
```

(3) For the ternary evaluation of the DNF φ the procedure FUNEVA of Table 10.1.10 is used, though not verbally. The relevant procedure has the name TREE.

Discussion. In the above step (2), for algorithmic simplicity no vertex(node)-wise optimization to limit the breadth of the decision tree is implemented. This yields suboptimal results, i.e. DDNF's which should be shortened by merging the terms according to

$$T_i X_j + T_i \bar{X}_j = T_i \ .$$

By applying the relevant procedures at the end of §10.5 this can be done additive to the procedure of Table 10.4.1. Beyond the intention to keep this text on a medium level of programming refinement, it should not be forgotten that nodewise optimization can cost a lot of time if FUNEVA would have to run many times.

It goes without saying that the DNF of φ should be as short as possible as to merging according to

$$T_i X_j \vee T_i \bar{X}_j = T_i \ , \tag{10.4.1}$$

Table 10.4.1 (continued)

```
PROCEDURE SUCCESSION (VAR ORDER: LITERALS);
{DETERMINING THE SEQUENCE OF VARIABLES TO BE SPECIFIED IN THE TREE}
    VAR ALL: SET OF 1 .. N;
        POINTER: LIST;
        I: INTEGER;     {INDEX OF VARIABLE}
        PLACE: INTEGER {SPECIFIES POSITION IN THE ARRAY 'ORDER'}
    BEGIN
      POINTER:= START;
      PLACE:= 1;
      ALL:= [1 .. DIMENSION];
      WHILE (POINTER < > NIL) AND (ALL < > [  ]) DO
          BEGIN
              FOR I:= 1 TO DIMENSION DO
                  IF ((I IN POINTER↑. TERM) OR       {XI IN TERM}
                      (I +N IN POINTER↑. TERM))  {NOT XI IN TERM}
                      AND (I IN ALL) THEN            {XI NOT YET IN ARRAY 'ORDER'}
                      BEGIN
                        ALL:= ALL−[I];†
                        ORDER[PLACE]:= I;
                        PLACE:= PLACE+1
                      END;
              POINTER:= POINTER↑. AFTER
          END
    END; {PROCEDURE SUCCESSION}
```

Table 10.4.1 (continued)

```
PROCEDURE TREE (ARGUMENT: LITERALS; DEPTH: INTEGER)
          {RECURSIVE TREE CONSTRUCTION}
  VAR POINTER: LIST;
      T, I, PHI: INTEGER; {T FOR THE VALUE OF A TERM}
      D: BOOLEAN; {D FOR DON'T CARE}
  BEGIN
    POINTER:= START;
    D:= FALSE;
    PHI:= 0;
    WHILE (POINTER < > NIL) AND (PHI = 0) DO
    {TERNARY EVALUATION IN THE SPIRIT OF FUNEVA OF SECTION 10.1}
      BEGIN
        T:= 1;
        FOR I:= 1 TO N DO
            BEGIN
              IF I IN POINTER↑. TERM THEN
                  T:= T*ARGUMENT[I];
              IF I+N IN POINTER↑. TERM THEN
                  T:= T*(1−ARGUMENT[I]);
            END;
        IF T=1 THEN PHI:= 1 ELSE
            BEGIN
              IF T < > 0 THEN D:= TRUE;
              POINTER:= POINTER↑. AFTER
            END
    END;
    IF (PHI=0) AND (D=TRUE) THEN PHI:= 3;
```

Table 10.4.1 (continued)

```
  {EVALUATION OF VALUE OF PHI; PHI=0 IS DISREGARDED}
  IF PHI=1 THEN
      BEGIN
          FOR I:=1 TO DIMENSION DO {PRINT TERM OF DDNF}
              IF ARGUMENT [I]=1
                THEN WRITE['+', I]
                ELSE IF ARGUMENT[I]=0 THEN WRITE[-I];
          WRITELN
      END;
  IF PHI=3 THEN {FURTHER DECISION IS NECESSARY}
      BEGIN
          ARGUMENT [ORDER[DEPTH]]:=0; {LEFT HAND SON}
          TREE (ARGUMENT, DEPTH+1);
          ARGUMENT [ORDER[DEPTH]]:=1; {RIGHT HAND SON}
          TREE (ARGUMENT, DEPTH+1)
      END
  END; {PROCEDURE TREE}
BEGIN {MAIN BODY OF PROGRAM DECISION_TREE}
  INREAD;
  FOR I:=1 TO N DO {INITIALIZATION}
      BEGIN
          ORDER[I]:=0;
          ARGUMENT[I]:=3
      END;
  SUCCESSION (ORDER);
  TREE (ARGUMENT, 1) {START OF RECURSION AT DEPTH 1}
END.
```

† 3 for don't care to denote a non-specified variable value, as in procedure FUNEVA in §10.1.
‡ Clearly, one could also search the array 'ORDER', if a new index I was put there earlier. However, searching the set ALL for novel I, which are not yet deleted from ALL is simpler (to formulate).

and to absorption according to

$$T_i \vee T_i T_j = T_i \ . \tag{10.4.2}$$

However, the program of Table 10.4.1 also works with sub-optimal DNFs. The following example will ease the understanding of the details of Table 10.4.1.

Example 10.4.1 (Example 4.4.2 redone)

from Eq. (4.4.21)

$$\varphi = X_6X_7 \vee X_1X_2X_3 \vee X_1X_4X_7 \vee X_3X_5X_6 \vee X_1X_2X_5X_7 \vee X_1X_3X_4X_5 \vee X_2X_3X_4X_6 \ . \tag{10.4.3}$$

The input file is shaped to become

+6+7o+1+2+3o+1+4+7o+3+5+6o+1+2+5+7o+1+3+4
+5o+2+3+4+6o

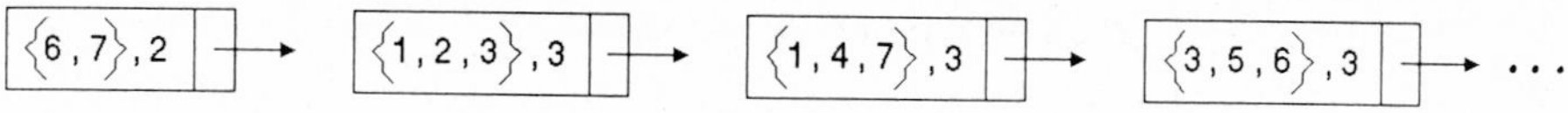

Fig. 10.4.1. Linked list of sets whose cardinalities are added as another parameter; see also last part of §10.5

By the procedure INREAD this is transformed according to Fig. 10.4.1. By the procedure SUCCESSION the following sequence of variables to be fixed to 0 or 1 is determined:

$$X_6, X_7, X_1, X_2, X_3, X_4, X_5 \ .$$

Note that the first four terms of the DNF suffice to fix this sequence. Details of the decision tree are given (in the notation of Fig. 3.2.1) in Fig. 10.4.2. The terms of the resulting DDNF, which correspond to the 1-nodes in Fig. 10.4.2, are printed as

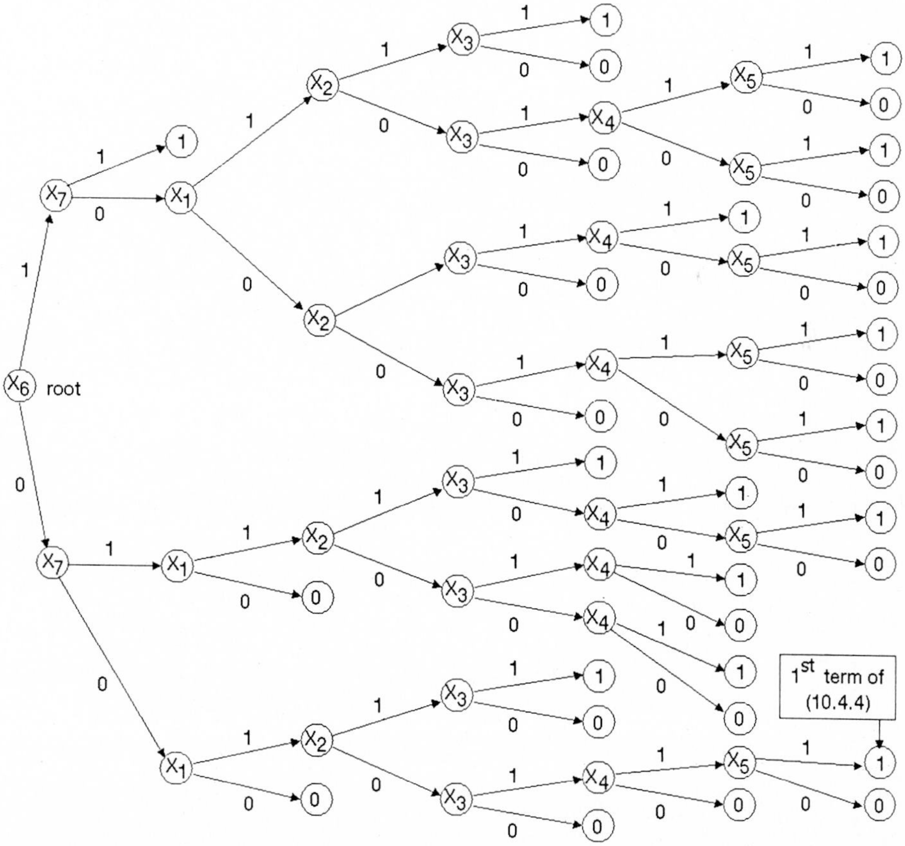

Fig. 10.4.2. Decision tree example according to the program DECISION_TREE

given in Table 10.4.2. The corresponding DDNF is

$$X_\varphi = X_1\bar{X}_2X_3X_4X_5\bar{X}_6\bar{X}_7 \leftarrow \text{first line of Tab. 10.4.1}$$
$$\vee \ldots \vee X_6X_7 \,. \tag{10.4.4}$$

Note that in Fig. 10.4.2 a 1-edge resp. an 0-edge originating in node X_i corresponds to X_i resp. $\bar{X}_i$ in the DDNF, if they lie on a path from the root (X_6) to a 1-type leaf. The price for the relative simple-mindedness of the program DECISION_TREE is an appreciable increase in the length of the DDNF. Equation (10.4.4) consists of 15 terms including 7 minterms, whereas the DDNF of Fig. 4.4.10 consists of only 9 terms including only 3 minterms.

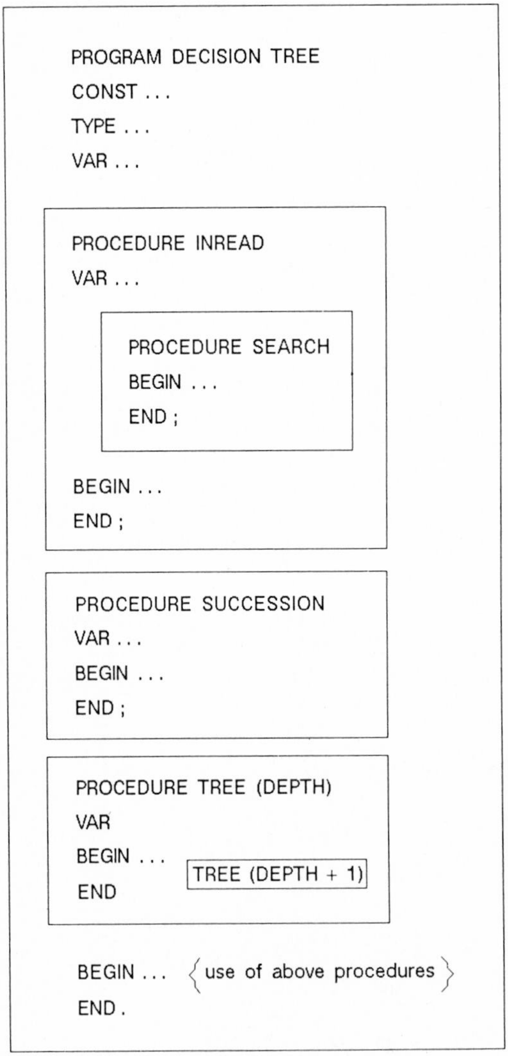

Fig. 10.4.3. Main blocks of Table 10.4.2

Table 10.4.2. Output of program DECISION_ TREE for example 10.4.1

+1	−2	+3	+4	+5	−6	−7	
+1	+2	+3	−6	−7			
+1	−2	−3	+4	−6	+7		
+1	−2	+3	+4	−6	+7		
+1	+2	−3	−4	+5	−6	+7	
+1	+2	−3	+4	−6	+7		
+1	+2	+3	−6	+7			minterms
−1	−2	+3	−4	+5	+6	−7	
−1	−2	+3	+4	+5	+6	−7	
−1	+2	+3	−4	+5	+6	−7	
−1	+2	+3	+4	+6	−7		
+1	−2	+3	−4	+5	+6	−7	
+1	−2	+3	+4	+5	+6	−7	
+1	+2	+3	+6	−7			
+6	+7						

In order to understand Table 10.4.1, its rough block structure is given in Fig. 10.4.3. The recursiveness of the procedure TREE is indicated by a little block with an incremented parameter value inside the definition block of this procedure.

10.5 Selected Set Manipulations

As a kind of an appendix to the rest of §10 this section is devoted to the implementation of several selected set operations. To get easy-to-use results we will use sets of integers throughout. Integers 1 through N will, usually, refer to the indices of $X_1, \ldots, X_n$ and $N+1$ through $2N$ to $\bar{X}_1, \ldots, \bar{X}_n$. Let us call $\{1, \ldots, 2N\}$ the "fundamental" set.

The Cardinality of a Given Set

Given a set S which could possibly contain some or all of the elements of the fundamental set. How big is card $\{S\}$? The little PASCAL procedure (formally a function) of Table 10.5.1 will produce card $\{S\}$. It is the final value of CARD.

Table 10.5.1. Function "CARDSET" to find the cardinality of a set

```
CONST N2 = . . . ; {N2 EQUALS 2N}
TYPE SETI=SET OF 1 . . N2; {SETI FOR SET OF INTEGERS}
FUNCTION CARDSET(S: SETI): INTEGER;
   VAR CARD, I: INTEGER;
   BEGIN CARD:= 0;
         FOR I:= 1 TO N2 DO
               IF I IN S THEN CARD:= CARD+1;
           CARDSET:= CARD
   END;
```

Finding a Set With the Smallest Number of Elements

Now, given a set of sets, e.g. the index sets of the terms of a DNF (with the index j of an $\bar{X}_j$ transformed to $N+j$), the (a) set with the smallest cardinality is found by the procedure SMALLEST of Table 10.5.2. Obviously the "new" $S[G]$ is a "smallest" set.

Table 10.5.2. Procedure "SMALLEST" to find a set of smallest cardinality in a given set of G sets

```
TYPE ISIS: ARRAY[1 .. G] OF SETI; {ISIS FOR INITIAL SET OF INDEX SETS}
PROCEDURE SMALLEST (S: ISIS; VAR SMALL: SETI);
  VAR I: INTEGER;
      T: ARRAY[1 .. G] OF INTEGER; {FOR CARDINALITIES OF ISIS SETS}
  BEGIN
    FOR I:= 1 TO G DO T[I]:= CARDSET(S[I]);
    FOR I:= 1 TO G-1 DO
      IF T[I] < T[I+1] THEN BEGIN T[I+1]:= T[I]; S[I+1]:= S[I] END;
    SMALL:=S[G]
  END;
```

Deletion of an Element of a Set and of a Set Which Contains a Certain Element

To delete an element I of a set of integers S is trivially described by:

IF I IN S THEN S:= S−[I] .

Also, the deletion of a set (of integers) which contains a given element I is very simple. If the set under consideration should be empty thereafter, this can be implemented by

IF I IN S THEN S:=[ ] .

Table 10.5.3. Procedure "DELSET" (delete a set)

```
PROCEDURE DELSET (J: INTEGER, VAR S: ISIS; VAR M: INTEGER);
  VAR K: INTEGER;
  BEGIN FOR K:= J TO M-1 DO S[K]:= S[K+1];
          S[M]:= [ ];
          M:= M-1
  END;
```

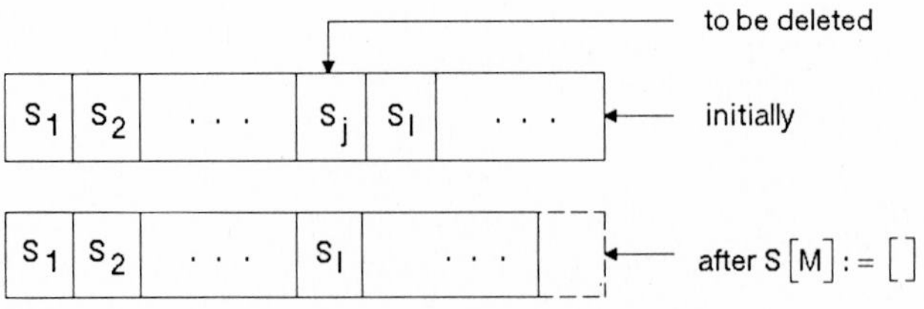

Fig. 10.5.1. Keeping the set of useful sets compact

Otherwise, if S_j is to vanish from an array of m sets $S_1, \ldots, S_m$ one can use procedure DELSET of Table 10.5.3. After the deletion of S_j the remaining sets are shifted one position to the left as shown in Fig. 10.5.1. Now, the deletion of S_j in case it contains i (as an element) is possible via IF I IN S[J] THEN DELSET (J,S,M).

Removing Sets Which Contain Others

Let $S[1\ ..\ M]$ be an array of M sets some of which may be contained in others. The "bigger" sets are to be "weeded out"; see Table 10.5.4. The procedure ABSORP simply clears those sets which contain others. To keep the set of "useful" sets compact and to clear right-hand sets, see Fig. 10.5.1, use can be made of procedure DELSET of Table 10.5.3.

Table 10.5.4. Procedure "ABSORP" to weed out sets which contain others

```
CONST G= ...
TYPE ISIS: ARRAY [1 .. G] OF SETI;
PROCEDURE ABSORB (VAR S: ISIS, VAR M: INTEGER);
  VAR I, J: INTEGER;
  BEGIN
    FOR I:= 1 TO M-1 DO
      FOR J:= I+1 TO M DO
        BEGIN IF (S[I]<=S[J]) AND (S[I]<>[  ]) THEN S[J]:=[  ];
              IF (S[J]<=S[I]) AND (S[J]<>[  ]) THEN S[I]:=[  ];
        END
  END;
```

Sets of Integers Inside Arrays

Some manipulations with sets of integers are studied next, where these sets are stored intelligently, namely in some ordered way in arrays. Specifically, in assembly language, where such an array may be a single word, set manipulations such as searching in sets, determining its cardinality, etc. become simple bit manipulations which can often be performed with a high degree of parallelism using the bitwise Boolean operators *and*, *or* or *exor*.

Keeping to high-level languages such as PASCAL, let the above $2N$ integers 1 through $2N$ be given in single cells of an INTEGER ARRAY of $2N$ components. A cell i contains 0, if i is not in the set to be modelled, i.e. if X_i is not part of the Boolean term to be modelled, and 1 if it is. (For $i=N+1, \ldots, 2N$ X_i is replaced by $\bar{X}_{i-N}$.)

For instance, for $N=5$ the term $X_1\bar{X}_2X_3$ corresponds to the integer set $\{1,3,7\}$ which is now modelled to be the array

$$(1,\ 0,\ 1,\ 0,\ 0,\ 0,\ 1,\ 0,\ 0,\ 0) \tag{10.5.1}$$

$X_1\ X_2\ X_3\ X_4\ X_5\ \bar{X}_1\ \bar{X}_2\ \bar{X}_3\ \bar{X}_4\ \bar{X}_5 \leftarrow$ corresponding literals

$\{1,\ 2,\ 3,\ 4,\ 5,\ 6,\ 7,\ 8,\ 9,\ 10\} \leftarrow$ numbers of positions

The procedure SEARCHAR of Table 10.5.5 describes the identification of the members of the set stored in array AR.

Table 10.5.5. Procedure "SEARCHAR" (search in an array) to identify and print the positions of 1s in an array of 1s and 0s

```
CONST N2 = . . . ; {N2 = 2N}
TYPE ARRAYI = ARRAY[1 . . N2] OF 0 . . 1;†
. . .
PROCEDURE (AR: ARRAYI);
  VAR I: INTEGER;
  BEGIN FOR I := 1 TO N2 DO
        IF AR[I] = 1 THEN WRITELN(I)
  END;
```

The 10-tuple (10.5.1), would yield the output

1
3
7

After this rather simple-minded introduction let us treat the problem of checking for pairs of Boolean functions' terms which can be merged according to

$$\hat{T}_i X_j \vee \hat{T}_i \bar{X}_j = \hat{T}_i \ . \tag{10.5.2}$$

The procedure TRYMERGE 1 assumes that two $2N$-tuples of 0s and 1s are given in the arrays ARRAYI1 and ARRAYI2, with the single elements being AR1 [I] and AR2 [I], respectively. The main point of interest is to stop the comparison as soon as possible. Looking at the example of merging according to

$$X_1 \bar{X}_2 X_3 \vee X_1 X_2 X_3 = X_1 X_3$$

with (for a Boolean function of five variables as in (10.5.1))

$$\mathrm{AR1} = (1, \underline{0}, 1, 0, 0, 0, \underline{1}, 0, 0, 0) \mathrel{\hat{=}} X_1 \bar{X}_2 X_3 \ ,$$

$$\mathrm{AR2} = (1, \underline{1}, 1, 0, 0, 0, \underline{0}, 0, 0, 0) \mathrel{\hat{=}} X_1 X_2 X_3 \ ,$$

it is obvious, how the comparison can be implemented. Details are given in Table 10.5.6. (It is assumed that no single term contains X_i and $\bar{X}_i$.)

Obviously, the variable COUNT counts the HAMMING distance of the two compared $2N$-tuples. For the desired merging this must not only be exactly two but also for a given i $\mathrm{AR1}_i{}^{\dagger} \neq \mathrm{AR2}_i$ and $\mathrm{AR1}_{i+n} \neq \mathrm{AR2}_{i+n}$ must be true. This i, if it exists, is identified by the variable DEL. Finally, the corresponding X_i or $\bar{X}_i$ is deleted from the term whose model is in AR1, and AR2 is cleared.

It is worth mentioning that the merging of two terms as in Eq. (10.5.2) can be accomplished nicely also without the ordering of an array. Let S1 and S2 contain the indices of the literals of $\hat{T}_i X_j$ and $\hat{T}_i \bar{X}_j$, respectively, in the numbering scheme

† Here this is, rather, a Boolean array, even though not Boolean in the PASCAL sense.

Table 10.5.6. Procedure "TRYMERGE1" (try to merge) to test two Boolean terms for mergability according to Eq. (10.5.2)

```
CONST N= . . . ; N2= . . . {N2=2N}
TYPE ARRAYI1, ARRAYI2=ARRAY[1 . . N2] OF INTEGER;
. . . .
PROCEDURE TRYMERGE1 (AR1:ARRAYI1, AR2: ARRAYI2);
  VAR I, COUNT, DEL: INTEGER;
  BEGIN
  I:= 1;
  COUNT:= 0;
  DEL:= 0;
  REPEAT
    IF (AR1[I] <> AR2[I]) THEN
    BEGIN
      COUNT:= COUNT+1;
      IF DEL=0 THEN DEL:= I ELSE IF DEL+N <> I THEN COUNT:= 3;
    END;
    I:= I+1
  UNTIL (COUNT=3) OR (I=N2+1);
  IF COUNT <> 2 THEN WRITELN ('TERMS ARE NOT MERGABLE');
            ELSE BEGIN
                    AR1[DEL] := 0;
                    AR1[DEL+N]:= 0;
                    FOR I: =1 TO N2 DO AR2[I]:= 0
                 END;
  END;
```

Table 10.5.7. Procedure TRYMERGE 2

```
CONST N= . . . ; N2= . . . ,
TYPE SETI=SET OF 1 . . N2;
PROCEDURE TRYMERGE2 (VAR S1: SETI; VAR S2: SETI);
  VAR I: INTEGER;
  BEGIN
    I:= 1;
    REPEAT
    IF (I IN S1) AND ((I+N) MOD N2 IN S2)
        AND (S1-[I]=S2-[(I+N)MOD N2])
      THEN BEGIN
              S1:= S1-[I,(I+N)MOD N2];
              S2:= [ ]
           END
      ELSE I:= I+1
    UNTIL (I > N2) OR (S2=[ ])
  END;
```

(positions) of Eq. (10.5.1). Then the procedure TRYMERGE2 of Table 10.5.7 is not difficult to understand. The vital points are

(1) that in S1 and S2 there must exist a pair of elements I and (I+N)MOD N2, respectively, and
(2) that, except for this pair, S1 and S2 are equal.

In what follows, use will be made of the POINTER type of PASCAL. Let us start with some arguments of motivation.

Because standard PASCAL does not allow for the definition:

TYPE ... = SET OF SET OF ...; sets of sets were introduced above as arrays of sets. But PASCAL offers a better way to realize the SET OF SET construction, viz. the realization as a linear list of sets via pointer types. There are two main reasons why this is advantageous:

(1) Because pointer types are dynamic, one need not specify a maximum number of data elements in the declaration part of the program, as one has to do for arrays. With pointer types one only needs the memory space that one really fills with relevant data.
(2) When inserting or deleting a set in the array of sets, one has to shift many sets, as explained above. With big arrays, this can be a very time-consuming operation, whereas, using pointer types, one only has to change a few pointers.

Regarding these two points and the definition of SETI in Table 10.5.1, the linear list is defined by Table 10.5.8 (see standard text books on PASCAL).

Table 10.5.8. Basic pointer type definition

```
TYPE SETLIST = ↑NODE;
        NODE = RECORD
                   ONESET: SET;
                   AFTER: SETLIST {REST OF THE CONNECTED LIST}
               END;
```

When working with the linear list, one always needs a pointer, which indicates the head of the list:

VAR START: SETLIST;

In the following, let us assume, that this pointer has the right value.
As an example, the set of sets:

$\{\{1,2\}, \{1,3\}, \{2,3\}\}$

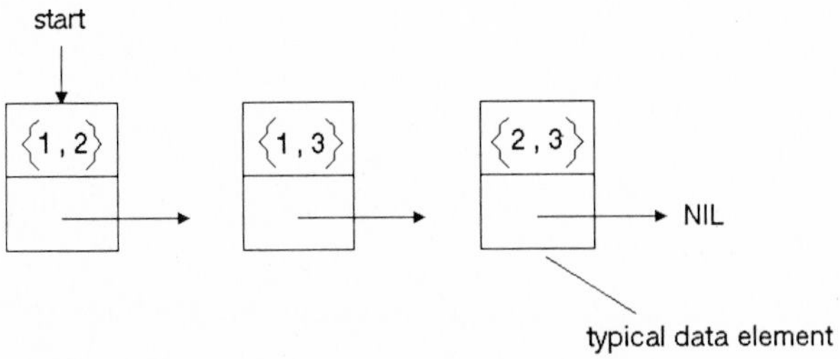

Fig. 10.5.2. Linking sets to become a list of sets

would look as given in Fig. 10.5.2. The single data element consists of the net set and the address of the following data element.

Searching a Special Set

Simulating the set of sets construction, we need procedures to insert, delete and search a single set. Given a linear list, the procedure SEARCH_SET searches for the set, which is specified as SEEK and returns with the pointer PLACE, which indicates the place in the list, where SEEK stands, or returns with NIL, if the set is not found. Details are given in Table 10.5.9. (As defined above, SETI = SET OF 1 . . 2N.)

Table 10.5.9. PASCAL implementation of the procedure "SEARCH_SET"

```
PROCEDURE SEARCH_SET (START: SETLIST; SEEK: SETI; VAR PLACE: SETLIST);
  VAR LOOP: SETLIST; {SEE TAB. 10.5.8}
  BEGIN
    PLACE:= NIL;
    LOOP:= START;
    WHILE (LOOP< > NIL) AND (PLACE = NIL) DO
    BEGIN
      IF LOOP↑.ONESET=SEEK THEN PLACE:= LOOP;
      LOOP:= LOOP↑.AFTER {POINTER IS MOVED TO NEXT LIST ELEMENT}
    END
  END;
```

Inserting a New Set in a Given Set of Sets

It is often useful to have the linear list sorted; with the smallest† set first and the biggest† set last. The procedure INSERT_SET inserts the set INS into the linear list at the appropriate place and changes START if necessary. For that the function CARDSET of Table 10.5.1 is used. For details see Table 10.5.10.

Comment. Using a sorted list is certainly advantageous as compared to searching a smallest‡ set (e.g. via procedure SMALLEST of Table 10.5.2), if several times in succession a smallest set is asked for.

Note that this algorithm is the basis for creating a linked list.

Deleting a Set from a Given List of Sets

The following procedure DELETE_SET deletes the set DEL from a given list. For that purpose the above procedure SEARCH_SET is used to find DEL in the list. It is assumed that a given set appears only once in the list. This is plausible whenever the given list of sets corresponds to the terms of a DNF. Details are noted in Table 10.5.11.

---

† ARj_i is component i (from left) in array j.

‡ As concerns cardinality.

Table 10.5.10. PASCAL version of the procedure "INSERT_SET"

```
PROCEDURE INSERT_SET (INS: SETI; VAR START: SETLIST);
  VAR ALLOCATE, BEFORE, LOOP: SETLIST;
  BEGIN
    NEW (ALLOCATE); {PASCAL STANDARD PROCEDURE}
  ALLOCATE↑. ONESET := INS;
  BEFORE := NIL; {BEFORE IS GOING TO POINT TO INS}
  LOOP := START;
  WHILE LOOP < > NIL DO
      BEGIN
      IF CARDSET (LOOP↑.ONESET) < =CARDSET (INS)
        THEN BEFORE := LOOP;
        LOOP := LOOP↑. AFTER
      END
  IF BEFORE=NIL {INS SHOULD BE THE FIRST LIST ELEMENT}
      THEN BEGIN
             ALLOCATE↑.AFTER:= START;
             START := ALLOCATE
             END
      ELSE BEGIN
             ALLOCATE↑.AFTER := BEFORE↑.AFTER;
             BEFORE↑.AFTER := ALLOCATE
             END
  END;
```

Table 10.5.11. PASCAL version of the procedure DELETE_SET

```
PROCEDURE DELETE_SET (DEL: SETI; VAR START: SETLIST);
  VAR AWAY, BEFORE: SETLIST;
  BEGIN
    AWAY := NIL;
    SEARCH_SET (START, DEL, AWAY);
    IF AWAY < > NIL {AWAY POINTS TO DEL, THE SET TO BE DELETED}
    THEN BEGIN
           IF AWAY = START
           THEN START := AWAY↑. AFTER
           ELSE BEGIN {ISOLATING DEL}
                  BEFORE := START;
                  WHILE BEFORE↑.AFTER < > AWAY DO BEFORE:=BEFORE↑.AFTER;
                  BEFORE↑. AFTER := AWAY↑. AFTER
                  END
           DISPOSE (AWAY) {PASCAL STANDARD PROCEDURE}
           END
    ELSE WRITELN ('DEL NOT IN LIST')
  END;
```

Comment. DELETE_SET should not be mixed-up with DELSET of Table 10.5.3. Beyond the difference between the data types, in DELSET the position (in an ARRAY) of the set to be deleted is supposed to be known.

Operations on the DNF

With the last three procedures some of the earlier algorithms of section 10.5 can be implemented in a more elegant and efficient way. For example, searching the smallest set is now trivial, because the list is sorted.

To employ the absorption-rule, the procedure ABSORB_SET is needed. Since a given set is to be absorbed by any of its subsets, here the procedure DELETE_SET of Table 10.5.11 is applied to delete a given set, if one of its subsets is found. More details are given in Table 10.5.12. Note that the typical one-step forward movement of a pointer is expressed by

POINTER := POINTER↑. AFTER (see Table 10.5.8 for AFTER) .

Mergability of terms in the DNF is examined by the procedure MERGE_SET, where N is (again) the number of Boolean variables, see Table 10.5.13.

Comment. The procedure MERGE_SET is, at its core, based on the procedure TRYMERGE 2 of Table 10.5.7. However, it is not only meant for a single pair of Boolean terms, but rather for a DNF. More details can be found in Table 10.5.13. After using ABSORB_SET AND MERGE_SET on a given DNF, all terms which can be absorbed or merged are removed from the DNF. Since both operations don't influence each other, they can be executed in any order.

Table 10.5.12. PASCAL version of the procedure ABSORB_SET

```
PROCEDURE ABSORB_SET (VAR START: SETLIST);
  VAR LOOP1, LOOP2, AWAY: SETLIST); {LOOP1 AND LOOP2 POINT TO A PAIR OF SETS}
  BEGIN                           {WHICH ARE COMPARED TO CHECK IF ONE IS}
    LOOP1 := START;               {A SUBSET OF THE OTHER!               }
    WHILE LOOP1 <> NIL DO
    BEGIN
      LOOP2 := START
      WHILE LOOP2 <> NIL DO
        IF (LOOP1↑.ONESET <= -LOOP2↑.ONESET) AND (LOOP1 <> LOOP2)
        THEN BEGIN
              AWAY := LOOP2; {AWAY IS POINTER FOR DELETION}
              LOOP2 := LOOP2↑.AFTER;
              DELETE_SET (AWAY↑.ONESET, START)
              END
        ELSE LOOP2 := LOOP2↑.AFTER;
      LOOP1 := LOOP1↑.AFTER
  END
END;
```

Table 10.5.13. PASCAL program for the procedure "MERGE_SET"

```
CONST N = . . . ;
PROCEDURE MERGE_SET (VAR START: SETLIST);
  VAR LOOP1, LOOP2, AWAY: SETLIST;
    X: INTEGER;
    FOUND, FINISH: BOOLEAN;
  BEGIN
    REPEAT
      FINISH:= TRUE;
      LOOP1:= START;
      WHILE LOOP1 < > NIL DO
      BEGIN
        LOOP2:=LOOP1↑. AFTER;
        WHILE LOOP2 < > NIL DO
        BEGIN
          X:= 1;
          FOUND:= FALSE;
          REPEAT
            IF ((X IN LOOP1↑.ONESET) AND (X+N IN LOOP2↑.ONESET)
               AND (LOOP1↑.ONESET-[X]=LOOP2↑.ONESET-[X+N]))
               OR
               ((X+N IN LOOP1↑.ONESET) AND (X IN LOOP2↑.ONESET)
               AND (LOOP1↑.ONESET-[X+N]=LOOP2↑.ONESET-[X]))
            THEN BEGIN
                   AWAY:= LOOP2;
                   FOUND:= TRUE; FINISH:= FALSE;
                   LOOP1↑.ONESET:= LOOP1↑.ONESET-[X,X+N];
                   LOOP2:= LOOP2↑.AFTER;
                   DELETE_SET (AWAY↑.ONESET, START)
                   END
            ELSE X:= -X+1
          UNTIL (X > N) OR (FOUND=TRUE);
          IF FOUND=FALSE THEN LOOP2:=LOOP2↑.AFTER
        END;
      LOOP1:=LOOP1↑. AFTER
      END
    UNTIL FINISH=TRUE
  END;
```

Exercises

10.1

Give a simple algorithm (and a PASCAL implementation for it) to check if two Boolean functions φ_1 and φ_2 of n variables are equal.

10.2

Find (and implement in PASCAL) an efficient[†] algorithm for the production of all the code words of a k-out-of-n equal weight code. More precisely:

[†] Not necessarily in the strict sense of algorithm complexity theory.

(a) Give a very simple program for the case $k=3$, $n=5$, including the printing of a table such as Table 10.1.9.
(b) Give an elegant solution on the basis of a recursive procedure for nesting FOR-loops for general k and n, without bothering about printing, just producing $\binom{n}{k}$ binary n-tuples with k 1s and $n-k$ 0s each, or, at least, addresses (positions) of the k 1s.

10.3

To factor out common literals in a polynomial (or a DNF) of the variables $X_1, \ldots, X_n$ the set S of common indices must be known.

(a) Determine S under the assumption that the m polynomial (DNF) terms $\hat{T}_1, \ldots, \hat{T}_m$ are stored as binary arrays of equal length $2n$.
(b) Give a simple hand-calculated example!

11 Appendix: Probability Theory Refresher

In this text only very elementary concepts of stochastics are used. Nevertheless, to make this text as self-contained as possible, a few basic results of the applied theory of stochastics (theories of random events, variables, and processes) are reviewed, mostly without proofs in the framework of measure theory, but just made plausible by the pseudo heuristics usual and useful in engineering and operations research. Good references for engineers are [26], [27], [28], [54].

As an addendum to section 11.4 the very last section, 11.5, contains a short refresher of the Laplace transform. Except for renewal theory the Laplace transform is also used to solve linear differential equations appearing in the Markov model.

In order to intensify this short discourse on basic stochastics, some exercises are also added to this section.

11.1 Boolean Algebra of Sets

In probability calculus sets of *random events* a, b, ... or a_1, a_2, ... are of some importance; e.g. in reliability theory a typical random event is

$$a_i \triangleq \text{“component } i \text{ is in a failure (faulty) state”.}$$

Combined events can be described by the operations of
union:

$$a_i \cup a_j \quad (\text{event } a_i \text{ OR event } a_j) \ ,$$

intersection:

$$a_i \cap a_j \quad (\text{event } a_i \text{ AND event } a_j) \ ,$$

and *complementation* (with respect to the “sure” event Ω):

$$\bar{a}_i \quad (\text{NOT event } a_i) \ .$$

For these 3 fundamental set operations the laws of Boolean algebra hold; specifically (see §2) the laws of

commutation:

$$a_i \cap a_j = a_j \cap a_i \ , \qquad a_i \cup a_j = a_j \cup a_i \ ,$$

idempotence (no axiom):

$$a_i \cap a_i = a_i \ , \qquad a_i \cup a_i = a_i \ ,$$

association (no axiom):

$$a_i \cap (a_j \cap a_k) = (a_i \cap a_j) \cap a_k \ ,$$

$$a_i \cup (a_j \cup a_k) = (a_i \cup a_j) \cup a_k \ ,$$

distribution:

$$a_i \cap (a_j \cup a_k) = (a_i \cap a_j) \cup (a_i \cap a_k) \ ,$$

$$a_i \cup (a_j \cap a_k) = (a_i \cup a_j) \cap (a_i \cup a_k) \ ,$$

existence of unit elements (events):

$$\Omega \cap a_i = a_i \ , \qquad \bar{\Omega} \cup a_i = a_i \ , \qquad \text{with} \qquad \bar{\Omega} = \phi \text{ being the “impossible” event} \ ,$$

existence of the inverse element (event):

$$a_i \cap \bar{a}_i = \phi \ , \qquad a_i \cup \bar{a}_i = \Omega \ .$$

De Morgan's law (no axiom)

$$\overline{a \cap b} = \bar{a} \cup \bar{b}$$

is very plausible: If the joint event $a \cap b$ does not happen, then either one of them (or both) do not happen. Using a venn diagram, the absorption law

$$a \cup (a \cap b) = a$$

is easily verified.

The above algebra of sets is very useful in connection with positive set functions

$$q(a_i) \geqq 0 \ .$$

The Venn diagram of Fig. 11.1.1 reveals that

$$q(a_i \cup a_j) = q(a_i) + q(a_j) - q(a_i \cap a_j) \ . \tag{11.1.1}$$

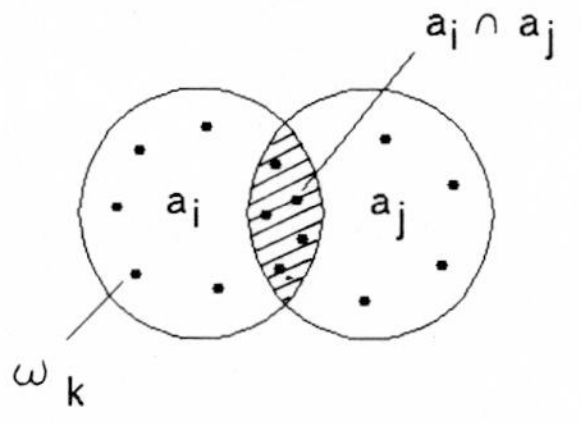

Fig. 11.1.1. Venn diagram of two intersecting sets of elements ω_k, e.g. points of the plane

Obviously, because $a_i \cap a_j$ is part of both a_i and a_j, $q(a_i \cap a_j)$ must be subtracted from $q(a_i)+q(a_j)$.

Figure 11.1.2 yields immediately, by the same principle of inclusion and exclusion of sets,

$$q(a_i \cup a_j \cup a_k) = q(a_i)+q(a_j)+q(a_k)-q(a_i \cap a_j)-q(a_i \cap a_k)-g(a_j \cap a_k) + q(a_i \cap a_j \cap a_k) \ . \qquad (11.1.2)$$

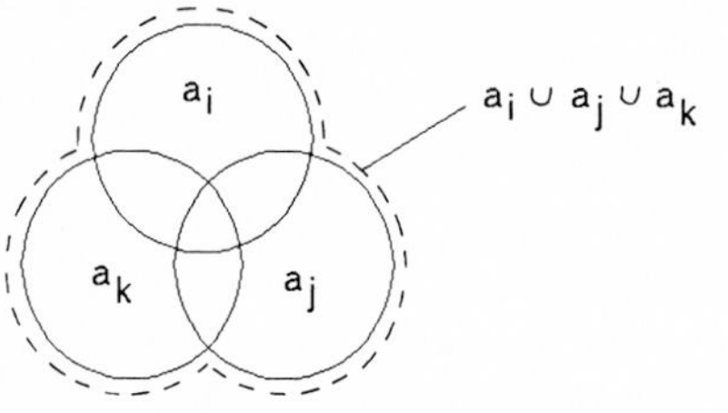

Fig. 11.1.2. Union of three sets

Hint. The interested reader will derive this result easily from Eq. (11.1.1) by replacing a_j by $a_j \cup a_k$.

11.2 Elementary Probability Calculus

For practical applications, especially with a finite set of random events *a*, *b*, *c*, . . . or a_i, $i=1, \ldots, n$; with

$$\Omega = \bigcup_{i=1}^{n} a_i \qquad (11.2.1)$$

the "sure" event, it is good engineering practice to "define" the *probability* $P\{a_i\}$ of a random event a_i as the limiting value of the relative frequency (count) of the appearance of a_i in N independent trials:

$$P\{a_i\} = \lim_{N\to\infty} \frac{C_N(a_i)}{N} \ , \qquad (11.2.2)$$

where $C_N(a_i)$ is the number of times the random experiments with the possible outcomes $a_1, a_2, \ldots, a_n$ produced the result (event) a_i. Clearly, the limiting operation is problematic; this is one of the central problems of mathematical statistics. In this appendix we will skip over this problem and define a set of axioms to deal with probability calculus. (For much deeper analyses see [26], [29], [30].)

$$0 \leqq P\{a_i\} \leqq 1 \ , \qquad (11.2.3)$$

$$P\{a_i \cup a_j\} = P\{a_i\} + P\{a_j\} \ , \quad \text{if} \quad a_i \cap a_j = \varnothing \ , \qquad (11.2.4)$$

$$P\{\bar{a}_i\} = 1 - P\{a_i\} \ , \quad \bar{a}_i := \Omega \backslash a_i \ , \qquad (11.2.5)$$

$$P(\Omega) = 1 \ , \quad P\{\varnothing\} = 0 \ . \qquad (11.2.6)$$

Since $P\{a\}$ is also a positive set function, Eq. (11.1.1) can be transcribed at once to become

$$P\{a_i \cup a_j\} = P\{a_i\} + P\{a_j\} - P\{a_i \cap a_j\} , \qquad (11.2.7)$$

and Eq. (11.1.2) yields

$$P\{a_i \cup a_j \cup a_k\} = P\{a_i\} + P\{a_j\} + P\{a_k\} - P\{a_i \cap a_j\} - P\{a_i \cap a_k\} - P\{a_j \cap a_k\} + P\{a_i \cap a_j \cap a_k\} . \qquad (11.2.8)$$

In general, as is easy to prove by induction, one has the (Poincaré–Sylvester) sum formula (also known as the *principle of inclusion–exclusion*)

$$P\left\{\bigcup_{i=1}^{m} a_i\right\} = \sum_{i=1}^{m} P\{a_i\} - \sum_{i=1}^{m-1} \sum_{j=i+1}^{m} P\{a_i \cap a_j\} + \sum_{i=1}^{m-2} \sum_{j=i+1}^{m-1} \sum_{k=j+1}^{m} P\{a_i \cap a_j \cap a_k\} - + \ldots + (-1)^{m-1} P\left\{\bigcap_{i=1}^{m} a_i\right\} . \qquad (11.2.9)$$

(A simple proof is by induction, using Eq. (8.3.5).)

For disjoint a_i (also as an extension of Eq. (11.2.4))

$$P\left\{\bigcup_{i=1}^{m} a_i\right\} = \sum_{i=1}^{m} P\{a_i\} . \qquad (11.2.10)$$

Conditional Probability

To explain the concept of *conditional probability* along the heuristic lines of the explanation of "simple" probability, imagine that out of original N trials (yielding event a† or not) those not yielding event b† are deleted. By this selection process, due to the "condition" b,

N changes to $C_N(b)$ and $C_N(a)$ to $C_N(a \cap b)$.

The last change is due to the fact that, under the condition b, only that part of a is "visible" that is within b. Hence the heuristic definition of the conditional probability of a under (the condition) b is

$$P\{a|b\} := \lim_{N\to\infty} \frac{C_N(a \cap b)}{C_N(b)} = \lim_{N\to\infty} \frac{C_N(a \cap b)/N}{C_N(b)/N} .$$

† a and b are supposed to be members of the set $a_1, a_2, \ldots$.

In case the limit of the quotient equals the quotient of the limits, by definition (11.2.2)

$$P\{a|b\}=\frac{P\{a\cap b\}}{P\{b\}}\ , \tag{11.2.11}$$

which is used in axiomatic probability theory as a defining equation for *conditional probability*, so that problems with the limits $N\to\infty$ are precluded.

Definition 11.2.1. If

$$P\{a|b\}=P\{a\}\ , \tag{11.2.12}$$

a and b are said to be *stochastically independent* (of each other)†. — —

An immediate consequence is, by Eq. (11.2.11), the *product rule* for independent events a and b:

$$P\{a\cap b\}=P(a\}\,P\{b\}\ . \tag{11.2.13}$$

Specifically, if for Boolean X_i and X_j, the events a and b are

$$a=(X_i=1)\ ,\quad b=(X_j=1)\ ,$$

we have

$$P\{a\cap b\}=P\{(X_i=1)\cap(X_j=1)\}=P\{X_iX_j=1\}\ .$$

Hence, for $(X_i=1)$ stochastically independent of $(X_j=1)$

$$P\{X_iX_j=1\}=P\{X_i=1\}\,P\{X_j=1\}\ , \tag{11.2.14}$$

and in general

$$P\left\{\prod_i X_i=1\right\}=\prod_i P\{X_i=1\}\ . \tag{11.2.15}$$

One of the most important applications of the concept of conditional probability is within the formula of *total probability*.

Theorem 11.2.1 (Law of total probability).
Let

$$a\subset\bigcup_{i=1}^{m} b_i\ ;\quad b_j\cap b_k=\varnothing\ ;\quad j\neq k\ ,$$

then

$$P\{a\}=\sum_{i=1}^{m} P\{a|b_i\}\,P\{b_i\}\ . \tag{11.2.16}$$

† It is common practice to omit "of each other" and to shorten stochastically to $s-$.

Proof. From the Boolean algebra of sets here

$$a \cap \left(\bigcup_{i=1}^{m} b_i \right) = \bigcup_{i=1}^{m} (a \cap b_i) = a \ .$$

Since

$$(a \cap b_i) \subset b_i \ ,$$

the disjointness of the b_i implies the disjointness of the different

$$a_i := (a \cap b_i) \ .$$

Hence, by Eq. (11.2.10)

$$P\{a\} = \sum_{i=1}^{m} P\{a \cap b_i\} \ .$$

Replacing $P\{a \cap b_i\}$ according to Eq. (11.2.11) results immediately in Eq. (11.2.16), q.e.d.

Remark. In practice the set of b_is of Eq. (11.2.16) is often a partition of Ω, such that

$$P\left\{ \bigcup_{i=1}^{m} b_i \right\} = 1 \ . \qquad (11.2.17)$$

Equation (11.2.17) is called the *normation rule* of probability calculus.

11.3 Random Variables and Random Processes

The sure event Ω and "ordinary" events a_i are the union of elements ω_i the so-called *elementary events*:

$$a_i = \{\omega_j : j \in I_i\} \ , \qquad (11.3.1)$$

where I_i is the proper index set. In the stochastic theory of Boolean functions it is often very useful to take elementary states i.e. Boolean n-tuples $\mathbf{X}_i$ as elementary events ω_i. Often the event $\mathbf{X} = \mathbf{X}_i$ is abbreviated as

$$\mathbf{X}_i = (X_{i,1}, X_{i,2}, \ldots X_{i,n}) =: \omega_i$$

Similarly minterms being 1 are useful elementary events:

$$(\hat{M}_i = 1) =: \omega_i \ .$$

Then the probability of a function φ being 1 is by Eq. (11.2.4) simply the sum of the probabilities of its minterms being 1 individually:

$$P\{\varphi = 1\} = \sum_{i=1}^{m} P\{\hat{M}_{l_i} = 1\} \ ; \quad \hat{M}_{l_i} \in \hat{M}(\varphi) \ ; \quad m = N_{\hat{M}}(\varphi) \ . \qquad (11.3.2)$$

Also the two states 0 and 1 of a single indicator variable X are useful as elementary events so long as only this one variable is investigated.

Definition 11.3.1a. Given Ω and a function $Z: \Omega \to \{z_i: i=1, \ldots, m\}$.
Then Z is a discrete *random variable*, and the linear expression

$$E\{Z\} = \sum_{i=1}^{m} z_i \, P\{Z = z_i\} \tag{11.3.3}$$

is the *expected* (*or mean*) *value* of Z. — —

Comment. By the heuristics introduced via Eq. (11.2.2) $E\{Z\}$ is the limiting value of the arithmetic mean of the N measured values z_i' of Z during the N trials. In fact, by (11.2.2)

$$\sum_{i=1}^{m} z_i P\{Z=z_i\} = \sum_{i=1}^{m} \left[z_i \lim_{N\to\infty} \frac{C_N(Z=z_i)}{N} \right]$$

$$= \lim_{N\to\infty} \frac{1}{N} \sum_{i=1}^{m} [z_i \, C_N(Z=z_i)] = \lim_{N\to\infty} \frac{1}{N} \sum_{j=1}^{N} z_j' \, ,$$

where, obviously,

$$\sum_{i=1}^{m} [z_i \, C_N(Z=z_i)]$$

is the sum of all the N measured values z_j' of Z, with the possible m different values $z_1, \ldots, z_m$ factored out.

Now, if Z is a Boolean random varialble X, then by Eq. (11.3.3)

$$E\{X\} = 0 \cdot P\{X=0\} + 1 \cdot P\{X=1\} = P\{X=1\} \, . \tag{11.3.4}$$

Example. Using Eq. (11.3.4), Eq. (11.2.14) becomes

$$E\{X_i X_j\} = E\{X_i\} \, E\{X_j\} \, . \tag{11.3.5}$$

Definition 11.3.1b. The continuous-case analogue of Eq. (11.3.3) is the expected value

$$E\{Z\} = \int_{-\infty}^{\infty} z \, d F_Z(z) \, , \tag{11.3.6}$$

where, typically, Z is the life (time) of a system and

$$F_Z(z) := P\{Z \leqq z\} \, ; \quad f_Z(z) := d F_Z(z) \, dz \tag{11.3.7}$$

are the *distribution function*† and the *probability density function*‡, respectively of the

† Often called the cumulative distribution function (cdf.).

‡ Often abbreviated by pdf.

random variable Z. In this text the following result is needed: For $Z =: L \geqq 0$ (L for life)

$$E\{L\} = \int_0^\infty [1 - F_L(t)]\, dt \;. \tag{11.3.8}$$

Proof. By Fig. 11.3.1, from Eq. (11.3.6), with stripes orthogonal to the stripes of the typical Riemann integral approximation:

$$E(L) = \sum_{i=1}^{\infty} i\Delta t[F_L(i\Delta t + \Delta t) - F_L(i\Delta t)] + 0(\Delta t) \;.$$

But this is the approximation of the size of the shaded area in Fig. 11.3.1, for which Eq. (11.3.8) is another integral formula.

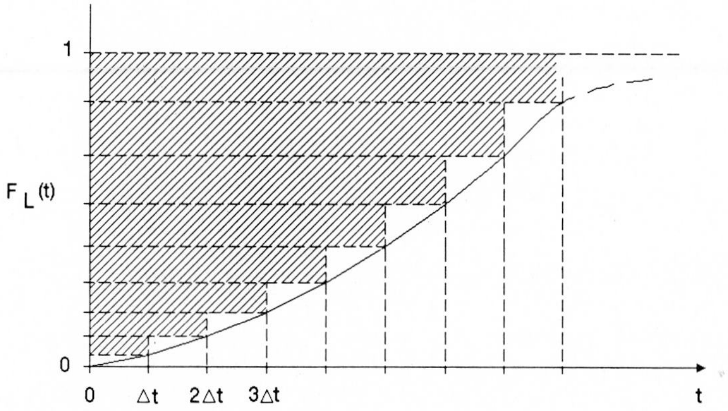

Fig. 11.3.1. Towards a special integral for the mean value

Frequently, several random variables are combined in a single cdf/pdf of several variables. In the case of two random variables Z_1 and Z_2 the combined cdf is

$$F_{Z_1,Z_2}(z_1, z_2) = P\{(Z_1 \leqq z_1) \cap (Z_2 \leqq z_2)\} \;. \tag{11.3.9}$$

Definition 11.3.2. Two random variables Z_1 and Z_2 are *stochastically* (*s*-) *independent*, if

$$F_{Z_1,Z_2}(z_1, z_2) = F_{Z_1}(z_1)F_{Z_2}(z_2) \Leftrightarrow f_{Z_1,Z_2}(z_1, z_2) = f_{Z_1}(z_1)f_{Z_2}(z_2) \;. \text{— —} \tag{11.3.10}$$

In many practical examples the pdf of $L_1 + L_2$ (for *s*-independent L_1 and L_2) is needed. In fact, for $L_1, L_2 \geqq 0$:

$$F_{L_1+L_2}(t) = \int_0^t f_{L_1}(\tau)F_{L_2}(t-\tau)\,d\tau \;. \tag{11.3.11}$$

Proof. Remember the law of total probability (theorem 11.2.1). Now, choose

$$P\{a\} = F_{L_1+L_2}(t) \ , \quad P\{b_i\} = f_{L_1}(i\Delta\tau)\Delta\tau \ , \quad P\{a|b_i\} = F_{L_2}(t-i\Delta\tau) \ .$$

Then the typical Riemann integral approximation of Eq. (11.3.11) for *s*-independent L_1 and L_2, i.e.

$$F_{L_1+L_2}(t) = \sum_{i=1}^{[t/\Delta\tau]} F_{L_2}(t-i\Delta\tau) f_{L_1}(i\Delta\tau)\Delta\tau + 0(\Delta\tau) \ , \tag{11.3.12}$$

is just a special case of Eq. (11.2.16), where the integer $[t/\Delta\tau]$ corresponds to m of Eq. (11.2.16). With ever increasing m the r.h.s. of Eq. (11.3.12) approaches the limiting value Eq. (11.3.11), q.e.d.

Differentiation of Eq. (11.3.11) yields

$$f_{L_1+L_2}(t) = \int_0^t f_{L_1}(\tau) f_{L_2}(t-\tau)\,d\tau \ . \tag{11.3.13}$$

Expected values of cardinal powers of random variables are called *moments*.

Definition 11.3.3.

$E\{Z^m\}$ is the *m-th moment* of Z ,

$E\{(Z-E\{Z\})^n\}$ is the *m-th central moment* of Z.

The 2nd central moment is called the *variance*, its square root the standard *deviation*. Usually both are denoted by σ_Z^2 and σ_Z, respectively. — —

Obviously,

$$\begin{aligned}\sigma_Z^2 &:= E\{(Z-E\{Z\})^2\} = E\{Z^2 - 2ZE\{Z\} + (E\{Z\})^2\} \\ &= E\{Z^2\} - 2E\{Z\}E\{Z\} - (E\{Z\})^2 = E\{Z^2\} - (E\{Z\})^2 \ .\end{aligned} \tag{11.3.14}$$

For two random variables Z_1 and Z_2 the parameter

$$\rho_{Z_1,Z_2} := E\{(Z_1 - E\{Z_1\})(Z_2 - E\{Z_2\}) = E\{Z_1 Z_2\} - E\{Z_1\}E\{Z_2\} \tag{11.3.15}$$

is called the *covariance* of Z_1 and Z_2.
In this appendix only a few intuitive and informal remarks on the concept of a *random process* in general, and that of a *Markov* process in particular can be made. More details can be found in many specialized books, e.g. (for applied workers) [27], [20].

Definition 11.3.4. A random process is an ordered set of random variables $\{Z(t), t \in T\}$. Both, the set of the different values any $Z(t)$ can assume, and the set of the different values of t can be finite (including denumerably infinite) or continuous.

Definition 11.3.5. If for a random process, where $t > t_1 > t_2 > \ldots$

$$P\{Z(t) = z_j \mid [Z(t_1) = z_i] \cap [(Z(t_2) = z_k] \cap \ldots\}$$
$$= P\{Z(t) = z_j \mid Z(t_1) = z_i\} =: p_{i,j}(t, t_1)$$

this random process is called a *Markov* process. Moreover, if $p_{i,j}(t, t_1)$ only depends on $t - t_1$ the Markov process is (time) stationary or *homogeneous.* If the number of different z_j (for any t) is finite (or countably infinite) the Markov process is often called a *Markov chain* [20].

11.4 Elementary Renewal Theory

Stochastic point processes are sequences of random points; mostly in time. An (*ordinary*) *renewal process* (RP) [35] is a stochastic *point process*, where the lengths of all the intervals between neighbouring points are random variables, which

- are all stochastically independent (of each other)
- have the same probability distribution with pdf f_1.

The renewal function $H(t)$ is by definition simply the expected number of points (renewals) between 0 and t:

$$H(t) := E\{N(0, t)\} \ . \tag{11.4.1}$$

Next we derive $H(t)$ in terms of the pdf f_1 of point distance. To this end we need the following lemma.

Lemma 11.4.1. For any *orderly* [32] point process

$$P\{N(0, t) = k\} = F_k(t) - F_{k+1}(t) \ , \tag{11.4.2}$$

where F_j is the (cumulative) probability distribution function (cdf) of the distance T_k of the k-th random point from the time origin, where – in the case of an ordinary renewal process – the 0-th renewal point is located.

Proof. By definition of the cdf

$$F_k(t) := P\{T_k \leqq t\} = P\{N(0, t) \geqq k\}$$
$$= P\{N(0, t) = k\} + P\{N(0, t) \geqq k + 1\}^\dagger$$
$$= P\{N(0, t) = k\} + F_{k+1}(t) \ ,$$

q.e.d.

---

† Notice that the random events $\{N(o, t) = k\}$ and $\{N(o, t) \geqq k + 1\}$ are disjoint

Hint. Renewal processes are orderly [32].

Now, by the general definition of an expected value (discrete case)

$$\begin{aligned} H(t) &= \sum_{k=1}^{\infty} kP\{N(0,t)=k\} \\ &= F_1(t)-F_2(t)+2F_2(t)-2F_3(t)+3F_3(t)-3F_4(t)+-\ldots \\ &= \sum_{k=1}^{\infty} F_k(t) \;. \end{aligned} \tag{11.4.3}$$

Since termwise differentiation is allowed here, the renewal density is

$$h(t) = \sum_{k=1}^{\infty} f_k(t) \;. \tag{11.4.4}$$

Since T_k is the sum of k stochastically-independent random variables, each with the pdf f_1, using the convolution theorem of the Laplace transform (notation of [35]), by Eq. (11.5.8)

$$f_k^*(s) = [f_1^*(s)]^k \;. \tag{11.4.5}$$

Hence the Laplace transformed renewal density is

$$h^*(s) = \sum_{k=1}^{\infty} [f_1^*(s)]^k = f_1^*(s)/[1-f_1^*(s)] \;. \tag{11.4.6}$$

If the above ordinary renewal process is an alternating one with alternating interval lengths L (for component life) and D (for component down time) then for the renewal process of the restarts

$$h^*(s) = f_L^*(s) f_D^*(s)/[1-f_L^*(s) f_D^*(s)] \;, \tag{11.4.7}$$

where use was made of Eq. (11.5.8) yielding for the renewal distance $L+D$:

$$f_{L+D}^*(s) = f_L^*(s) f_D^*(s) \;. \tag{11.4.8}$$

Equation (11.4.7) refers to the (ordinary) renewal process of restarts. When looking at the moments of failure, one does no longer have an ordinary renewal process but rather a specially *modified* one [35]. Specifically, the time to the first "renewal" is D and the renewal distance is (as before) $L+D$. Hence, for failures as "renewals", (11.4.5) has to be changed to

$$\tilde{f}_k^*(s) := \begin{cases} f_L^*(s) \;, & k=1 \\ f_L^*(s)[f_L^*(s) f_D^*(s)]^{k-1} \;, & k \geqq 2 \;. \end{cases} \tag{11.4.9}$$

This result is plausible, since, compared to Eq. (11.4.4), in $\tilde{h}(t)$ each random variable under consideration is shortened by D.

Next we derive the Laplace transform of the availability $A(t)$ of a component, whose changes of state (good or bad) occur according to an ordinary alternating renewal process. Since it is plausible to assume that the component under consideration was "good" at $t = 0$, there must be an even number of renewals (of any kind) between 0 and t to have the component in a "good" state at t:

$$A(t) = \sum_{i=0}^{\infty} P\{N(0, t) = 2i\} . \quad (11.4.10)$$

By (11.4.2) and the integration rule of the Laplace transform[†]

$$\begin{aligned} A^*(s) &= \sum_{i=0}^{\infty} \frac{1}{s} [f_{2i}^*(s) - f_{2i+1}^*(s)] \\ &= \frac{1}{s} \sum_{i=0}^{\infty} \{[f_L^*(s)]^i [f_D^*(s)]^i - [f_L^*(s)]^{i+1} [f_D^*(s)]^i\} \\ &= \frac{1}{s} [1 - f_L^*(s)] \sum_{i=0}^{\infty} [f_L^*(s) f_D^*(s)]^i \\ &= \frac{1}{s} [1 - f_L^*(s)] / [1 - f_L^*(s) f_D^*(s)] . \end{aligned} \quad (11.4.11)$$

Asymptotical results for the case $t \to \infty$, i.e. the stationary case can easily be gained by using the limit theorem Eq. (11.5.9) yielding

$$A(\infty) := \lim_{t \to \infty} A(t) = \lim_{s \to 0} \{[1 - f_L^*(s)] / [1 - f_L^*(s) f_D^*(s)]\} . \quad (11.4.12)$$

To gain stationary values for the renewal density and for the availability we use the definition of the Laplace transform

$$\begin{aligned} g^*(s) &:= \int_0^{\infty} g(t) \exp(-st) \mathrm{d}t \\ &= \int_0^{\infty} g(t) \mathrm{d}t - s \int_0^{\infty} t\, g(t) \mathrm{d}t + 0(s) . \end{aligned} \quad (11.4.13)$$

If g is a pdf f_Z of a non-negative Z, then by normation

$$\int_0^{\infty} f_Z(t) \mathrm{d}t = 1 , \quad (11.4.14)$$

and by general definition of the mean value (continuous case)

$$\int_0^{\infty} t f_Z(t) \mathrm{d}t = E(Z) . \quad (11.4.15)$$

Hence, by (11.4.13)

$$f_Z^*(s) = 1 - sE(Z) + 0(s) = 1 + 0(s) . \quad (11.4.16)$$

[†] Division by s corresponds to time domain integration; see section 11.5

Now, at last, we can calculate $h(\infty)$ and $A(\infty)$. From Eq. (11.4.7)

$$h(\infty) = \lim_{s\to 0}\left\{[1+0(s)]\Big/\left[E(L+D)+\frac{0(s)}{s}\right]\right\}$$

$$= \frac{1}{E(L)+E(D)} \,. \tag{11.4.17}$$

Likewise, from Eq. (11.4.13)

$$A(\infty) = \lim_{s\to 0}\frac{E(L)+0(s)/s}{E(L+D)+0(s)/s} = \frac{E(L)}{E(L)+E(D)} \,. \tag{11.4.18}$$

11.5 Laplace Transform Refresher

This short review of applications aspects of the Laplace transform is possibly needed to fully understand elementary renewal theory; e.g. that of section 11.4. (Because of its brevity and its minor importance this section was not made into another appendix.)

The Laplace transform is defined as the following integral transform of the real function $g(t)$:

$$g^*(s) := \mathscr{L}\{g(t)\} := \int_0^\infty g(t)\exp(-st)\mathrm{d}t \,, \tag{11.5.1}$$

where $s=x+iy$ is a complex (number) variable. Note that $g(t)$ is only of interest for non-negative arguments. Trivially,

$$g^*(0) = \int_0^\infty g(t)\mathrm{d}t \,, \tag{11.5.2}$$

$$\mathscr{L}\{1\} = \int_0^\infty \exp(-st)\mathrm{d}t = 1/s \,. \tag{11.5.3}$$

Differentiation (roughly) corresponds to multiplication by s:

$$\mathscr{L}\left\{\frac{\mathrm{d}}{\mathrm{d}t}g(t)\right\} = sh^*(s) - h(+0) \,. \tag{11.5.4}$$

Proof. By partial integration

$$\mathscr{L}\left\{\frac{\mathrm{d}}{\mathrm{d}t}g(t)\right\} = [g(t)\exp(-st)\,]_{+0}^{\infty} - \int_{+0}^{\infty} g(t)(-s)\exp(-st)\mathrm{d}t = -g(+0) + sg^*(s) \,.$$

Integration corresponds to division by s:

$$\mathscr{L}\left\{\int_0^t g(\tau)\mathrm{d}\tau\right\}=g^*(s)/s \ . \qquad (11.5.5)$$

Proof. For s from the right half (complex) plane; by partial integration:

$$\mathscr{L}\left\{\int_{+0}^t g(\tau)\mathrm{d}\tau\right\}=[-\frac{1}{s}\exp(-st)\int_0^t g(\tau)\mathrm{d}\tau\ ]_{+0}^{\infty}+\frac{1}{s}\int_{+0}^{\infty} g(t)\exp(-st)\mathrm{d}t=\frac{1}{s}g^*(s) \ .$$

Convolution corresponds to multiplication:

$$\mathscr{L}\left\{\int_0^\infty g_1(\tau)g_2(t-\tau)\mathrm{d}\tau\right\}=g_1^*(s)g_2^*(s) \ . \qquad (11.5.6)$$

Proof. By (11.5.1), with

$$g(t):=\int_0^\infty g_1(\tau)g_2(t-\tau)\mathrm{d}\tau \ ,$$

being the convolution (integral) of g_1 and g_2:

$$\mathscr{L}\left\{\int_0^\infty g_1(\tau)g_2(t-\tau)\mathrm{d}\tau\right\}=\int_0^\infty\left[\int_0^\infty g_1(\tau)g_2(t-\tau)\exp(-st)\mathrm{d}\tau\right]\mathrm{d}t$$

$$=\int_0^\infty g_1(\tau)\exp(-s\tau)\left\{\int_0^\infty g_2(t-\tau)\exp[-s(t-\tau)]\mathrm{d}t\right\}\mathrm{d}\tau \ , \qquad (11.5.7)$$

where $\exp(-s\tau)$ was factorized according to

$$\exp(-st)=\exp(-s\tau)\exp[-s(t-\tau)] \ .$$

If $g_2(t)=0$ for $t<0$, then for $\tau>0$

$$\int_0^\infty g_2(t-\tau)\exp[-s(t-\tau)]\mathrm{d}t=\int_0^\infty g_2(t)\exp(-st)\mathrm{d}t=g_2^*(s) \ .$$

This is independent of τ, and can therefore be extracted from the inner integral of the last line of Eq. (11.5.7). The remaining integral is $g_1^*(s)$. Thus, Eq. (11.5.7) equals Eq. (11.5.6), q.e.d.

As a trivial application of Eq. (11.5.6) one can transform Eq. (11.3.13) yielding

$$f^*_{\mathrm{L}_1+\mathrm{L}_2}(s)=f^*_{\mathrm{L}_1}(s)f^*_{\mathrm{L}_2}(s) \ . \qquad (11.5.8)$$

From $g^*(s)$ near $s=0$ one can derive $g(\infty)$ via the following limit theorem:

$$\lim_{t\to\infty} g(t)=\lim_{s\to 0}[sg^*(s)] \ . \qquad (11.5.9)$$

Proof. From the differentiation rule Eq. (11.5.4) and the expansion of the exponential function there follows

$$sg^*(s)=g(+0)+\int_0^\infty \frac{dg(t)}{dt}[1-st+\tfrac{1}{2}(st)^2-+\ldots]dt \ .$$

Invoking results of calculus

$$sg^*(s)=g(+0)+\int_0^\infty \frac{d}{dt}g(t)dt+0(s)=\lim_{t\to\infty} g(t)+0(s) \ .$$

Hence, Eq. (11.5.9) is true, q.e.d.

Example 11.5.1 Transforming the negative exponential function

Let

$$g(t)=\exp(-\alpha t) \ , \quad \alpha>0 \ . \tag{11.5.10}$$

Then by Eq. (11.5.1)

$$g^*(s)=\int_0^\infty \exp(-\alpha t)\exp(-st)dt=\int_0^\infty \exp[-(\alpha+s)t]dt=\frac{1}{\alpha+s} \ . \tag{11.5.11}$$

Example 11.5.2

Let

$$g(t)=t\exp(-\alpha t) \ , \quad \alpha>0 \ . \tag{11.5.12}$$

Then by (11.5.1)

$$g^*(s)=\int_0^\infty t\exp[-(\alpha+s)t]dt \ .$$

This can be solved by partial integration. However, using Eq. (11.3.8) makes partial integration superfluous. Defining the formal pdf and cdf, respectively:

$$f_L(t):=(\alpha+s)\exp[-(\alpha+s)t] \ ; \quad F_L(t)=1-\exp[-(\alpha+s)t] \ ,$$

one has

$$g^*(s)=\frac{1}{\alpha+s}E\{L\}$$

Now, by Eq. (11.3.8)

$$g^*(s)=\frac{1}{\alpha+s}\int_0^\infty \exp[-(\alpha+s)t]dt=\frac{1}{(\alpha+s)^2} \ . \tag{11.5.13}$$

Exercises

11.1

Show that in Eq. (11.1.2) always

(a) $q(a_i \cup a_j \cup a_k) \leqq q(a_i)+q(a_j)+q(a_k)$, (E11.1)

(b) $q(a_i \cup a_j \cup a_k) \geqq q(a_i)+q(a_j)+q(a_k)$

$$-q(a_i \cap a_j)-q(a_i \cap a_k)-q(a_j \cap a_k) \; . \tag{E11.2}$$

11.2

Draw a Venn diagram to help explain the law of total probability. Show that $b_j \cap b_k = \phi$ is only necessary for those parts of b_j and b_k which are parts of a.

11.3

Let there be m (well defined) indicator variables $X_1, \ldots, X_m$. Let $E(X_i)=U_i$; $i=1, \ldots, m$.

(a) How many of them are in 1-state, i.e. assume the value 1 (at a given time which need not be specified here)?

(b) How big is the expected value of the number specified in question a?

11.4

Determine the pdf of the distance of neighbouring points of an ordinary renewal process with a constant renewal density function.

11.5

Given the pdf's of two s-independent random variables Z_1 and Z_2. How big is $P\{Z_1 < Z_2\}$?

Solutions of the Exercises for §§2 through 11

2.1

By Eq. (2.2.8) for $X_i \leftarrow \bar{X}_i$

$$\bar{X}_i \vee \bar{\bar{X}}_i = 1 \ .$$

Applying Eq. (2.2.1) yields

$$\bar{\bar{X}}_i \vee \bar{X}_i = 1 \ .$$

Since X_i is the unique inverse of $\bar{X}_i$, by (2.2.8) $\bar{\bar{X}}_i = X_i$, q.e.d.

2.2

$$\varphi_{13} = X_1 \ , \quad \varphi_4 = \bar{X}_1 \ , \quad \varphi_{11} = X_2 \ , \quad \varphi_6 = \bar{X}_2 \ .$$

2.3

(a) Exchanging the columns of X_1 and X_2 in Table 2.1.2 has the same effect as exchanging the inner pair of lines; but this has no effect on both φ_9 (AND) and φ_{15} (OR), q.e.d.

(b) Other commutative functions (which really depend on both X_1 and X_2) are φ_2, φ_7, φ_8, φ_{10}.

2.4

For n variables there are 2^n argument n-tuples of functions $\varphi_i(X_1, \ldots, X_n)$. Any φ_i is given by its column with 2^n entries. Hence there can be 2^{2^n} such different columns, viz. functions of n variables, q.e.d.

2.5

From Table 2.1.2

$$\bar{\varphi}_7=\overline{X_1 \not\equiv X_2}=\varphi_{10}=X_1X_2 \vee \bar{X}_1\bar{X}_2 \ . \tag{S1.1}$$

From Eq. (2.1.1), replacing X_2 by $\bar{X}_2$ (and observing $\bar{\bar{X}}_2=X_2$):

$$X_1 \not\equiv \bar{X}_2=X_1X_2 \vee \bar{X}_1\bar{X}_2 \ ,$$

q.e.d.

2.6

By the distributive law Eq. (2.2.3) with $X_i \leftarrow X_i \vee X_j$, $X_j \leftarrow X_k$, $X_k \leftarrow X_l$:

$$(X_i \vee X_j)(X_k \vee X_l)=(X_i \vee X_j)X_k \vee (X_i \vee X_j)X_l$$

By the commutative law (2.2.1) and again by (2.2.3)

$$(X_i \vee X_j)X_k=X_iX_k \vee X_jX_k$$

and

$$(X_i \vee X_j)X_l=X_iX_l \vee X_jX_l \ .$$

By the associative law Eq. (2.2.20) the results of the last two formulas can be combined to give the desired result, q.e.d. .

2.7

Proof by induction: (E2.2) is true for $n=2$ (de Morgan's law). Let (for some $m>2$)

$$\overline{\bigvee_{i=1}^{m} X_i}=\bigwedge_{i=1}^{m} \bar{X}_i \ . \tag{S2.1}$$

Now, by de Morgan's law

$$\overline{\bigvee_{i=1}^{m+1} X_i}=\overline{\left(\bigvee_{i=1}^{m} X_i\right) \vee X_{m+1}}=\overline{\left(\bigvee_{i=1}^{m} X_i\right)}\,\bar{X}_{m+1} \ ,$$

and by (S2.1)

$$\overline{\bigvee_{i=1}^{m+1} X_i}=\bigwedge_{i=1}^{m+1} \bar{X}_i \ .$$

Hence, if (E2.2) is true for $n=m$, it is also true for $n=m+1$, i.e. by the principle of induction, it is true for all $n \in \mathbb{N}$.

2.8

First

$$\bar{X}=\overline{X \vee X}=: X \veebar X \ .$$

By this result and by de Morgan's law

$$X_i \wedge X_j=\overline{\overline{X_i \wedge X_j}}=\overline{\bar{X}_i \vee \bar{X}_j}=(X_i \veebar X_i) \veebar (X_j \veebar X_j) \ ,$$

and

$$X_i \vee X_j=\overline{\overline{X_i \vee X_j}}=\overline{X_i \veebar X_j}=(X_i \veebar X_j) \veebar (X_i \veebar X_j) \ .$$

Hence, NOT, AND, and OR can be expressed by NOR, q.e.d.

2.9

There are $\binom{n}{k}$ combinations of k variables $\varphi(X_1, \ldots, X_n)$ may not depend on explicitly. Hence, the number of functions depending truly on $X_1, \ldots, X_n$ is

$$N(n)=2^{2^n}-n\,N(n-1)-\binom{n}{2}N(n-2)- \ \ldots \ -N(0) \ .$$

$N(0)=2$, since $\varphi=0$ and $\varphi=1$ do not depend on any variables. Specifically

$$N(1)=2^{2^1}-N(0)=4-2=2 \ ,$$

$$N(2)=2^{2^2}-2N(1)-N(0)=16-4-2=10 \ ,$$

$$N(3)=2^{2^3}-3N(2)-3N(1)-N(0)=256-30-6-2=218 \ .$$

2.10

(a) $$\begin{bmatrix} 0 & 1 & 1 \\ 1 & 0 & 1 \\ 0 & 1 & 0 \end{bmatrix} \begin{bmatrix} 0 & 0 & 1 & 1 \\ 1 & 1 & 1 & 0 \\ 0 & 1 & 0 & 1 \end{bmatrix} = \begin{bmatrix} 1 & 1 & 1 & 1 \\ 0 & 1 & 1 & 1 \\ 1 & 1 & 1 & 0 \end{bmatrix} .$$

(b) $$\begin{bmatrix} 0 & 1 & 1 \\ 1 & 0 & 1 \\ 0 & 1 & 0 \end{bmatrix} (\wedge) \begin{bmatrix} 0 & 0 & 1 & 1 \\ 1 & 1 & 1 & 0 \\ 0 & 1 & 0 & 1 \end{bmatrix} = \begin{bmatrix} 1 & 0 & 1 & 1 \\ 0 & 1 & 1 & 0 \\ 1 & 1 & 1 & 0 \end{bmatrix} .$$

3.1

In every row (line or column) two variables have constant values. For every neighbouring pair of squares a third variable has a constant value. Hence, since every pair of neighbouring squares is trivially in a row, it can differ in values of at most one variable, and this it does, because every square corresponds uniquely to a quadruple of variable values. An alternative proof uses Fig. 3.3.4, especially the fact that two adjacent nodes of a hypercube have Hamming distance 1.

3.2

Look at Table S3.1!

Table S3.1. States of an SR-flipflop

$X_1(t)$	$X_2(t)$	$\varphi_1(t)$	$\varphi_2(t)$	$\varphi_1(t+\Delta t)$	$\varphi_2(t+\Delta t)$
0	0	0	1	0	1
0	0	1	0	1	0
0	1	0	1	1	0
0	1	1	0	1	0
1	0	0	1	0	1
1	0	1	0	0	1
1	1	—	—	0	0

For $X_1(t)=0$, $X_2(t)=1$ and vice versa there are definite stable patterns of the pair $\varphi_1(t+\Delta t)$ and $\varphi_2(t+\Delta t)$; i.e. these inputs are "stored".

3.3

The set $\bar{A}_1 \cap A_2$ is that part of A_2 that is disjoint with A_1. By Fig. 3.5.1 $A_1 \cup A_2$ also consists of A_1 united with that part of A_2 that is not in A_1. But that part is found by intersecting $\bar{A}_1$ and A_2, q.e.d.

3.4

Look at Fig. S3.1!

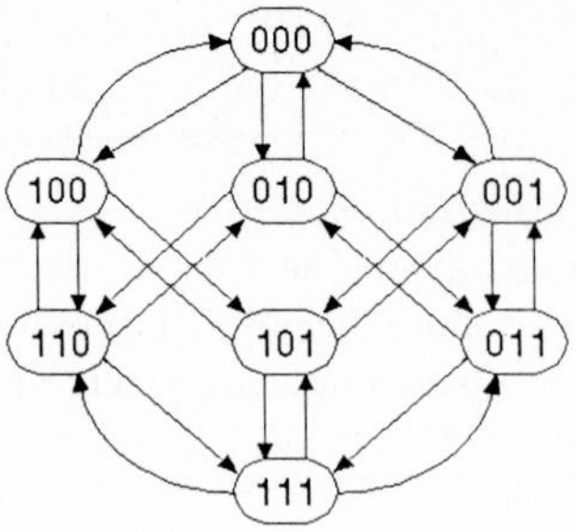

Fig. S3.1. Transition graph of a three-components system

3.5

Fig. 3.8.2(a): $\varphi_a = (X_1 \vee X_2)(X_2 \vee X_3)(X_3 \vee X_1)$

(E2.1): $= (X_1 X_2 \vee X_1 X_3 \vee X_2 \vee X_2 X_3)(X_3 \vee X_1)$

(2.2.14): $= (X_2 \vee X_1 X_3)(X_3 \vee X_1)$

(2.2.12): $= X_2 X_3 \vee X_1 X_2 \vee X_1 X_3$,

and this is φ_b of Fig. 3.8.2b., q.e.d.

3.6

First the binary tree of Fig. S3.2 is found. It can be "simplified" to a non-tree digraph as shown in Fig. S3.3.

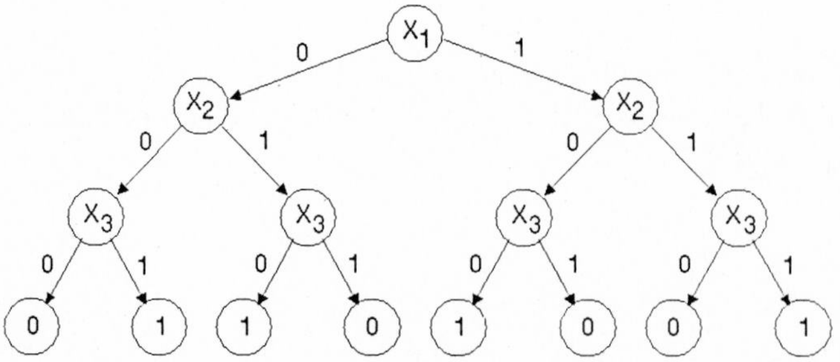

Fig. S3.2. Binary decision tree for $\varphi = X_1 \not\equiv X_2 \not\equiv X_3$

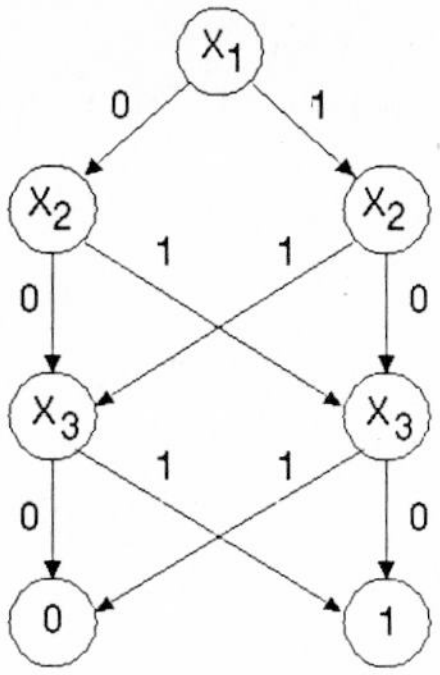

Fig. S3.3. Non-tree version of Fig. S3.2

4.1

With $\varphi=\bar{\psi}$, Eq. (4.3.3) is changed upon negation via Shannon's inversion rule (theorem 4.1.1) to

$$\bar{\bar{\psi}}=\psi=[\bar{X}_i \vee \psi(X_i=1)][X_i \vee \psi(X_i=0)] , \qquad \text{(S4.1)}$$

q.e.d., since, as a general function, ψ is just as good as φ.

4.2

According to de Morgan's law, by Eq. (4.3.30)

$$\begin{aligned}\bar{X}_\varphi=&(\bar{X}_1 \vee \bar{X}_2 \vee \bar{X}_3)(\bar{X}_1 \vee \bar{X}_2 \vee \bar{X}_5 \vee \bar{X}_7)\\ &\wedge (\bar{X}_1 \vee \bar{X}_3 \vee \bar{X}_4 \vee \bar{X}_5)(\bar{X}_1 \vee \bar{X}_4 \vee \bar{X}_7) \wedge (\bar{X}_6 \vee \bar{X}_7)\\ &\wedge (\bar{X}_2 \vee \bar{X}_3 \vee \bar{X}_4 \vee \bar{X}_6)(\bar{X}_3 \vee \bar{X}_5 \vee \bar{X}_6)\\ =&(\bar{X}_1 \vee \bar{X}_2 \vee \bar{X}_3\bar{X}_5 \vee \bar{X}_3\bar{X}_7) \wedge (\bar{X}_1 \vee \bar{X}_4 \vee \bar{X}_3\bar{X}_7 \vee \bar{X}_5\bar{X}_7)\\ &\wedge (\bar{X}_6 \vee \bar{X}_7) \wedge (\bar{X}_3 \wedge \bar{X}_6 \vee \bar{X}_2\bar{X}_5 \vee \bar{X}_4\bar{X}_5) .\end{aligned}$$

Combining the upper and the lower pair of parentheses, respectively, yields

$$\begin{aligned}\bar{X}_\varphi=&(\bar{X}_1 \vee \underline{\underline{\bar{X}_3\bar{X}_7}} \vee \bar{X}_2\bar{X}_4 \vee \underline{\underline{\bar{X}_2\bar{X}_5\bar{X}_7}} \vee \bar{X}_3\bar{X}_4\bar{X}_5)\\ &\wedge (\bar{X}_6 \vee \underline{\underline{\bar{X}_3\bar{X}_7}} \vee \underline{\underline{\bar{X}_2\bar{X}_5\bar{X}_7}} \vee \bar{X}_4\bar{X}_5\bar{X}_7)\\ =&\bar{X}_3\bar{X}_7 \vee \bar{X}_2\bar{X}_5\bar{X}_7 \vee \bar{X}_1\bar{X}_6 \vee \bar{X}_1\bar{X}_4\bar{X}_5\bar{X}_7\\ &\vee \bar{X}_2\bar{X}_4\bar{X}_6 \vee \bar{X}_3\bar{X}_4\bar{X}_5\bar{X}_6 ,\end{aligned} \qquad \text{(S4.2)}$$

which equals (but for a permutation of terms), Eq. (4.3.31).

4.3

(a) Because the first two terms of (E4.2) are disjoint, initially

$$\varphi=X_1X_4 \vee \bar{X}_1\bar{X}_4 \vee \ldots$$

The term $\bar{X}_2 X_5$ is replaced by $\bar{X}_1 \bar{X}_2 X_5 \vee X_1 \bar{X}_4 \bar{X}_2 X_5$ to get that part which is disjoint with $X_1 X_4$. Since $\bar{X}_1 \bar{X}_2 X_5$ is not disjoint with $\bar{X}_1 \bar{X}_4$, it must be changed to $\bar{X}_1 \bar{X}_2 X_4 X_5$. This yields

$$\varphi = X_1 X_4 \vee \bar{X}_1 \bar{X}_4 \vee \bar{X}_1 \bar{X}_2 X_4 X_5 \vee X_1 \bar{X}_2 \bar{X}_4 \bar{X}_5 \vee \ldots$$

The term $X_1 X_3 X_4$ is redundant, since it is absorbable by $X_1 X_4$. The term $X_2 X_3 \bar{X}_4$ is first changed to $X_1 X_2 X_3 \bar{X}_4$ to make it disjoint with $\bar{X}_1 \bar{X}_4$. Fortunately, it is now also disjoint with the other terms of the provisional result. So, now

$$\varphi = X_1 X_4 \vee \bar{X}_1 \bar{X}_4 \vee \bar{X}_1 \bar{X}_2 X_4 X_5 \vee X_1 \bar{X}_2 \bar{X}_4 X_5 \vee X_1 X_2 X_3 \bar{X}_4 \,. \qquad \text{(S4.3)}$$

The term $X_1 X_2 X_3 X_5$ is first changed to $X_1 X_2 X_3 \bar{X}_4 X_5$ and is absorbed by the last term of (S4.3). The final result is therefore (S4.3) without further terms.

(b)
$$\begin{aligned}\varphi &= X_1 X_4 \vee \bar{X}_1 \bar{X}_4 \vee \bar{X}_2 X_5 \vee X_2 X_3 \bar{X}_4 \vee X_1 X_2 X_3 X_5 \\ &= X_1 (X_4 \vee \bar{X}_2 X_5 \vee X_2 X_3 \bar{X}_4 \vee X_2 X_3 X_5) \\ &\quad \vee \bar{X}_1 (\underline{\underline{\bar{X}_4}} \vee \bar{X}_2 X_5 \vee \underline{\underline{X_2 X_3 \bar{X}_4}}) \\ &= X_1 [X_4 \vee \bar{X}_4 (\bar{X}_2 X_5 \vee X_2 X_3)] \vee \bar{X}_1 (\bar{X}_4 \vee X_4 \bar{X}_2 X_5) \\ &= X_1 X_4 \vee X_1 \bar{X}_2 \bar{X}_4 X_5 \vee X_1 X_2 X_3 \bar{X}_4 \vee \bar{X}_1 \bar{X}_4 \vee \bar{X}_1 \bar{X}_2 X_4 X_5 \,,\end{aligned}$$

which happens to be identical with (S4.3) but for the sequencing of terms.

(c) See Fig. S4.1. The underlined states yield (S4.3).

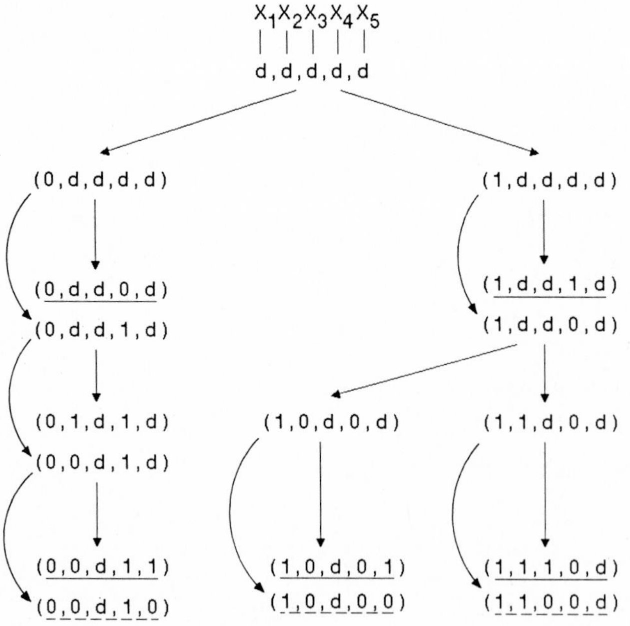

Fig. S4.1. Decision tree for (E4.2).

4.4

The CDNF of φ is here

$$\varphi = X_1 X_2 X_3 \vee X_1 X_2 \bar{X}_3 \vee X_1 \bar{X}_2 X_3 \vee \bar{X}_1 X_2 X_3 \ .$$

(a) To achieve $\varphi(\mathbf{0})=1$ we must add the minterm $\bar{X}_1 \bar{X}_2 \bar{X}_3$.
(b) To achieve $\varphi(\mathbf{1})=0$ we must delete the minterm $X_1 X_2 X_3$.
(c) In general a coherent function of $\mathbf{X}=(X_1, \ldots, X_n)$ must contain in its CDNF the minterm $X_1 \ldots X_n$ and must not contain the minterm $\bar{X}_1 \ldots \bar{X}_n$.

4.5

Replace in corollary 4.4.2 X_i by $\bar{\hat{T}}_i$. The desired results follow immediately from Eqs. (4.4.13) and (4.4.13a) and from lemma (4.4.2).

A proof by induction is also quite simple: From

$$\varphi_1 \vee \varphi_2 = \varphi_1 \vee \bar{\varphi}_1 \varphi_2$$

there follows

$$\hat{T}_1 \vee \hat{T}_2 = \hat{T}_1 \vee \bar{\hat{T}}_1 \hat{T}_2 \ ,$$

and further, by a de Morgan rule,

$$\hat{T}_1 \vee \hat{T}_2 \vee \hat{T}_3 = (\hat{T}_1 \vee \hat{T}_2) \vee (\overline{\hat{T}_1 \vee \hat{T}_2})\hat{T}_3 = \hat{T}_1 \vee \bar{\hat{T}}_1 \hat{T}_2 \vee \bar{\hat{T}}_1 \bar{\hat{T}}_2 \hat{T}_3 \ ,$$

$$\hat{T}_1 \vee \ldots \vee \hat{T}_{k+1} = (\hat{T}_1 \vee \ldots \vee \hat{T}_k) \vee (\overline{\hat{T}_1 \vee \ldots \vee \hat{T}_k})\hat{T}_{k+1}$$
$$= \underbrace{\hat{T}_1 + \bar{\hat{T}}_1 \hat{T}_2 \vee \ldots \vee \bar{\hat{T}}_1 \ldots \bar{\hat{T}}_{k-1} \hat{T}_k}_{\text{by induction assumption}} \vee \bar{\hat{T}}_1 \ldots \bar{\hat{T}}_k \hat{T}_{k+1} \ ,$$

q.e.d.

4.6

Any term can either only contain X_i, or $\bar{X}_i$ (not both) or neither of them. Hence,

$$N_n = 3^n - 1$$

since the case where a term contains no literals yields no real term.

4.7

Let $\mathbf{X}_i$ the binary n-tuple corresponding to $\hat{M}_i$ by replacing all X_j of $\hat{M}_i$ by 1 and all $\bar{X}_j$ by 0. Then by Table S4.1 Eq. (4.3.10) is true.

Table S4.1. Truth table of two minterms, their disjunction and their antivalence

$X_1 \ldots X_n$	$\hat{M}_i$	$\hat{M}_j$	$\hat{M}_i \not\equiv \hat{M}_j$	$\hat{M}_i \vee \hat{M}_j$
$0 \ldots 0$	0	0	0	0
	⋮	⋮	⋮	⋮
	0	0	0	0
$\leftarrow \mathbf{X}_i \rightarrow$	1	0	1	1
	0	0	0	0
	⋮	⋮	⋮	⋮
	0	0	0	0
$\leftarrow \mathbf{X}_j \rightarrow$	0	1	1	1
	0	0	0	0
	⋮	⋮	⋮	⋮
$1 \ldots 1$	0	0	0	0

4.8

Let us start with Eq. (4.3.3) in the abbreviated notation

$$\varphi = X_i \varphi(X_i = 1) \vee \bar{X}_i \varphi(X_i = 0) \ .$$

Assume that the following decomposition of $\varphi(X_i = 1)$ and $\varphi(X_i = 0)$ is possible:

$$\varphi(X_i = 1) = \varphi'(X_i = 1) \vee \varphi_i \ , \qquad \varphi(X_i = 0) = \varphi'(X_i = 0) \vee \varphi_i \ ,$$

where φ_i is independent of X_i. Then Eq. (E4.3) is a trivial consequence of $X_i \vee \bar{X}_i = 1$. In case φ is given as a DNF, then, obviously, φ_i is the DNF of those terms, which do not contain X_i or $\bar{X}_i$.

4.9

(a) See Fig. S4.2.
(b) See Fig. S4.3.

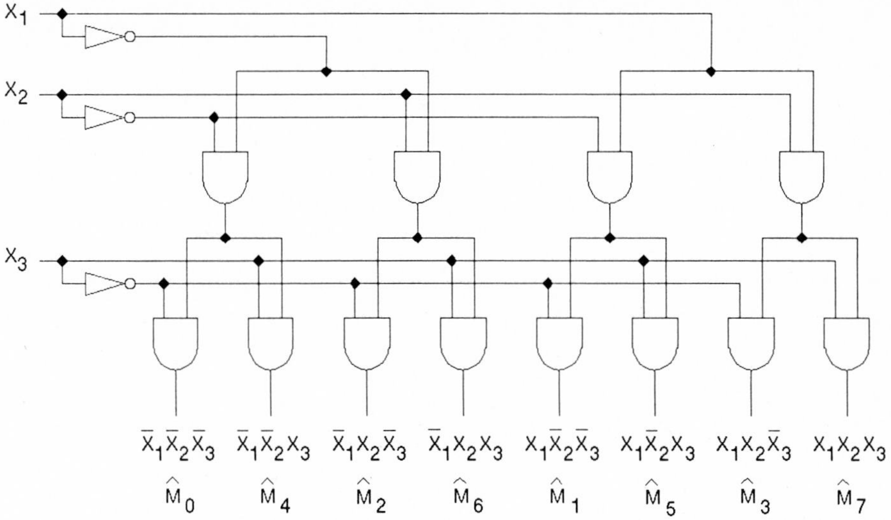

Fig. S4.2. Tree-type demultiplexer using 2-inputs AND gates

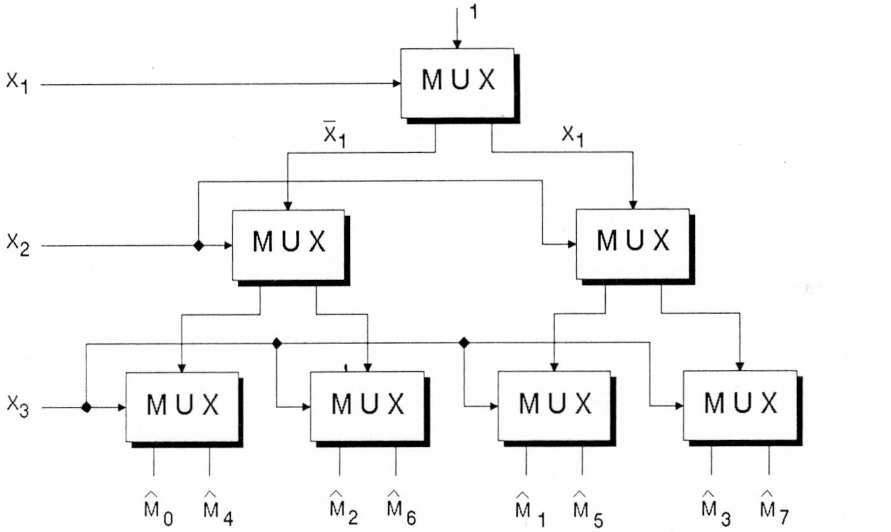

Fig. S4.3. Tree-type demultiplexer using 2-outputs demultiplexers. X_1, X_2, and X_3 are control inputs

5.1

From Fig. 5.1.1(a)

$$\varphi = X_1 X_2 \vee \bar{X}_2 X_3 \tag{S5.1}$$

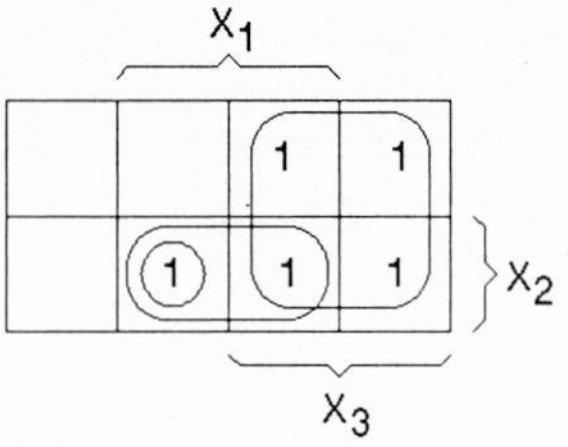

Fig. S5.1. K-map of a function whose shortest DDNF is longer than its shortest DNF

is a shortest DNF which is at the same time a DDNF. However, for the function defined by the K-map of Fig. S5.1 this is different. There the shortest DNF is

$$\varphi = X_3 \vee X_1 X_2 \ ,$$

and the shortest DDNF is

$$\varphi = X_3 \vee X_1 X_2 \bar{X}_3 \ .$$

5.2

Table S5.1. Determination of all the PIs of φ of Fig. E5.1

i	$X_3X_2X_1$		i, j	$X_3X_2X_1$	PI
4	1 0 0	✓	4, 5	1 0 –	$\bar{X}_2X_3$
3	0 1 1	✓	3, 7	– 1 1	X_1X_2
5	1 0 1	✓	5, 7	1 – 1	X_1X_3
7	1 1 1	✓			

5.3

(a) $\varphi = \hat{M}_3 \vee \hat{M}_4 \vee \hat{M}_5 \vee \hat{M}_7 = \bar{X}_3X_2X_1 \vee X_3\bar{X}_2\bar{X}_1 \vee X_3\bar{X}_2X_1 \vee X_3X_2X_1$. Obviously $C(\hat{M}_3, \hat{M}_7) = X_2X_1$, $C(\hat{M}_4, \hat{M}_5) = X_3\bar{X}_2$, $C(\hat{M}_5, \hat{M}_7) = X_3X_1$. This checks with Table S5.1.

(b) In the case $\hat{T}_i = \hat{T}_j$, then by $X_k \vee \bar{X}_k = 1$,

$$C(\hat{T}_iX_k, \hat{T}_i\bar{X}_k) = \hat{T}_i\hat{T}_i^{\dagger} = \hat{T}_i = \hat{T}_iX_k \vee \hat{T}_i\bar{X}_k \ .$$

† Here $\hat{T}_i^2$ would not be a well-defined notation.

5.4

Negating (S5.1) twice:

$$\bar{\varphi}=(\bar{X}_1 \vee \bar{X}_2)(X_2 \vee \bar{X}_3)=\bar{X}_1 X_2 \vee \bar{X}_1 \bar{X}_3 \vee \bar{X}_2 \bar{X}_3 \ ,$$

$$\bar{\bar{\varphi}}=\varphi=(X_1 \vee \bar{X}_2)(X_1 \vee X_3)(X_2 \vee X_3)=(X_1 \vee \bar{X}_2 X_3)(X_2 \vee X_3)$$

$$=X_1 X_2 \vee X_1 X_3 \vee \bar{X}_2 X_3 \ ,$$

see Eq. (5.1.5).

5.5

$\hat{T}_2$ and $\hat{T}_4$ are core PIs. $\hat{T}_3$ is superfluous because it implies $\hat{T}_4$, i.e.

$\hat{M}(\hat{T}_3) \subset \hat{M}(\hat{T}_4)$.

Since

$$\hat{M}(\hat{T}_1) \subset [\hat{M}(\hat{T}_2) \cup \hat{M}(\hat{T}_4)] \ ,$$

there is only one minimal DNF consisting of the two core PIs:

$$\varphi=\hat{T}_1 \vee \hat{T}_4 \ .$$

This follows directly from observing that the core PIs cover all relevant minterms.

5.6

From Fig. 5.3.2b there follows the K-map of Fig. S5.2.

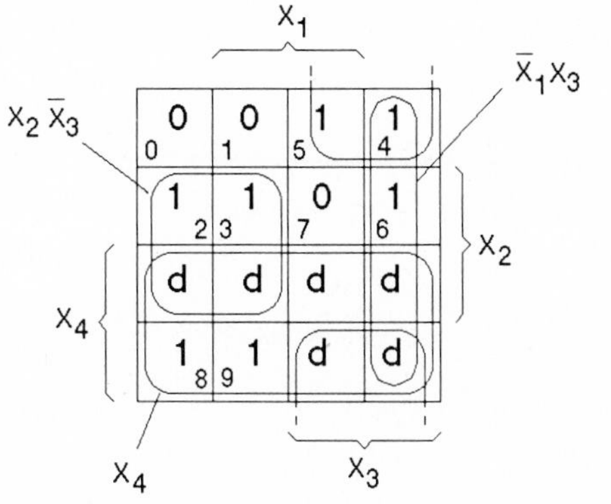

Fig. S5.2. K-map of φ_g

Again it is best to set all *d*s to 1, giving

$$\varphi_g = X_4 \vee X_2\bar{X}_3 \vee \bar{X}_1 X_3 \vee \bar{X}_2 X_3 \ .$$

5.7

The desired minimized function φ certainly must contain as many minterms as possible. In the *K*-map φ must be given as a chess-board pattern; see Fig. S5.3 for one of two choices in the case of 4 variables. Hence, in general the desired maximum minimal length is for *n* variables and 2^{n-1} minterms

$$n2^{n-1} + 2^{n-1} - 1 = (n+1)2^{n-1} - 1 \ .$$

It corresponds to 2^{n-1} terms of length *n* each and the $\vee$ between any two terms.

1		1	
	1		1
1		1	
	1		1

Fig. S5.3. *K*-map of minimized DNF of maximum length

6.1

(a) If in a modulo 2 (1-Bit) addition one operand is complemented (0 instead of 1 or vice versa) the result is complemented.
(b) The addition modulo 2 is obviously associative.

6.2

By Eq. (6.1.1), using $X_i X_i = X_i$, and $T_i \not\equiv T_i = 0$ several times,

$$\begin{aligned}(X_1X_2 \vee X_1X_3) \vee X_2X_3 &= (X_1X_2 \not\equiv X_1X_3 \not\equiv X_1X_2X_3) \not\equiv X_2X_3 \\ &\not\equiv (X_1X_2 \not\equiv X_1X_3 \not\equiv X_1X_2X_3)X_2X_3 \\ &= X_1X_2 \not\equiv X_1X_3 \not\equiv X_1X_2X_3 \\ &\not\equiv X_2X_3 \not\equiv X_1X_2X_3 \not\equiv X_1X_2X_3 \not\equiv X_1X_2X_3 \\ &= X_1X_2 \not\equiv X_1X_3 \not\equiv X_2X_3 \ ,\end{aligned}$$

This result checks with Eq. (6.1.16).

6.3

By Eqs. (6.2.2), (6.1.1) and (6.1.15)

$$\Delta_{X_1}\varphi=(X_2 \vee X_3)\not\equiv X_3=X_2\not\equiv X_3\not\equiv X_2X_3\not\equiv X_3=X_2\not\equiv X_2X_3=X_2\bar{X}_3 \ . \tag{S6.1}$$

By Eq. (6.3.5) with

$$\psi=X_1X_2 \ , \quad \Delta_{X_1}\psi=X_2$$

and

$$\Delta_\psi\tilde{\varphi}=(1 \vee X_3)\not\equiv(0 \vee X_3)=1\not\equiv X_3=\bar{X}_3$$

we find

$$\Delta_{X_1}\varphi=\bar{X}_3X_2 \ ,$$

as in (S6.1).

6.4

Here, with α corresponding to $X_2X_3=0$,

$$\varphi=(X_1 \vee X_2X_3)X_4 \ , \quad \varphi_\alpha=X_1X_4 \ .$$

i.e. the TGF

Hence the logical test condition i.e. the TGF is

$$\psi_\alpha=\varphi\not\equiv\varphi_\alpha=(X_1 \vee X_2X_3)X_4\not\equiv X_1X_4.$$

By Eq. (6.1.1)

$$X_1 \vee X_2X_3 = X_1\not\equiv X_2X_3\not\equiv X_1X_2X_3 \ .$$

Hence

$$\begin{aligned}\psi_\alpha&=X_4(X_1\not\equiv X_2X_3\not\equiv X_1X_2X_3\not\equiv X_1)\\&=X_4[X_2X_3(1\not\equiv X_1)]=\bar{X}_1X_2X_3X_4 \ .\end{aligned}$$

Interpretation. By $X_2X_3=1$ the correct value 1 would appear at the output of the left-hand AND gate, instead of the faulty value 0. Furthermore, $X_1=0$ and $X_4=1$ would sensitize a path for the fault to "propagate" from its origin to the output of the circuit.

7.1

(a) By Eq. (7.1.6)

$$\begin{aligned}\varphi &= (X_1+X_2-2X_1X_2)+X_3-2(X_1+X_2-2X_1X_2)X_3\\ &= X_1+X_2+X_3-2X_1X_2-2X_1X_3-2X_2X_3+4X_1X_2X_3\ .\end{aligned} \tag{S7.1}$$

(b) By aiming at the form (7.2.9) one has here

$$\begin{aligned}
&a_0=\varphi(0,0,0)=(0\not\equiv 0)\not\equiv 0=0\ ,\\
&a_0+a_1=\varphi(1,0,0)=(1\not\equiv 0)\not\equiv 0=1\Rightarrow a_1=1,\\
&a_0+a_2=\varphi(0,1,0)=(0\not\equiv 1)\not\equiv 0=1\Rightarrow a_2=1,\\
&a_0+a_3=\varphi(0,0,1)=(0\not\equiv 0)\not\equiv 1=1\Rightarrow a_3=1,\\
&a_0+a_1+a_2+a_{1,2}=\varphi(1,1,0)=(1\not\equiv 1)\not\equiv 0=0\Rightarrow a_{1,2}=-2\ ,\\
&a_0+a_1+a_3+a_{1,3}=\varphi(1,0,1)=(1\not\equiv 0)\not\equiv 1=0\Rightarrow a_{1,3}=-2\ ,\\
&a_0+a_2+a_3+a_{2,3}=\varphi(0,1,1)=(0\not\equiv 1)\not\equiv 1=0\Rightarrow a_{2,3}=-2\ ,\\
&a_0+a_1+a_2+a_3+a_{1,2}+a_{1,3}+a_{2,3}+a_{1,2,3}=\varphi(1,1,1)\\
&\qquad =(1\not\equiv 1)\not\equiv 1=1\\
&\qquad \Rightarrow a_{1,2,3}=4\ ,
\end{aligned}$$

which is equivalent to (S7.1).

(c) Since interchanging X_1 with X_3 does not change (S7.1), (E7.1) is true. (Of course, the commutativeness of the EXOR operation is used as well.)

7.2

By (7.3.10), (7.2.8), (7.1.6)

$$\begin{aligned}\Delta_{X_1}(X_1X_2 \vee X_1X_3 \vee X_2X_3) &= \Delta_{X_1}(X_1X_2+X_1X_3+X_2X_3-2X_1X_2X_3)\\ &= X_2+X_3-2X_2X_3=X_2\not\equiv X_3\ .\end{aligned}$$

This checks with Eq. (6.2.11).

7.3

Because otherwise the necessary condition $\varphi(\mathbf{0})=0$ would be violated.

$a_0=0$. Hence, because of

7.4

Because of $\varphi(0)=0$: $c_0=0$ and $\varphi(\mathbf{1})=1$

$$c_1+c_2+\ldots+c_m=1 \ ,$$

and

$$a_1+\ldots+a_{1,\ldots,n}=1 \ .$$

8.1

Obviously $p_i, p_\varphi \in [0, 1]$. Now, for certain representations of φ, typically polynomials, the argument $\mathbf{X}$ can be replaced by $\mathbf{p}=(p_1, \ldots, p_n)$ yielding p_φ instead of X_φ.

8.2

From Eq. (11.2.8)

$$P\{a_1 \cup a_2 \cup a_3\} = P\{a_1\}+P\{a_2\}+P\{a_3\}-P\{a_1 \cap a_2\} \\ -P\{a_1 \cap a_3\}-P\{a_2 \cap a_3\}+P\{a_1 \cap a_2 \cap a_3\} \ . \tag{S8.1}$$

Clearly, since $P\{a_1 \cap a_2 \cap a_3\} \geqq 0$

$$P\{a_1 \cup a_2 \cup a_3\} \geqq P\{a_1\}+P\{a_2\}+P\{a_3\}-P\{a_1 \cap a_2\} \\ -P\{a_1 \cap a_3\}-P\{a_2 \cap a_3\} \ .$$

Further, since, with $P\{\cdot\}$ a non-negative set function,

$$P\{a_1 \cap a_2\} \geqq P\{a_1 \cap a_2 \cap a_3\} \ ,$$

the sum of the last two lines of the r.h.s. of (S8.1) is not positive. Hence

$$P\{a_1 \cup a_2 \cup a_3\} \leqq P\{a_1\}+P\{a_2\}+P\{a_3\} \ ,$$

q.e.d.

8.3

(a) The Markov model differential equations here are

$$\dot{P}_0(t) = -\lambda P_0(t)+\mu P_1(t) \ ,$$

$$\dot{P}_1(t) = \lambda P_0(t)-(\lambda+\mu)P_1(t)+\mu P_2(t) \ ,$$

$$\dot{P}_2(t) = \lambda P_1(t)-(\lambda+\mu)P_2(t)+\mu P_3(t) \ ,$$

$$\ldots$$

In case $\dot{P}_1(\infty)=0$, $P_i(\infty)=: P_i$ these equations change to

$$P_1=\frac{\lambda}{\mu}P_0 , \tag{S8.2}$$

$$P_k=P_{k-1}+\frac{\lambda}{\mu}(P_{k-1}-P_{k-2}) , \quad k=2, 3, \ldots \tag{S8.3}$$

So, given P_0, a recursive solution is possible. However, in general, only the normation condition

$$P_0+P_1+ \ldots =1 \tag{S8.4}$$

is given.

Writing Eq. (S8.2, 3) as

$$P_1-\frac{\lambda}{\mu}P_0=0 ,$$

$$P_k-\frac{\lambda}{\mu}P_{k-1}=P_{k-1}-\frac{\lambda}{\mu}P_{k-2} , \quad k \geqq 1 ,$$

it becomes obvious that a possible solution is

$$P_k=\frac{\lambda}{\mu}P_{k-1}=\left(\frac{\lambda}{\mu}\right)^k P_0 ,$$

and, with (S8.4),

$$P_0\left[1+\frac{\lambda}{\mu}+\left(\frac{\lambda}{\mu}\right)^2+ \ldots \right]=1 ,$$

so that for $\lambda < \mu$

$$P_0=1-\frac{\lambda}{\mu} ,$$

whence

$$P_k=\left(\frac{\lambda}{\mu}\right)^k\left(1-\frac{\lambda}{\mu}\right) . \tag{S8.5}$$

(b) The queue length is (in state K) $L=K-1$. Hence

$$E(L)=P_2+2P_3+3P_4+ \ldots =\left(1-\frac{\lambda}{\mu}\right)\sum_{i=1}^{\infty} i\left(\frac{\lambda}{\mu}\right)^{i+1} .$$

Now (for $x < 1$)

$$\sum_{i=1}^{\infty} i x^{i-1} = \frac{d}{dx} \sum_{i=1}^{\infty} x^i = \frac{d}{dx} \frac{x}{1-x} = \frac{1}{(1-x)^2} .$$

Hence

$$E(L) = \left(1 - \frac{\lambda}{\mu}\right)\left(\frac{\lambda}{\mu}\right)^2 \frac{1}{\left(1 - \frac{\lambda}{\mu}\right)^2} = \frac{(\lambda/\mu)^2}{1 - \lambda/\mu} . \quad \text{(S8.6)}$$

8.4

By the definition of states 1 and 2 of Fig. 8.2.5

$$P_1 = p_1 p_2 , \quad \text{(S8.7)}$$

$$P_2 = p_1(1-p_2) + p_2(1-p_1) = p_1 + p_2 - 2p_1 p_2 . \quad \text{(S8.8)}$$

By Eq. (S8.7)

$$p_2 = P_1/p_1 .$$

Hence, by Eq. (S8.8)

$$p_1 + P_1/p_1 = P_2 + 2P_1 ,$$

$$p_1^2 - (P_2 + 2P_1)p_1 = -P_1 ,$$

$$p_1 = P_1 + \tfrac{1}{2}P_2 \pm |\sqrt{\tfrac{1}{4}(P_2 + 2P_1)^2 - P_1}| . \quad \text{(S8.9)}$$

Both solutions are real, since

$$\tfrac{1}{4}(P_2 + 2P_1)^2 - P_1 = \tfrac{1}{4}(p_1 + p_2)^2 - p_1 p_2$$

$$= \tfrac{1}{4}(p_1 - p_2)^2 \geqq 0 .$$

Yet only the solution with the positive sign is, generally, correct, since formally Eq. (S8.9) equals

$$p_1 = \tfrac{1}{2}(p_1 + p_2) \pm \tfrac{1}{2}(p_1 - p_2) = \begin{cases} p_1 \\ p_2 \end{cases} .$$

8.5

By the differentiation rule Eq. (11.5.4) the differential equations (8.2.31) and (8.2.32) are transformed to the linear algebraic equations

$$sP_1^*(s) = 1 - a\lambda P_1^*(s) + \mu P_2^*(s) , \quad \text{(S8.10)}$$

$$sP_2^*(s) = a\lambda P_1^*(s) - (\lambda + \mu)P_2^*(s) + b\mu P_3^*(s) . \quad \text{(S8.11)}$$

Equation (8.2.33) yields

$$P_1^*(s)+P_2^*(s)+P_3^*(s)=1/s \; . \tag{S8.12}$$

In order to shorten expressions, $P_i^*(s)$ is reduced to P_i^*. From Eq. (S8.10)

$$P_2^*=-\frac{1}{\mu}+\frac{s+a}{\mu}P_1^* \; .$$

Insertion in Eq. (S8.12) yields

$$P_3^*=\frac{1}{s}-P_2^*-P_1^*=\frac{1}{s}+\frac{1}{\mu}-\frac{s+a\lambda}{\mu}P_1^*-P_1^*=\frac{1}{s}+\frac{1}{\mu}-\frac{s+a\lambda+\mu}{\mu}P_1^* \; .$$

Replacing in Eq. (S8.11) P_2^* and P_3^* by the above functions of P_1^*:

$$P_1^*=\frac{s+\lambda+\mu}{a\lambda}P_2^*-\frac{b\mu}{a\lambda}P_3^*$$

$$=-\frac{s+\lambda+\mu}{a\lambda\mu}+\frac{(s+\lambda+\mu)(s+a\lambda)}{a\lambda\mu}P_1^*-\frac{b\mu(s+\mu)}{sa\lambda\mu}+\frac{(s+a\lambda+\mu)b}{a\lambda}P_1^* \; .$$

Factoring out P_1^* yields

$$P_1^*=\frac{s^2+(b'\mu+\lambda)s+b\mu^2}{s[s^2+(a'\lambda+b'\mu)s+a\lambda^2+ab\lambda\mu+b\mu^2} \; ; \qquad a':=a+1,\; b':=b+1 \tag{S8.13}$$

Check. Applying the limit theorem, Eq. (11.5.9)

$$\lim_{s\to 0} s[P_1^*(s)]=P_1(\infty)=b\mu^2/(a\lambda^2+ab\lambda\mu+b\mu^2) \; .$$

This checks with the stationary value P_1 of Eq. (8.2.34).

8.6

Under the condition that at τ there is the first restart, i.e.

$$\tau=L+D \; ,$$

one can use τ as a new point to count time from. Hence, under this condition availability at t equals $A_i(t-\tau)$. Replacing the integral in Eq. (8.1.6) by a proper approximating sum (in the sense of Riemann integration), as was exemplified in the derivation of Eq. (11.3.11), answers for the integral in Eq. (8.1.6). Now, under the condition that there is no down time between 0 and t, $X_i(t)=1$, and the corresponding conditional availability is 1. The probability of this condition is

$$P\{L>t\}=1-F_L(t) \; .$$

This answers for the last two terms of the r.h.s. of Eq. (8.1.6), q.e.d.

9.1

By Eq. (8.2.15)

$$\lambda_{i,i} = -\sum_{\substack{j=1 \\ j \neq i}}^{m} \lambda_{i,j}$$

and by Eq. (9.1.1)

$$E\{D_i\} = -1/\lambda_{i,i}.$$

From Fig. E8.1

$$\lambda_{i,i} = -(\lambda + \mu) \ , \qquad i \geqq 1 \ .$$

Hence for $i > 0$

$$E\{D_i\} = \frac{1}{\lambda + \mu} \ ,$$

which means that all queue lengths prevail for the same mean time.

9.2

From example 4.4.3

$$\varphi = X_1[X_2 \vee X_3] + \bar{X}_1 X_2 X_3 = X_1[X_2 + \bar{X}_2 X_3] + \bar{X}_1 X_2 X_3 \ .$$

By theorem 8.2.1

$$p_\varphi = p_1[p_2 + \bar{p}_2 p_3] + \bar{p}_1 p_2 p_3 \ .$$

Hence, to conform with what theorem 9.2.2 would yield from the fully developed polynomial form, Eq. (9.2.44)

$$v_\varphi = p_1[p_2(\lambda_1 + \lambda_2) + \bar{p}_2 p_3(\lambda_1 - \mu_2 + \lambda_3)] + \bar{p}_1 p_2 p_3(-\mu_1 + \lambda_2 + \lambda_3) \ .$$

Obviously, in this very simple example the gain is only one multiplication.

9.3

For equal components (of a coherent technical system) Eq. (8.2.5) would be specialized to

$$p_\varphi = \sum_i c_i p^{i_1} \bar{p}^{i_2}; \qquad \bar{p} := 1 - p \ .$$

Correspondingly Eq. (9.2.28) would become

$$\nu_\varphi = \sum_{\mathrm{i}} c_{\mathrm{i}} p^{\mathrm{i}_1} \bar{p}^{\mathrm{i}_2} (i_1 \lambda - i_2 \mu) \ .$$

9.4

Any of the three components can be the first one to fail. From the moment of this first failure to the moment of the second component failure the residual system life L_r is measured. Clearly, if L_i ends at t, then $L_r > \tau$ iff both L_j and L_k end after $t+\tau$, where $i, j, k \in \{1, 2, 3\}$, $i \neq j \neq k \neq i$. Since these events are disjoint, their probabilities can be added. Hence similar to the proof of (11.3.11)

$$\begin{aligned} P\{L_r > \tau\} &= P\{(L_2 > L_1 + \tau) \cap (L_3 > L_1 + \tau)\} \\ &\quad + P\{(L_1 > L_2 + \tau) \cap (L_3 > L_2 + \tau)\} \\ &\quad + P\{(L_1 > L_3 + \tau) \cap (L_2 > L_3 + \tau)\} \\ &= \int_0^\infty f_{L_1}(t) \bar{F}_{L_2}(t+\tau) \bar{F}_{L_3}(t+\tau)\, dt + \int_0^\infty f_{L_2}(t) \bar{F}_{L_1}(t+\tau) \bar{F}_{L_3}(t+\tau)\, dt \\ &\quad + \int_0^\infty f_{L_3}(t) \bar{F}_{L_1}(t+\tau) \bar{F}_{L_2}(t+\tau)\, dt \ . \end{aligned}$$

which equals Eq. (9.4.16).

9.5

From Eqs. (9.1.8)–(9.1.10)

$$P_1(t) + P_2(t) = \frac{2\lambda + \mu - \gamma_1}{\gamma_2 - \gamma_1} \exp(-\gamma_1 t) - \frac{2\lambda + \mu - \gamma_2}{\gamma_2 - \gamma_1} \exp(-\gamma_2 t) \ .$$

Now, by Eqs. (9.1.8) and (9.1.7)

$$E\{D_{\varphi,1}\} = \frac{1}{\gamma_2 - \gamma_1} \left(\frac{2\lambda + \mu - \gamma_1}{\gamma_1} - \frac{2\lambda + \mu - \gamma_2}{\gamma_2} \right) = \frac{2\lambda + \mu}{\gamma_1 \gamma_2} \ .$$

By Eq. (9.1.11) and the binomial formula $(a+b)(a-b) = a^2 - b^2$:

$$\gamma_1 \gamma_2 = \tfrac{1}{4}(9\lambda^2 + 6\lambda\mu + \mu^2 - \lambda^2 - 6\lambda\mu - \mu^2) = 2\lambda^2.$$

Hence

$$E\{D_{\varphi,1}\} = \frac{2\lambda + \mu}{2\lambda^2} = \frac{1}{\lambda} + \frac{\mu}{2\lambda^2} \ .$$

10.1

A conceptually very simple algorithm can be based on the concurrent development of the truth tables of φ_1 and φ_2. The relevant procedure should be left as soon as $\varphi_1 \neq \varphi_2$ is obvious. A technically very simple kernel of the desired algorithm is given in Table S10.1.

Table S10.1. Kernel of the algorithm "COMPTWOFUN" (compare two functions)

```
LABEL 1;
FOR X1:=FALSE TO TRUE DO
  FOR X2:=FALSE TO TRUE DO
   ⋱
       .............................
       FOR XN:=FALSE TO TRUE DO
          BEGIN
          PHI1:= ....; {INSERT BOOLEAN FORM OF FUNCTION PHI1}
          PHI2:= ....; {INSERT BOOLEAN FORM OF FUNCTION PHI2}
             IF PHI1<>PHI2 THEN
               BEGIN WRITELN ('PHI1 NOT EQUAL TO PHI2');
                       GOTO 1
               END
       END;
  WRITELN ('PHI1 EQUALS PHI2')
1:END
```

Clearly, this can be rendered more elegant, when using the procedure of Table 10.1.8, which is not pursued here.

10.2

(a) The final result should be Table S10.2.

Table S10.2. Table of the codewords of the three-out-of-five code

No.	12345	No.	12345
1	11100	6	10011
2	11010	7	01110
3	11001	8	01101
4	10110	9	01011
5	10101	10	00111

Because $\binom{5}{2}=10$, there are 10 code words. Table S10.3 shows a very simple implementation of the desired algorithm.

Table S10.3. Calculating and printing Table S10.2

```
PROGRAM CODE_3_OUT_OF_5 (OUTPUT);
VAR L, I1, I2, I3, J: INTEGER;
BEGIN
  L:=O; {L IS LINE-NO. OF TABLE}
  WRITE ('NO.12345');
  FOR I1:=1 TO 3 DO {START OF POSITIONING 1'S}
    FOR I2:=I1+1 TO 4 DO
      FOR I3:=I2+1 TO 5 DO
        BEGIN {START OF PRINTING PROGRAM}
          L:=L+1
          WRITE (L:3);
          FOR J:=1 TO 5 DO
                BEGIN
                  IF (J=I1) OR (J=I2) OR (J=I3)
                      THEN WRITE ('1')
                      ELSE WRITE ('0')
                END;
          WRITELN
        END
END.
```

(b) In extended notation the implementation would run as in Table S10.4; see also Table 10.1.9. However, in PASCAL N and K must be input as constants.

Table S10.4. Main body of a lengthy program to fill Table 10.1.9

```
FOR I1:=1 TO N-K+1 DO
  BEGIN I [1]:=I1;
      FOR I2:=I[1]+1 TO N-K+2 DO
          BEGIN I[2]:=I2;
              ...............................
                FOR IK:=I[K-1]+1 TO N DO
                  BEGIN I[K]:=IK;
                      FOR J:=1 TO N DO X[J]:=0;
                        FOR J:=1 TO K DO X[IJ]]:=1;
                        .... {USE OF CODE WORD}
                END
                ............
      END
END
```

Now, similar to Table 10.1.8 the nesting of the K FOR loops concerning I [1] through I [K] is implemented by a short recursive procedure; see Table S10.5. Note that K is now a variable.

Table S10.5. Procedure "NESTFOR" for the nesting of a variable number of FOR loops of variable length each

```
CONST N= . . . ;
PROCEDURE RECURSION (K: INTEGER);
  VAR I: ARRAY [O . . . N] {VARIABLE LENGTH IS NOT ALLOWED}
  PROCEDURE NESTFOR (DEPTH: INTEGER);
    VAR J: INTEGER;
    BEGIN
      FOR J:=I [DEPTH−1]+1 TO N−K+DEPTH DO
        BEGIN
          I[DEPTH]:=J;
          IF DEPTH<=K  THEN NESTFOR (DEPTH+1)
                       ELSE . . . {USE I[1], . . , I[K]}
        END
    END; {NESTFOR}
  BEGIN {RECURSION}
    I[O]:=O;
    NESTFOR (1)
  END;
```

10.3

(a) Let us begin our investigation with only two terms, $\hat{T}_1$ and $\hat{T}_2$. Once the index sets S_1 and S_2 of $\hat{T}_1$ and $\hat{T}_2$, respectively, are known one gets $S_1 \cap S_2$ trivially as

$$S12 := S1 * S2 \ .$$

Table S10.6. Determining the common factor of a polynomial, given as a two-dimensional array of terms $\hat{T}_i$ of literals $\tilde{X}_j . X_{k+n} = \bar{X}_k$; $k \in \{1, \ldots, n\}$

```
PROGRAM FACTOR (INPUT, OUTPUT);
CONST N2= . . . ; M= . . . ; {N2=2N.N/M ARE NUMBERS OF VARIABLES/TERMS}
VAR I, J: INTEGER;
   T: ARRAY [1 . . M, 1 . . N2] OF INTEGER; {T[I, J]=1, IF TERM I CONTAINS
      LITERAL XJ}
   S: SET OF 1 . . N2;
   AS: ARRAY [1 . . N2] OF S;
BEGIN
  T[1, 1]:= . . . ; . . . , T[M, N2]:= . . . ;
  S:=[1 . . N2];
  FOR I:=1 TO M DO
    BEGIN
      AS[I]:=[  ];
      FOR J:=1 TO N2 DO
        IF T[I, J]=1 THEN AS [I]:=AS[I]+[J];
      S:=S * AS[I]
    END;
IF S=[  ] THEN WRITE ('NO COMMON FACTOR')
          ELSE FOR J:=1 TO N2 DO
                  IF J IN S THEN WRITELN(J)
END.
```

To find S_i from $\hat{T}_i$ use is made of procedure SEARCHAR of Table 10.5.5. If the array AR contains $\hat{T}$, then the corresponding index set S is found as follows:

```
S:=[O];
FOR I:=1 TO 2*N DO
    IF AR[I]=1 THEN S:=S+[I]
```

The complete program is given in Table S10.6. Note, as concerns initialization, that any subset is the intersection of itself with the original set.

(b) Example: Let $\varphi(X_1, X_2, X_3) = X_1X_2X_3 \vee X_1X_2X_3\bar{X}_4 \vee X_1X_2X_4X_5$.[†] Clearly X_1X_2 can be factored out. From φ there follows

$$(\hat{T}_{i,j})_{m,n} = \begin{bmatrix} 1 & 1 & 1 & 0 & 0 & 0 & 0 & 0 & 0 & 0 \\ 1 & 1 & 1 & 0 & 0 & 0 & 0 & 0 & 1 & 0 \\ 1 & 1 & 0 & 1 & 1 & 0 & 0 & 0 & 0 & 0 \end{bmatrix} .$$

Initially, $S = \{1, 2, \ldots, 10\}$.
Then, since $S[1] = \{1, 2, 3\}$, one gets $S = \{1, 2, 3\}$.
Since $S[2] = \{1, 2, 3, 9\}$, the next value of S is $\{1, 2, 3\}$.
Finally, since $S[3] = \{1, 2, 4, 5\}$, the final S is $\{1, 2\}$.

11.1

(a) Since

$$(a_i \cap a_j \cap a_k) \subseteq (a_i \cap a_j) ,$$

by the definition of $q(\cdot)$

$$q(a_i \cap a_j \cap a_k) \leqq q(a_i \cap a_j) .$$

Hence

$$-q(a_i \cap a_j) - q(a_i \cap a_k) - q(a_j \cap a_k) + q(a_i \cap a_j \cap a_k) \leqq 0$$

and (E11.1) is true.

(b) Since $q(a_i \cap a_j \cap a_k) \geqq 0$, (E11.2) is trivially true.

11.2

See Fig. S11.1.

[†] Don't mind here that the first term can absorb the second one.

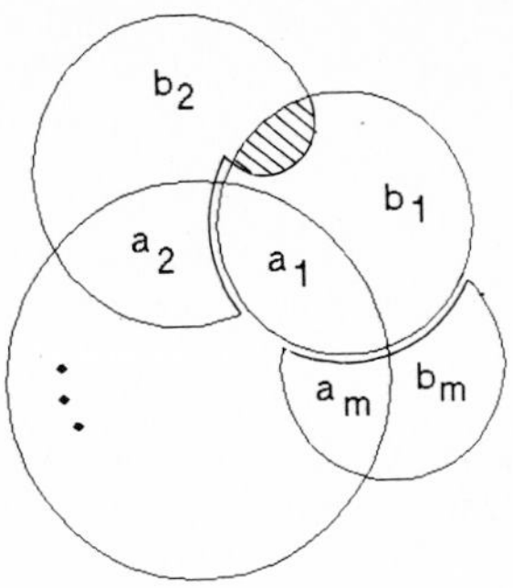

Fig. S11.1. Visualizing the law of total probability

So long as – in the example of Fig. S11.1 –

$$a_1 = a \cap b_1 \ , \qquad a_2 = a \cap b_2 \ ; \qquad a_1 \cap a_2 = \varnothing \ ,$$

it does not matter, if $b_1 \cap b_2 \neq \varnothing$. Only the $a_1, \ldots, a_m$ must be a partition of a.

11.3

a) Let N_1 be the number asked for. Obviously

$$N_1 = \sum_{i=1}^{m} X_i \ .$$

b) $E\{N_1\} = \sum_{i=1}^{m} E\{X_i\} = m\,U$.

11.4

Let $h(t) = c$. Then by Eq. (11.5.3)

$$h^*(s) = c/s \ .$$

Hence, by Eq. (11.4.6)

$$f_1^*(s)/[1 - f_1^*(s)] = c/s$$

or

$$f_1^*(s) = \frac{c}{s+c} \ .$$

Finally, by example 11.5.1:

$$f_1(t) = c \exp(-ct) \ .$$

11.5

From $Z_1 \in (z-\Delta z, z]$ in order to satisfy $Z_2 > Z_1$ there follows $Z_2 > z$. Obviously

$$P\{Z_1 \in (z-\Delta z, z]\} = f_{Z_1}(z)\Delta z + o(\Delta z), \quad P\{Z_2 > z\} = 1 - F_{Z_2}(z) .$$

Hence, the desired probability is, for *s*-independent Z_1, Z_2:

$$\int_{-\infty}^{\infty} f_{Z_1}(z)[1 - F_{Z_2}(z)]\, dz .$$

References

[1] Gschwind H., McCluskey E.: Design of digital computers. Heidelberg: Springer, 1975.

[2] Mendelson E.: Boolean algebra and switching circuits. New York: McGraw-Hill, 1970.

[3] Rudeanu S.: Boolean functions and equations. Amsterdam: North-Holland, 1974.

[4] Hulme B., Worrell R.: A prime implicant algorithm with factoring. Trans. IEEE, vol. C-24 (1975), 1129–1131.

[5] Kim K.: Boolean matrix theory and applications. New York: Marcel Dekker Inc., 1982.

[6] Abraham J.: An improved algorithm for network reliability. Trans. IEEE, vol. **R-28** (1979), 58–61.

[7] Tiwari R., Verma M.: An algebraic technique for reliability evaluation. Trans. IEEE, **R-29** (1980), 311–313.

[8] Aggarwal K., Misra K., Gupta J.: A fast algorithm for reliability evaluation. Trans. IEEE, **R-24** (1975), 83–85.

[9] Thayse A.: Boolean calculus of differences. Berlin: Springer, 1981.

[10] Quine W.: The problem of simplifying truth functions. Am. Math. Mon., **59** (1952), 521–531.

[11] Quine W.: A way to simplify truth functions. Am. Math. Mon., **62** (1955), 627–631.

[12] Samson W., Mills B.: Circuit minimization: algebra and algorithms for new Boolean canonical expression. AFCRC. Techn. Rep. 54–21 (1954).

[13] Nelson R.: Simplest normal truth functions. J. symb. logic, **20** (1954), 105–108.

[14] Akers S.: On a theory of Boolean functions. SIAM J., **7** (1959), 487–498.

[15] Barlow R., Proschan F.: Mathematical theory of reliability. New York: Wiley 1965.

[16] Schneeweiss W.: On a special Boolean algebra for indicator variables. NTZ-Archiv, **3** (1981), 53–56.

[17] Schneeweiss W.: Zuverlässigkeits-Systemtheorie (Reliability systems theory). Köln: Datakontext-Verlag 1980.

[18] Enzmann W.: Ein Algorithmus zur Berechnung von Zuverlässigkeitsdaten komplexer redundanter Systeme. Angewandte Informatik (1973), 493–499.

[19] Satyaryanana A., Prabhakar A.: New topological formula and rapid algorithm for reliability analysis of complex networks. Trans. IEEE, **R-27** (1978), 82–100.

[20] Bharucha-Reid A.: Elements of the theory of Markov processes and their applications. New York: McGraw-Hill, 1960.

[21] Barlow R., Proschan F.: Statistical theory of reliability and life testing (probability models). New York: Holt, Rinehart & Winston, 1975.

[22] Schneeweiss W.: Computing failure frequency, MTBF & MTTR via mixed products of availabilities and unavailabilities. Trans. IEEE, **R-30** (1981), 362–363.

[23] Jensen K., Wirth N.: PASCAL user manual and report (2nd edn). New York: Springer, 1978.

[24] Grogono P.: Programming in PASCAL Reading, Mass.: Addison-Wesley, 1978.

[25] Peterson W., Weldon E.: Error-correcting codes (2nd edn). Cambridge, Mass.: MIT Press 1972.

[26] Feller W.: An introduction to probability theory and its applications. Vol. I (2nd edn). New York: Wiley, 1957.

[27] Papoulis A.: Probability, random variables, and stochastic processes. New York: McGraw-Hill, 1965.

[28] Parzen E.: Modern probability theory and its applications. New York: Wiley 1960.

[29] Loève M.: Probability theory I (4th edn) : New York: Springer, 1977.

[30] Loève M.: Probability theory II (4th edn) : New York: Springer, 1978.

[31] Huang X.: Calculating the failure frequency of a repairable system. Micro-electronics & Reliability **22** (1982), 945–947.

[32] Khintchine A.: Mathematical methods in the theory of queueing. London: Griffin, 1969.
[33] Schneeweiss W., Schulte M.: Sy Re Pa '86—a package of programs for systems reliability evaluation. Fernuniversitaet, D-58 Hagen, Informatik—Ber. 62 (Apr. 1987).
[34] Cox D., Isham V.: Point processes. London: Chapman and Hall, 1980.
[35] Cox D.: Renewal theory. London: Methuen, 1962.
[36] Singh C.: Calculating the time-specific frequency of system failure. Trans. IEEE, **R-28** (1979), 124–126.
[37] Akers S.: Binary decision diagrams. Trans. IEEE, **C-27** (1978), 509–516.
[38] De Luca A.: On some representations of Boolean functions. Kybernetik, **9** (1971), 1–10.
[39] Reingold E., Nievergelt J., Deo N.: Combinatorial algorithms: theory and practice. Englewood Cliffs: Prentice Hall, 1977.
[40] Kemp R.: Fundamentals of the average case analysis of particular algorithms. Stuttgart, New York: Teubner/Wiley, 1984.
[41] Corynen G.: A fast procedure for the exact computation of the performance of complex probabilistic systems. Proc. Internat. ANS/ENS Meeting on Probabilistic Risk Assessment, Port Chester N.Y. 1981.
[42] Schneeweiss W.: Approximate fault tree analysis with prescribed accuracy. Trans. IEEE-R **36** (1987) 250–254.
[43] Neuschwander J., Beister J.: On the reasons for the costs of implementing switching functions. Informationstechnik **28** (1986) 89–96.
[44] Beister J.: A unified approach to combinational hazards. Trans. IEEE, **C-23** (1974) 566–575.
[45] McNaughton R.: Unate truth functions. Trans. IRE, **EC-10** (1961) 1–6.
[46] Dowsing R., Rayword-Smith V., Walter C.: A first course in formal logic and its applications in computer science. London: Blackwell 1986.
[47] South G.: Boolean Algebra and its uses. London: Van Nostrand Reinhold 1974.
[48] Grätze G.: Lattice theory. San Francisco : Freeman 1971.
[49] Dietmeyer D.: Logic design of digital systems (2nd edn). Boston: Allyn & Bacon 1978.
[50] Avizienis A.: The N-version approach to fault-tolerant software Trans. IEEE. Software Eng. (1985), 1491–1501.
[51] Bryant R.: Graph-based algorithms for Boolean function manipulation. Trans. IEEE **C-35** (1986) 677–691.
[52] Cutler, R., Muroga S.: Derivation of minimal sums for completely specified functions. Trans. IEEE, **C-36** (1987) 277–292.
[53] Blahut R.: Theory and practice of error control codes. Reading Mass.: Addison-Wesley 1983.
[54] Trivedi K.: Probability and statistic with reliability, queueing, and computer science applications. Englewood Cliffs: Prentice Hall 1982.
[55] Harary F.: Graph theory. Reading Mass.: Addison-Wesley 1969.
[56] Borge C.: The theory of graphs and its applications. London: Methuen 1962 (1st edn).
[57] Schneeweiss W.: Distribution of computer life after tolerable faults. Proc. 3rd German Conference on Fault-Tolerant Computing Systems. Berlin: Springer 1987, 304–313.
[58] Wegener I.: The complexity of Boolean functions. Stuttgart: Teubner/Wiley 1987.

Subject Index

Note: Basic terms such as Boolean or graph are omitted, others are listed only in selected cases.